Piper: A Model Genus for Studies of
Phytochemistry, Ecology, and Evolution

Piper: A Model Genus for Studies of Phytochemistry, Ecology, and Evolution

Edited by

Lee A. Dyer
Tulane University, New Orleans, Louisiana

and

Aparna D. N. Palmer
Mesa State College, Grand Junction, Colorado

Kluwer Academic/Plenum Publishers
New York   Boston   Dordrecht   London   Moscow

Library of Congress Cataloging-in-Publication Data

Piper: a model genus for studies of phytochemistry, ecology, and evolution / edited by Lee
 A. Dyer and Aparna D.N. Palmer.
   p. cm.
 Includes bibliographical references and index.
 ISBN 0-306-48498-6
 1. Piper (Genus)   I. Dyer, Lee A.   II. Palmer, Aparna D. N.
 QK495.P67P56 2004
 583'.25–dc22   2004042122

ISBN 0-306-48498-6

© 2004 by Kluwer Academic/Plenum Publishers, New York
233 Spring Street, New York, New York 10013

http://www.kluweronline.com

10 9 8 7 6 5 4 3 2 1

A C.I.P. record for this book is available from the Library of Congress

All rights reserved

No part of this work may be reproduced, stored in a retrieval system, or transmitted in any form
or by any means, electronic, mechanical, photocopying, microfilming, recording, or otherwise,
without written permission from the Publisher, with the exception of any material supplied specifically for the
purpose of being entered and executed on a computer system, for exclusive use by the purchaser of the work

Permissions for books published in Europe: permissions@wkap.nl
Permissions for books published in the United States of America: permissions@wkap.com

Printed in the United States of America

Contributors

Donald P. Briskin
Department of Natural Resources and
  Environmental Sciences
University of Illinois at Urbana/Champaign
Urbana, IL 61801 USA

Ricardo Callejas
Departamento de Biología
Universidad de Antioquia
Apartado Aereo 1226
Medellin, Colombia

Rodolfo Antônio de Figueiredo
Centro Universitário Central Paulista
Rua Sebastião de Abreu Sampaio, 1234
13083-470 São Carlos, Brazil

Craig D. Dodson
Deparment of Chemistry
Mesa State College
Grand Junction, CO 81501 USA

Lee A. Dyer
Deparment of Ecology and Evolutionary
  Biology
310 Dinwiddie Hall
Tulane University
New Orleans, LA 70118 USA

Theodore H. Fleming
Department of Biology
University of Miami
Coral Gables, FL 33124 USA

Karin R. Gastreich
Organization for Tropical Studies
Apdo. 676-2050
San Pedro, Costa Rica

Margaret Gawienowski
Department of Natural Resources and
  Environmental Sciences
University of Illinois at Urbana/Champaign
Urbana, IL 61801, USA

Grant L. Gentry
Deparment of Ecology and Evolutionary
  Biology
310 Dinwiddie Hall
Tulane University
New Orleans, LA 70118 USA

Nancy Greig
Cockrell Butterfly Center
Houston Museum of Natural Science
One Hermann Circle Drive
Houston, TX 77030-1799 USA

M. Alejandra Jaramillo
Departmento de Bioquímica Médica,
CCS-ICB, Universidade Federal do
Rio de Janeiro, C. P. 68041, 2194-590
Rio de Janeiro, RJ, Brazil

Hideka Kobayashi
Department of Natural Resources and
  Environmental Sciences
University of Illinois at Urbana/Champaign
Urbana, IL 61801 USA

Deborah K. Letourneau
Department of Environmental Studies
Interdisciplinary Sciences Building
University of California
Santa Cruz, CA 95064 USA

Mary Ann Lila
Department of Natural Resources and
  Environmental Sciences
University of Illinois at Urbana/Champaign
Urbana, IL 61801 USA

Robert J. Marquis
Department of Biology
University of Missouri-St. Louis
8001 Natural Bridge Road
St. Louis, Misssouri 63121-4499 USA

Aparna D.N. Palmer
Department of Biology
Mesa State College
Grand Junction, CO 81501 USA

Joe Richards
Deparment of Chemistry
Mesa State College
Grand Junction, CO 81501 USA

Marlies Sazima
Departamento de Botânica
Universidade Estadual de Campinas
13083-970 São Paulo, Brazil

Eric J. Tepe
Department of Botany
Miami University
Oxford, OH USA

Michael A. Vincent
Department of Botany
Miami University
Oxford, OH USA

Linda E. Watson
Department of Botany
Miami University
Oxford, OH USA

Preface

This book offers a glimpse into the world that exists within, and revolves around, a remarkable group of tropical plants. Although it does not synthesize all of the scientific work that has been gathered on *Piper*, it is a significant advance in our understanding. This advance represents only the tip of the iceberg in terms of the scientific inquiry that will undoubtedly continue to accumulate on this group of organisms. Not unlike other model genera (e.g., *Drosophila*, *Escherichia*, *Arabidopsis*, *Neurospora*), *Piper* possesses several attributes that are essential for successful scientific work: it is readily accessible, easily manipulable, and strikingly diverse. For these reasons, it is a model system that will become important to many areas of science beyond chemistry, ecology, and evolutionary biology. Regrettably, as we compiled this volume, we could not cover all aspects of *Piper* biology that have been studied to date; most notably, this book lacks insight into the developmental genetics, physiology, anatomy, and invasion ecology of *Piper*. And, although it makes a substantial contribution to our knowledge of the Neotropical pipers, it does not do justice to the Paleotropical members of the genus. Nevertheless, this book strives to create a resource for investigators searching for a body of information with which they can plan future studies on and around this extraordinary group of organisms.

We would like to thank many people and organizations for making this book possible. Helpful comments and thorough reviews were provided by a skilled group of anonymous reviewers. We are especially grateful to Christine Squassoni for reading the entire book several times and helping to edit the final version. Other thoughtful comments and editing remarks were kindly provided by the authors of the chapters in this book and by the following individuals: J. Stireman, R. Matlock, T. Walla, I. Rodden, M. Singer, W. Carson, M. Olson, T. Floyd, R. M. Fincher, A. Smilanich, C. Pearson, M. Tobler, and D. Bowers. LAD thanks N. Greig, R. Marquis, O. Vargas, and D. Letourneau for first introducing him to *Piper* in Costa Rica. ADNP thanks LAD for infecting her with the same enthusiasm he has for *Piper*, asking her to collaborate with him, and inviting her to share in the editing of this book. Finally, we would like to thank W. Kelley, who has been a source of inspiration to both of us with his continual enthusiasm for *Piper* anatomy and evolution.

We are grateful for an enormous amount of help in the field and laboratory for our own research on *Piper*, especially from G. Vega Chavarria, H. Garcia Lopez, and M. Tobler. In addition to the authors in this book, we also thank the following individuals for their collaborative work in the field and laboratory: A. Barberena, R. Krach, C. Squassoni, O. Vargas, D. Rath, H. Rosenberg, A. Smilanich, R. M. Fincher, A. Hsu, T. Walla, H. Greeney, W. Williams, M. Rathbone, A. Schaefer, J. Sorenson, H. Kloeppl, J. Searcy, Z. Wright, J. Jay, T. Brenes, G. Brehem, N. Bishop, hundreds of Earthwatch volunteers, and many undergraduate students at Tulane University and Mesa State College. Funding for the editors' research on *Piper* and some of the expenses incurred in compiling this book came from Earthwatch Institute, the National Science Foundation (DEB 9318543 and DEB 0074806), the United States Department of Agriculture (CSREES 35316-12198), the Department of Energy (Southcentral Regional Center of NIGEC), Tulane University, and Mesa State College.

We dedicate this book to the *Eois* caterpillars that are the true *Piper* specialists. The book is also dedicated to the Antisana–Sumaco conservation axis in Northeast Ecuador in hopes that the ecosystems there will always offer a safe haven for wild pipers. *Viva la abundancia de las bosques del Piper.*

LEE. A. DYER
APARNA D. N. PALMER

Contents

CHAPTER 1. Introduction 1
                N. Greig

CHAPTER 2. **Mutualism, Antiherbivore Defense, and Trophic Cascades:** *Piper* **Ant-Plants as a Mesocosm for Experimentation** 5
                D. K. Letourneau

    2.1. Introduction 5
    2.2. Study Sites 8
    2.3. Plants 8
    2.4. Herbivores 9
    2.5. Ant Mutualists 10
    2.6. Top Predators 11
    2.7. Other Endophytic Arthropods, Nematodes, and Annelids 12
    2.8. Mutualism Experiments 13
        2.8.1. Evidence for Nutrient Procurement by *Pheidole bicornis* Plant-Ants 15
        2.8.2. Evidence for Defense against Folivores by *Pheidole bicornis* Plant-Ants 16
        2.8.3. Evidence for Higher Fitness in Establishing Fragments via Defense against Folivores 17
        2.8.4. Evidence for Additional Plant Fitness Advantages Afforded by *Ph. Bicornis* Plant-Ants 18
    2.9. Tritrophic Interactions and Antiherbivore Defense 18
    2.10. Trophic Cascades 22
        2.10.1. Experimental Test I: Do Trophic Cascades Operate on the Four Trophic Level System Associated with *P. cenocladum* Ant-Plants? 23

	2.10.2. Experimental Test II: Can Top–Down and Bottom–Up Forces Affect Animal Diversity in the Endophytic Community of *P. Cenocladum* Ant-Plants?	24
	2.10.3. Experimental Test III: Can Indirect Effects of Top Predators Extend to Other Plants in the Understory Community of *Piper* Ant-Plants?	26
2.11.	Conclusions	27
2.12.	Acknowledgments	29

CHAPTER 3. Pollination Ecology and Resource Partitioning in Neotropical *Pipers* 33

Rodolfo Antônio de Figueiredo and Marlies Sazima

3.1.	Introduction	33
3.2.	Pollination and Resource Partitioning in *Piper*	34
	3.2.1. Study Site and Species of the Brazilian Study	35
	3.2.2. Habit and Habitat Utilization	36
	3.2.3. Vegetative Reproduction	37
	3.2.4. Reproductive Phenology	38
	3.2.5. Pollination and Visitors	41
3.3.	Conclusions: Pollination and Resource Partitioning of Pipers in Light of Evolutionary and Conservative Ecology	49
3.4.	Guidelines for Future Research on the Pollination of Pipers	52
3.5.	Acknowledgments	52

CHAPTER 4. Dispersal Ecology of Neotropical *Piper* Shrubs and Treelets 58

Theodore H. Fleming

4.1.	Introduction	58
4.2.	The *Piper* Bats	59
4.3.	*Piper* Fruiting Phenology and Dispersal Ecology	62
	4.3.1. Fruiting Phenology	62
	4.3.2. Patterns of Seed Dispersal	64
	4.3.3. Fates of Seeds	65
	4.3.4. Postdispersal Distribution Patterns	67
4.4.	Coevolutionary Aspects of Bat–*Piper* Interactions	72
4.5.	Conclusions	74
4.6.	Acknowledgments	75

CHAPTER 5. Biogeography of Neotropical *Piper* 78

Robert J. Marquis

5.1.	Introduction	78
5.2.	Methods	79

	5.3.	Results	84
		5.3.1. Biogeographic Affinities and Regional Species Pools	84
		5.3.2. Correlates of Local Species Richness	86
		5.3.3. Variation in Growth Form and Habitat Affinity	88
	5.4.	Discussion	91
	5.5.	Acknowledgments	94
CHAPTER 6.	**Faunal Studies in Model *Piper* spp. Systems, with a Focus on Spider-Induced Indirect Interactions and Novel Insect–*Piper* Mutualisms**		**97**

Karin R. Gastreich and Grant L. Gentry

	6.1.	Introduction	97
	6.2.	The Case of *Piper obliquum*	99
	6.3.	The Case of *Piper urostachyum*	102
		6.3.1. Plant Characteristics that Encourage Mutualism	103
		6.3.2. Resident Arthropods	103
		6.3.2a. Herbivores	103
		6.3.2b. Mutualist predators	105
		6.3.2c. Parasites of the mutualism?	106
		6.3.2d. Top predators	106
		6.3.3. Possible Mutualisms and the Effects of Spiders	107
	6.4.	Summary and Conclusions	112
	6.5.	Acknowledgments	114
CHAPTER 7.	**Isolation, Synthesis, and Evolutionary Ecology of *Piper* Amides**		**117**

Lee Dyer, Joe Richards, and Craig Dodson

	7.1.	Introduction to *Piper* Chemistry	117
	7.2.	Isolation and Quantification of *Piper* Amides	120
	7.3.	Synthesis of *Piper* Amides and Their Analogs	121
	7.4.	Ecology of *Piper* Chemistry	128
	7.5.	Evolution of *Piper* Chemistry	131
	7.6.	Applied *Piper* Chemistry	132
	7.7.	Future Research on *Piper* Chemistry	133
	7.8.	Acknowledgments	134
CHAPTER 8.	**Kava (*Piper methysticum*): Growth in Tissue Culture and *In Vitro* Production of Kavapyrones**		**140**

Donald P. Briskin, Hideka Kobayashi, Mary Ann Lila, and Margaret Gawienowski

	8.1.	Introduction	140
	8.2.	Origins of Kava Use and Discovery by Western Cultures	141

8.3.	Description of Kava (*Piper methysticum*) and Its Growth for Use in Kava Production	141
8.4.	Active Phytochemicals Present in Kava Extracts	142
8.5.	Issues Regarding the Potential Hepatotoxicity of Kava Extracts	143
8.6.	Significance of Tissue Culture Growth in Kava Production and Phytochemical Research	144
8.7.	Establishment of Kava Cell Cultures and the Determination of *In Vitro* Kavapyrone Production	146
8.8.	Regeneration of Viable Kava Plants from Kava Cell Cultures	151
8.9.	Summary and Perspective	151

CHAPTER 9. Phylogenetic Patterns, Evolutionary Trends, and the Origin of Ant–Plant Associations in *Piper* Section *Macrostachys*: Burger's Hypotheses Revisited **156**

Eric. J. Tepe, Michael A. Vincent, and Linda E. Watson

9.1.	Introduction	156
9.2.	Taxonomic History of *Piper* sect. *Macrostachys* (MIQ.) C.DC.	158
9.3.	Natural History of *Piper* sect. *Macrostachys*	159
9.4.	Phylogenetic Relationships in *Piper* sect. *Macrostachys*	162
9.5.	Burger's Hypotheses Revisited	165
	9.5.1. Systematic Relationships	165
	9.5.2. Evolutionary Trends	167
9.6.	Ant–Plant Associations in *Piper* sect. *Macrostachys*	168
	9.6.1. Origins and Evolutionary Trends	168
	9.6.2. Evolution of the Mutualism	169
	9.6.2a. Obligate associations and hollow stems	169
	9.6.2b. Petiolar domatia and facultative associations	172
	9.6.2c. Pearl Bodies	173
	9.6.2d. Origin of ant-associated plant structures	173
9.7.	Conclusions	174
	Appendix 9.1	174

CHAPTER 10. Current Perspectives on the Classification and Phylogenetics of the Genus *Piper* L. **179**

M. Alejandra Jaramillo and Ricardo Callejas

10.1.	Introduction	179
10.2.	Classification	180
	10.2.1. Getting Cluttered	180
	10.2.2. Getting Articulated	181
10.3.	Phylogeny	181

	10.3.1.	Phylogenetic Relationships of the Piperales: A Test of *Piper*'s Monophyly	181
	10.3.2.	Infrageneric Relationships of *Piper*	183
		10.3.2a. Neotropical Taxa	183
		10.3.2b. South Pacific and Asian taxa	188
10.4.	Evolutionary Aspects		189
	10.4.1.	Flower Morphology	189
	10.4.2.	Plant Architecture	191
10.5.	Acknowledgments		192
	Appendix 10.1		194

CHAPTER 11. Future Research in *Piper* Biology **199**

M. Alejandra Jaramillo and Robert Marquis

11.1.	Introduction	199
11.2.	Plant–Animal Interactions	199
11.3.	Abiotic Factors	201
11.4.	Geographical Distribution	201
11.5.	Summary	202
	Index	**205**

Piper: A Model Genus for Studies of Phytochemistry, Ecology, and Evolution

1
Introduction

N. Greig
Cockrell Butterfly Center, Houston Museum of Natural Science, Houston, Texas

Why *Piper*? Costa Rican naturalist Isidro Chacon asked me that very question in 1985, when I told him of my interest in the genus as a subject for my dissertation research. "Why such ugly and boring plants?" he said. "There are so many of them, and they all look just the same. Why not choose something with prettier flowers, or more diverse growth forms and habits?"

In fact, the contributors to this volume have found *Piper*'s diversity in species, coupled with its lack of morphological diversity, to be compelling reasons to be fascinated with these admittedly less than spectacularly beautiful plants; they have chosen them as tools with which to address a broad range of biological questions. Thus this book, instead of treating a field of study such as plant chemistry, or plant–animal interactions, or physiology, has taken as its unifying principle this genus of plants, which over the past two decades or so has become a favorite research system for tropical biologists investigating topics from demographics and reproductive biology to chemistry, and from phylogeny to ecology.

Piper is the nominate genus of the family Piperaceae, a pantropical family composed of five to eight or more genera, depending upon the treatment. The two largest genera are *Piper* and *Peperomia*, each containing about 1,000 species. Peperomias are mostly small, succulent, often epiphytic herbs; pipers are woody and more diverse in habit, including shrubs (the great majority), climbing vines, and small trees (treelets). The other piperaceous genera are small, and their boundaries are uncertain (at least some have often been subsumed into *Piper*).

The greatest diversity of *Piper* species is in the Neotropics, where about two-thirds of the described species are found. Some 300 species are endemic to Southeast Asia, including the East Indian islands and northern Australia. Only two species are native to Africa. Most *Piper* species grow in wet, warm, lowland rain forests. Both diversity and abundance of *Piper* typically decrease with increasing elevation or with decreasing precipitation. Thus in Costa Rica, La Selva has some 50+ species, Sirena on the Osa Peninsula (with more annual rainfall than La Selva, but with a stronger dry season) has about 42 species, Monteverde has about 11 species, and Santa Rosa has about 5 species.

Piper species are rather uniform morphologically, with simple, alternate leaves and jointed stems with enlarged nodes. Branches break easily at these nodes, either when snapped or when in decay; one often encounters *Piper* "bones"—the separated internodes of the stem, having an uncanny resemblance to chicken leg or thigh bones—in the leaf litter on the forest floor. Their stem anatomy is unusual for dicots; they retain scattered vascular bundles in the mature tissues. Most species of *Piper* are aromatic, some highly so, because of ethereal oil cells in their tissues. Many (perhaps most) produce pearl bodies on the leaves or stems, or lining the petioles. In some species, these bodies are used as food by resident mutualist ants.

The inflorescences are particularly distinctive. Dozens (e.g., *P. garagaranum*) to thousands (e.g., *P. auritum*) of tiny, reduced flowers are packed into upright or pendant spikes or spadices. Each flower matures into a tiny, one-seeded drupelet, which together with the other drupelets, form a multiple fruit; the fruit becomes fleshy and, in some species, very fragrant when ripe. New World *Piper* species have bisexual (hermaphroditic) flowers, whereas many Old World species are dioecious.

Few of the species are economically important. An exception is *Piper nigrum*, which is the source of black pepper, the world's most widely used spice. Originally from India, black pepper was important in early trading between Europe and Asia; today it is grown throughout the humid tropics. The roots of *P. methysticum* are the source of Kava, used traditionally by the Polynesians as a narcotic beverage; today, it is finding increasing popularity as a sedative in the phytomedicine industry. In Southeast Asia, *P. betle* leaves are chewed along with lime (calcium) and the fruit of the betel palm after meals as a digestive aid. A few other species (e.g., *P. cubeba*, cubeb pepper; *P. longum*, long pepper; *P. auritum*, hoja santa) are used locally as condiments or medicinals.

Although the vast majority of species have no economic value, they are important in natural habitats throughout the humid tropics. In lowland rain forests in particular, *Piper* species are a dominant component of the understory flora; some (e.g., *P. arieianum* at La Selva in Costa Rica) may be very abundant. A suite of insect herbivores feeds on the leaves of *Piper*; the ripening fruits are attacked by a variety of seed predators. Ripe fruits provide food for frugivorous bats and birds, and other animals surely use *Piper* fruits as food at least occasionally (I've seen hermit crabs eating the fruits of *P. pseudofuligineum* along the beach on the Osa Peninsula in Costa Rica, and doubtless some of the small-statured pipers in primary forest are dispersed by rodents or marsupials, or even ants).

Prior to 1980, most work on the genus *Piper* came out of India and concerned its chemical constituents and the potential these chemicals had as pesticidal agents (e.g., piperine); horticultural treatises on the cultivation of *Piper nigrum* (black pepper) were also popular. But with the burst of ecological research in the tropics over the last two decades, particularly in the Neotropics, *Piper* began to find favor with scientists in a number of other fields. *Piper* species have been used to study herbivory, distribution and abundance, phenological patterns, seed dispersal by bats and birds, photosynthetic capacity, ant mutualisms, phytochemistry, and more.

Piper has many advantages as a research system. In lowland Neotropical sites in particular, *Piper* is frequently within the top five genera in terms of species diversity. The genus is one of the easiest of understory plants to recognize, with its "knobby knees" (the swollen nodes), often oblique leaf bases, characteristic "little candle" inflor/infructescences, and the typical "spicy" or aromatic smell of the crushed leaves or broken stems.

INTRODUCTION

Most species are relatively small so that many of their important parts (leaves, branches, flowers, and fruits) can easily be observed and are accessible for manipulation. Although much remains to be determined with regard to hard and fast species concepts and infrageneric relationships in *Piper*, the different "morphospecies" at a given site can easily be learned and identified in the field. *Piper* species occur in a range of successional habitats, from highly disturbed areas to pristine primary forest, offering opportunities to compare "behavior" or physiology of closely related species under different environmental conditions. The genus is particularly rich in chemical compounds with economic potential and/or ecological implications, and several species are inhabited by obligate ant mutualists. Finally, many *Piper* species are easy to propagate, especially from stem cuttings, so clones can be produced on which to test the effects of herbivory, different soil types, ant mutualists, and different nutritional regimes on growth, survival, or phytochemistry. Indeed any aspect in which it is useful to reduce "noise" by eliminating the possibility of interindividual differences can involve the use of pipers.

To facilitate future investigations using pipers, this book both synthesizes and adds to the previous body of knowledge on this fascinating group of plants. Its major themes include (1) *Piper* as a model genus for ecological and evolutionary studies, (2) important ecological roles of *Piper* species in lowland wet forests, and (3) evolution of distinctive *Piper* attributes. Each of the following chapters develops one or more of these themes in the context of previous research and future directions. Chapter 2 is a sweeping overview of three decades of work on *Piper–Pheidole* ant mutualistic interactions, demonstrating the complementary contributions of both ant and plant chemical defense to plant fitness, and explaining how this system has been useful in examining multitrophic interactions and in assessing theoretical models of trophic cascades. Chapter 3 adds to the heretofore meager knowledge of *Piper* pollination biology, by comparing and contrasting phenologies and pollination syndromes of the suite of species growing in a semideciduous forest in Brazil. Chapter 4 describes distribution patterns of *Piper* species at another Neotropical site, and offers hypotheses on how those patterns, and ultimately genetic diversity and speciation, may be influenced by the foraging ecology of frugivorous bats. Chapter 5 considers the biogeographic distribution of *Piper* in the Neotropics, correlating patterns of species richness and growth form with climatic and geographical features. Chapter 6 assesses the role of indirect interactions in antiherbivore mutualisms by testing the effect of predatory spiders on levels of herbivory in both ant-associated and non–ant-associated *Piper* species. Chapter 7 surveys the chemical diversity found in *Piper*, especially the ecologically and economically important amides so abundant in the genus, and anticipates community-level consequences of chemical versus biotic plant defenses. In Chapter 8, the authors examine the chemical makeup of Kava, the narcotic phytomedicine derived from *P. methystichum*; they also discuss the potential role of tissue culture in amplifying Kava production. Chapter 9 elaborates on the phylogenetic relationships within a single section (*Macrostachys*) of the genus, surveying character suites of both facultative and obligate myremecophytes (ant-plants) found in the group, and suggesting how these morphological characters relate to the evolution of both obligate and facultative ant mutualisms. Chapter 10 is a welcome review of our current understanding of *Piper* phylogeny and infrageneric relationships; it provides an excellent starting place for further developments in this much needed area of investigation.

The final chapter offers a succinct summary of our knowledge of *Piper* biology to date, with suggestions on "where to go from here." The authors state that the contributions of the current volume, while augmenting and consolidating our understanding of these plants and their importance, bring up as many questions as they resolve. Further investigation is needed in almost every aspect of *Piper* biology; perhaps the most critical goals to achieve are to gain a better understanding of *Piper* phylogeny and species-level relationships, and to learn more about the Paleotropical flora.

Thus the genus *Piper*, with its diversity, its ease of recognition, and its amenability to observation and manipulation, will likely remain a productive research system for students of ecology and evolutionary biology for years to come; it has indeed become the "*E. coli* of tropical biology" as Chris Fields predicted 20 years ago (personal communication). The authors who have contributed herein argue that there is still much to be learned. We hope that this book will serve as both an inspiration and a useful reference for those future studies.

2
Mutualism, Antiherbivore Defense, and Trophic Cascades: *Piper* Ant-Plants as a Mesocosm for Experimentation

D. K. Letourneau
Department of Environmental Studies, Interdisciplinary Sciences Building, University of California, Santa Cruz, California

2.1. INTRODUCTION

The investigation of *Piper* ant-plants extends back through half of my lifetime, beginning in Corcovado National Park on the Osa Peninsula of Costa Rica in the summer of 1978 (as a graduate student in the Organization for Tropical Studies, Fundamentals of Tropical Biology course). At a loss to help me find an independent study topic on agricultural systems (my dogged interest) at this site of rich lowland forest, mangroves, and sandy beaches, ecologists Doyle McKee and David Janos urged me to consider examining these intriguing, forest understory plants, and their associated arthropods. Three *Piper* species occurring in the park at La Llorona Point had been described (Burger 1971), but were as yet unstudied as ant–plant associations. The previous year, Risch *et al.* (1977) had published the first study of *Piper* ant-plants, on the closely related *Piper cenocladum* C.DC. (Piperaceae), with its resident *Pheidole bicornis* ants in the Caribbean lowlands of Costa Rica. These studies set the stage for expanding the investigation of these ant-inhabited understory plants, and suggested ways in which ant occupants may benefit their host plant. Over the years this spectacular and complex system of plants, herbivores, inquilines, predators, and parasites has become a very useful mesocosm of interacting species for testing an array of ecological questions.

In the 1970s, when I was introduced to *Piper* ant-plants as a potential insect–plant mutualism, studies of mutualistic interactions among species were gaining ground. Questions about mutualism signified a distinct change from the strong focus on competition

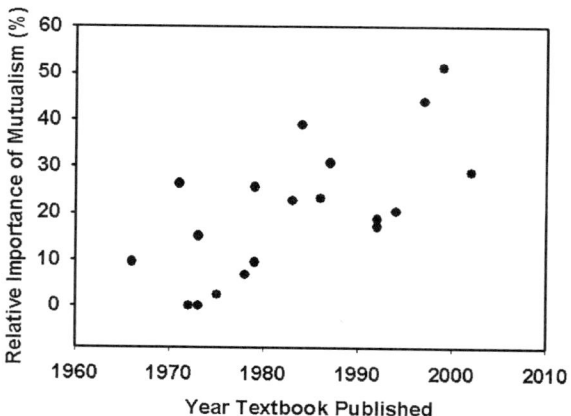

FIGURE 2.1. Significant change over time (GLM simple regression, $r^2 = 0.3916$, $P = 0.042$) in the relative interest in mutualistic interactions vs. competitive interactions, calculated using the number of pages listed in popular textbook indices (Smith 1966; Odum 1971, 1983; Krebs 1972, 1978, 1994; Spurr and Barnes 1973; Ricklefs 1973, 1979; Price, 1975, 1984, 1997; McNaughton and Wolf 1979; Begon et al. 1986; Erlich and Roughgarden 1987; Stiling 1992, 2002; Smith 1992; Speight et al. 1999) as mutualism/(mutualism + competition).

that had dominated the field of ecology for almost two decades (Boucher 1985, Price 1997). The context of changing views about interspecific interaction studies in ecology is reflected in a comparison of the ecology textbooks I used as a student and as an instructor (Fig. 2.1). Whereas the amount of coverage on competitive interactions in biotic communities has not changed over the years in a sample of textbooks used (GLM simple regression, $r^2 = 0.0000$, $P = 0.9968$), the number of pages devoted to mutualistic interactions among species has increased (GLM simple regression, $r^2 = 0.2728$, $P = 0.0218$). In index listings from the ecology textbooks used commonly at the University of California, the relative importance of mutualism, as a topic, compared with competition, shifted from an average of 10% before 1980 to 30% after 1980 (Fig. 2.1).

The study of a potential mutualistic interaction was my initial approach to the *Piper* ant–plant system, whereas my dissertation research at University of California—Berkeley involved biological control in agricultural systems, specifically, the role of plants in facilitating biological control by natural enemies. Standing with one "foot" in an agroecosystem and the other in a rain forest, I could not help but notice the linkages between basic and applied ecology. The potential for *Pheidole* ants to act as plant mutualists rested on their effectiveness as biological control agents against damaging pests to the *Piper* host plant. Before the 1980s, however, trophic ecology featured two-species interactions—not interactions among three species. Price et al. (1980) challenged the ecological community to include the effects of predators on herbivore–plant interactions and the effects of plants on predator–prey interactions. After many years of empirical and theoretical focus on two-species, plant–herbivore or predator–prey interactions, trophic studies began to take on a broader scope, involving at least three trophic levels. Indeed, research on multi–trophic-level interactions was much more prominent after Price's appeal (Price et al. 1980), with a doubling of the number of such articles appearing in the journal *Ecology* in 1986 versus 1976 (Letourneau 1988). Among these were studies of ant mutualists that protect their host plants from attack by herbivores. The *Piper–Pheidole* system became, in this way, a useful mesocosm for investigating not only mutualism as an interaction type, but the multi–trophic-level

MUTUALISM, ANTIHERBIVORE DEFENSE, AND TROPHIC CASCADES

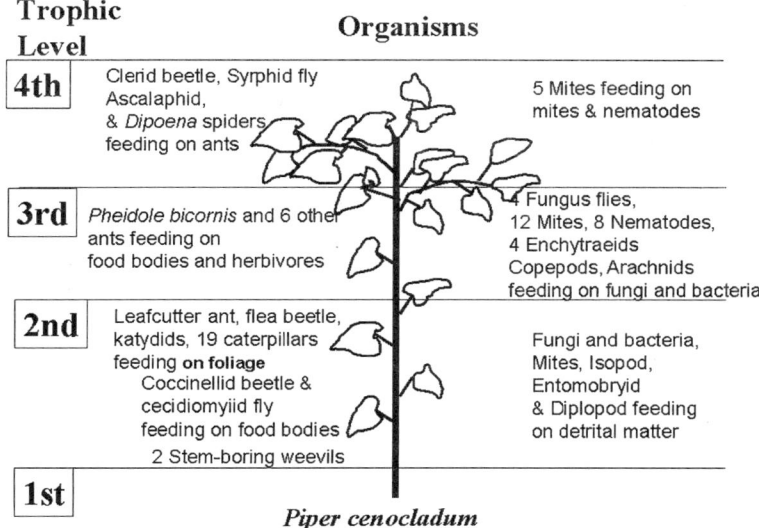

FIGURE 2.2. Sketch of the multi–trophic-level, largely endophytic, arthropod community supported by *Piper cenocladum* shrubs at La Selva Biological Station, Pto. Viejo, Heredia, Costa Rica.

interactions among plants and insects that define the underlying evolutionary mechanisms of ecological costs and benefits.

Because *Piper* ant-plants are highly manipulable, terrestrial systems supporting multiple trophic levels, studies of tri–trophic-level interactions evolved into tests of top–down and bottom–up effects in the context of "trophic cascades." This notion that changes in one trophic level can indirectly affect a nonadjacent trophic level was resurrected in the 1990s from theory originally promulgated in the 1960s and 1970s (Hairston, Smith, and Slobodkin 1960, Fretwell 1977, Power 1992). With multiple species interacting antagonistically or beneficially across four trophic levels, *Piper* ant-plants support a complex arthropod community (Fig. 2.2). Because many of the species are endophytic, this unique *Piper* minicommunity allows for the experimental exclusion or addition of organisms in four trophic levels.

My aim in this chapter is to highlight 25 years of field study on *Piper* ant-plants, using the system as a mesocosm to test emerging ecological hypotheses as the field progressed through three decades of intellectual advances. The term *microcosm*, and sometimes the term *mesocosm*, is used specifically for functional biological systems that are enclosed artificially and removed or isolated from the surrounding environment (Beyers and Odum 1993). Instead, I use the term *mesocosm* in the sense of studies on phytotelmata such as pitcher plants and tree holes (Maguire 1971, Addicott 1974), as systems that are naturally self-contained, but that interact freely with the external, natural environment. The following summary describes *Piper* ant–plant mesocosms in different study sites in Costa Rica, including surrounding plant communities in understory rain forest, three trophic levels of animals supported by living plant tissue, and organisms on three trophic levels supported by detrital resources found inside hollow stems and petiole chambers of these ant-plants.

2.2. STUDY SITES

The majority of my empirical and experimental studies on *Piper* ant-plants have been conducted in two lowland, tropical wet forests (La Selva Biological Station and Corcovado National Park), a tropical moist forest (Carara Biological Reserve), and a tropical premontane wet forest (Hacienda Loma Linda). The La Selva study sites at the Organization for Tropical Studies' La Selva Biological Station, at ~100 m elevation on the Caribbean side of Costa Rica, receive an average of 4,200 mm annual rainfall. Study sites in Corcovado National Park, Osa Peninsula, between the Pacific Ocean and the Golfo Dulce, at <100 m elevation, receive ~3,500 mm annual precipitation. The Carara Biological Reserve sites, on the Pacific coast near Jaco (elevation ~ 15 m), receive ~2,000 mm precipitation annually. Experiments at Hacienda Loma Linda near the Panama border at Agua Buena, Coto Brus, were conducted in a forest at 1,185 m elevation that receives ~3,300 mm annual precipitation. These forest preserves decrease in size from Corcovado (54,039 ha), La Selva (~1,500 ha, with an adjacent 45,899 ha in Braullio Carillo National Park), Carara (4,700 ha), to Loma Linda (~75 ha).

2.3. PLANTS

Piper ant-plants are understory shrubs with a central, nodulated stem from which large lanceolate leaves or thin branches arise (Figs. 2.2 and 2.3). Depending on the species and the location, *Piper* shrubs begin to produce flowers when they are 1–15 m in height and reproduce vegetatively through layering (fallen treelets root adventitiously) or fragmentation (petioles and twigs break off and root) (Greig 1993). The leaf blades can reach 40 cm when fully expanded and remain on the plant for at least 2 years. The stems are hollow when inhabited by *Pheidole bicornis* Forel ant colonies, because minor workers remove the pith. Single-celled, opalescent food bodies, rich in lipids and proteins, are produced on the adaxial side of the sheathing leaf bases (Risch and Rickson 1981). The sheaths serve as hollow petiole-like chambers that house ants (Fig. 2.3). *Piper sagittifolium* C.DC., *Piper obliquum* Ruiz & Pavon, and *Piper fimbriulatum* C.DC. occur on the Pacific slopes of Costa Rica, and the range of *Piper cenocladum* C.DC. extends throughout the Atlantic lowlands. All but *P. sagittifolium* are closely related, as part of the *P. obliquum* complex (Burger 1971, also see Chapter 9).

Piper ant-plants, depending on the species and location in Costa Rica, can be found in patches (especially *P. obliquum* and *P. cenocladum*) or as single shrubs (*P. sagittifolium*), throughout lowland to mid-elevation primary and late-successional secondary forests receiving at least 2,000 mm rainfall per year. *P. cenocladum* is typically found in low-lying sites that occasionally flood, but the other three species are just as common upland. The density of *Piper* ant-plants ranges from less than 10 per ha at Carara Biological Reserve, Puntarenas Province, to over 500 per ha near the Panama border at Agua Buena, Coto Brus Province (Letourneau and Dyer 1998a). The canopy over *Piper* ant-plants is made up of some of the most diverse forest tree, liana, and palm species in the world. Neighboring understory plants are even more diverse, including juveniles of all upper canopy species as well as a plethora of herbs, ferns, vines, shrubs, and understory trees (Dyer and Letourneau 1999a, Letourneau *et al.* 2004). In our study sites at La Selva Biological Station

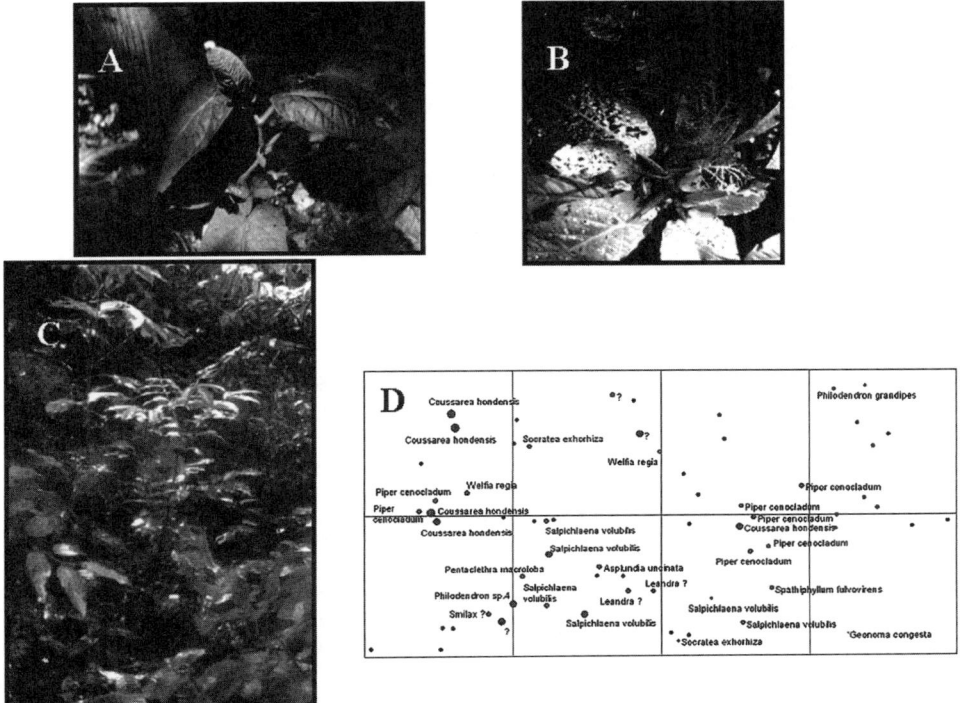

FIGURE 2.3. Level of herbivory found typically on *Piper cenocladum* shrubs with (A) a healthy colony of *Pheidole bicornis* ants and (B) a weakened colony and predaceous clerid beetle present; photograph (C) and GIS map (D) showing understory plant community associated with the ant-shrubs in a 2m × 4m plot at La Selva Biological Station.

in Costa Rica, hundreds of plant species are nearby neighbors, living within several meters of *Piper* ant-plants. Using 2-m by 4-m vegetation survey plots at La Selva Biological Station (Fig. 2.3), we occasionally found zero overlap among the plant species accompanying *P. cenocladum* in any two plots within the same hectare of forest. Overall, we found a total of almost 150 species in 30 of these 8-m^2 plots (Letourneau et al. 2004). Thus, the composition of the local vegetation surrounding these ant-plants is extremely variable.

2.4. HERBIVORES

The herbivores most commonly found feeding on *Piper* ant-plants at our study sites in Costa Rica are lepidopterans, hymenopterans, orthopterans, and coleopterans, including both specialists on the genus *Piper* and generalists that feed on many plant families (Fig. 2.4). Specialist Lepidoptera feeding on *P. cenocladum* foliage at La Selva include geometrids (e.g., *Eois dibapha*, *Eois apyraria*, *Epimecis* sp.) and skippers (e.g., *Quadrus cerealis*). Specialist stem-boring weevils (Coleoptera: Curculionidae: *Ambates* spp.) are common, as are leaf-feeding flea beetles (Coleoptera: Chrysomelidae: *Physimera* spp.)

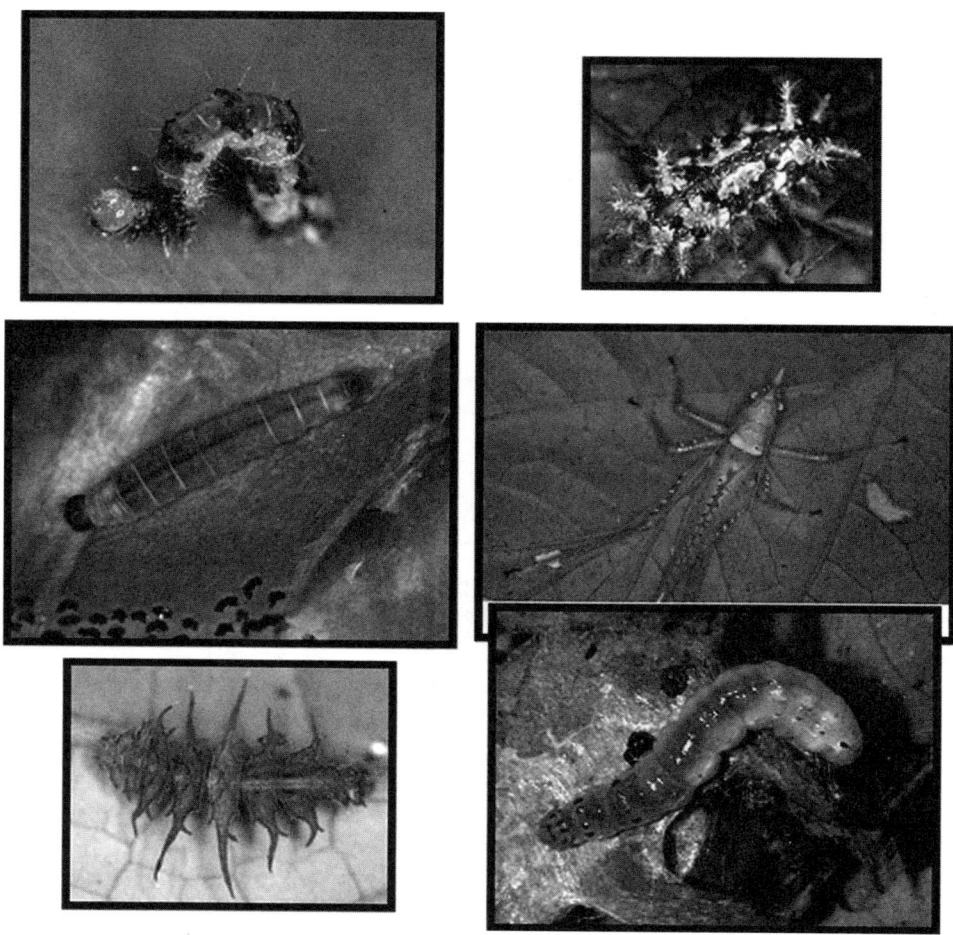

FIGURE 2.4. Examples of specialist and generalist herbivores that exploit the leaves and stems of *P. cenocladum* shrubs.

(Dyer, pers. obs., Marquis 1991, Letourneau, pers. obs.). Generalist lepidopteran folivores include an apatelodid, *Tarchon* sp., a limacodid, *Euclea plugma*, and a saturniid, *Automeris postalbida* (Dyer, Gentry, Garcia, pers. obs.). Feeding damage on foliage from leafcutter ants (Hymenoptera: Formicidae: *Atta* spp.) and orthopterans (primarily Orthoptera: Tettigoniidae) is also common. Standing leaf area lost to arthropods is usually between 5 and 13% for *Piper* ant-shrubs.

2.5. ANT MUTUALISTS

Pheidole bicornis is a small, dimorphic species (Formicidae: Myrmicinae) that occupies all the *Piper* ant-plant species in Costa Rica (Wilson 2003). This ant both stimulates

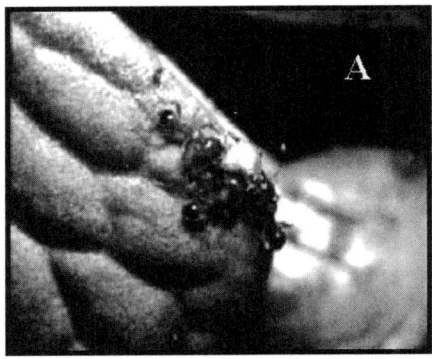
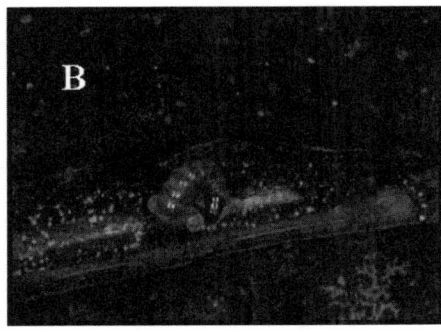

FIGURE 2.5. (A) *Pheidole bicornis* workers, tossing egg bait off the leaf and (B) *Tarsobaenus letourneauae* (= *Phyllobaenus* sp.) beetle larva, which feeds on single-celled food bodies produced on the adaxial surface of the sheathing leaf bases (hollow petiole chambers).

the production of food bodies by the plant and harvests these nutrient-rich, swollen epidermal cells for the brood (Fig. 2.5). The ants feed primarily on food bodies produced by the plant, and carry into the plant interior soft-bodied insects, insect eggs, and spores found on the leaves and flowers (Risch *et al.* 1977; Letourneau 1983; Letourneau 1998; Fischer *et al.* 2002). As omnivorous ants, the colonies kill insect eggs deposited on the plant and small larvae (second–fourth trophic level), harvest plant-produced food bodies (first trophic level), excavate and remove the pith tissue, but do not feed on photosynthetic or vascular plant tissue. We have found 0–7 *Ph. bicornis* queens in a single small *Piper* plant, but most often only one queen is present, accompanied by minor and major workers, brood, and some alates. Occasionally, other species of small ants occur in some of the petiolar cavities of *Piper* ant-plants. Seven species (*Pheidole bicornis*, *Ph. campanae*, *Ph. susannae*, *Ph. specularis*, *Ph. ruida*, *Hypoponera* sp., and *Neostruma myllorhapha*) have been found inhabiting *P. cenocladum* shrubs at La Selva. *Ph. bicornis* occupies 85–95% of *Piper* ant-plants in most forests studied in Costa Rica (Letourneau and Dyer 1998a).

2.6. TOP PREDATORS

Tarsobaenus letourneauae Barr = *Phyllobaenus* sp. (Coleoptera: Cleridae) occur as larvae inside the sheathing leaf bases of *P. cenocladum*, where they feed on both ant brood and food bodies (Letourneau 1990, Fig. 2.5). *Dipoena* sp. spiders (Arachnida: Aranae), including *D. banksii* = *D. schmidti* (Araneae: Theridiidae), specialize on ants, and occur almost everywhere *Piper* ant-plants are found. They position themselves and capture ants at the entrance/exit hole on the outside of the hollow petiole chamber. Different species of congeneric clerids occur in *Piper* ant-plants on the Caribbean and Pacific sides of the Central Cordillera of Costa Rica, and several other species of predatory larvae or nymphs occur occasionally on *Piper* ant-plants, including ant lions (Neuroptera: Ascalaphidae), and syrphids (Diptera: Syrphidae: *Microdon* spp.).

2.7. OTHER ENDOPHYTIC ARTHROPODS, NEMATODES, AND ANNELIDS

Animals in a number of phyla (Arthropoda, Nematoda, and Annelida) inhabit *Piper* ant-plants, along with an unknown array of fungi and bacteria. According to detailed dissections of 77 naturally occurring *P. cenocladum* plants at La Selva Biological Station in Costa Rica, the ant-plants contain a diverse community of endophytic organisms, most of which are species that have not yet been described (Fig. 2.6). Endophagous herbivores feed on stem and food body tissue. The latter include tiny, ubiquitous larval flies (Cecidiomyiidae) and beetle larvae (Coccinellidae). Arthropods that feed on decomposing plant material include collembolans, diplopods, lohmanniid mites, and isopods. Fly larvae

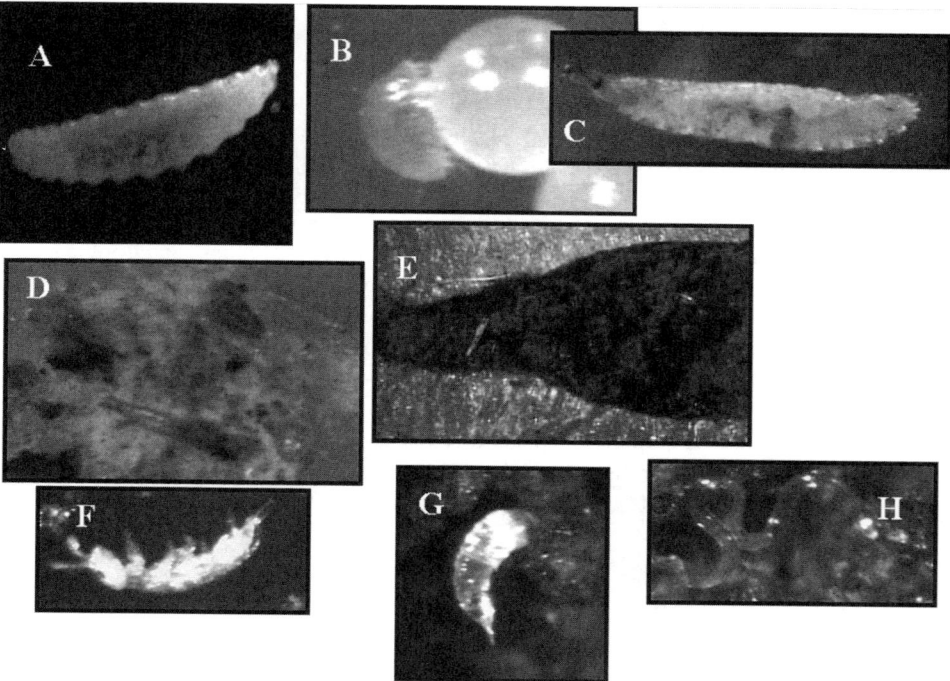

FIGURE 2.6. Endophytic resources and organisms dwelling in the stem and hollow petiole chambers of *P. cenocladum* shrubs (A = midrib-mining fly larva, B = food body–feeding mite, C = food body–feeding fly larva, D = detrital cache from hollow petiole chamber, E = frass stored in base of hollow stem, F = collembolan, G = copapod, and H = nematode).

(species of Cecidiomyiidae, Sciaridae, Mycetophilidae, and Chironomidae), enchytraeid worms, bacteria-feeding nematodes (*Eucephalobus, Rhabditidae, Diploscapter*, and *Plectus*) and fungi-feeding nematodes in the genera *Epidorylaimus, Eudorylaimus, Aporcelaimium*, and *Thonus* were found associated with debris caches and caches of decaying food bodies located in the stems and petioles of ant-shrubs as were a number of mites (Acarina), such as Damaeoidea (= Belboidea), Polyaspidoidea, *Tegeozetes* sp. (Tectocepheidae), *Xylobates* sp. (Haplozetidae), Ascidae, Phthiracaroidea, Euphtheracaroidea, Carabodoidea, Stigmaidae, Archipteriidae, *Podothrombium* sp. (Trombidiidae), and other oribatids. Other mites are likely to be predacious on the nematodes and arthropods in the detrital food web; these include members of Galumnoidea, *Mesostigmata* (= *Gamasida*), *Cosmolaelaps* sp., *Asca* sp. (Ascidae), *Uropodina*, and *Laelaspis* sp. (Hypoaspididae).

Using the best information available on the feeding preferences of the taxa represented by these species, it was possible to divide the endophytic community into different trophic levels within two-component food webs. For a single shrub, approximately two species and 1,083 organisms, on average, were members of the living resources food web (Table 2.1), depending on stem and petiole chamber tissues such as vascular tissue, pith, and food bodies. Primary detrital resources within the plant, including frass from ants and stem borers, discarded plant tissue from the stem, pith, and food bodies cached in the stem, supported from 0 to 82 individuals per shrub. The detrital component food web within *P. cenocladum* comprised an average of three species and approximately 13 individuals per shrub (Table 2.1).

2.8. MUTUALISM EXPERIMENTS

Because of the influence (lifelong it seems) of John H. Vandermeer and Daniel H. Janzen, two dedicated and inspiring faculty advisors at the University of Michigan, my fieldwork on insect–plant interactions has been divided between agricultural and tropical forest ecosystems. My approach, in crop–pest–enemy and forest plant–herbivore–predator interactions, has been to assess the effects of plants, as provisioners of food for natural enemies, on the regulation of plant-feeding insects. As a tightly interacting, symbiotic association, *Piper* ant-plants and their in-house *Pheidole bicornis* predatory ants suggested a possible mutualistic interaction for regulating *Piper* herbivores. Food resources and shelter provided by the plant could restrict ant-foraging bouts to the host plant itself, for maximum plant protection. On the other hand, as suggested by Risch *et al.* (1977), the primary benefit of these ants could, instead, be the provisioning of shrubs with vital nutrients, which enhance plant fitness.

My introduction to *Piper* ant-plants occurred during a rise in interest in mutualism spurred, in part, by a now famous study of an ant–plant interaction in the Neotropics. Janzen's (1966, 1967, 1969) studies on the Bull's Horn *Acacia* and its *Pseudomyrmex* ant-mutualists were catalytic during this transition in emphasis from competition and predation to mutualism, with, for example, five of the eight pages on mutualism in Krebs' (1972) ecology text devoted to this ant-plant example. Disturbance of ant-*Acacia* causes rapid and abundant recruitment of ants that sting and otherwise cause the retreat of most herbivores, perching birds, or even biologists. Ant colonies living in the trees not only protect the plants

TABLE 2.1
Mean Abundance (±SE) per *P. cenocladum* Shrub of the Major Taxonomic/Functional Groups in the Living and Detrital Resources Component Food Webs, Listed by Factorial Treatment (Tmt) Design and by Trophic Level (1st–4th)

Tmt			N	1st	1st	1st	2nd	2nd	2nd/3rd	3rd	4th	2nd	3rd	3rd	3rd	3rd	4th
F	P	L		# Petioles	FBD[1]	Stem Ht[2]	SBs[3]	FBC[4]	*Ph. bicornis*	Ants[5]	TPs[6]	Dec[7]	FF[8]	Enc[9]	Nem[10]	Mites[11]	Mites[12]
N	N	H	10	13(1)	83(12)	99(18)	0.4(0.2)	5.0(1.6)	1075(287)	6(5)	0.0(0)	0.4(0.4)	0.6(0.4)	5.5(2.0)	3.0(1.8)	1.9(0.9)	2.8(1.2)
N	N	L	10	12(1)	87(7)	97(11)	0.3(0.2)	5.1(2.3)	824(138)	6(5)	0.0(0)	0.7(0.5)	0.8(0.5)	3.8(1.8)	3.9(2.6)	4.3(2.5)	5.8(3.5)
N	Y	H	9	15(2)	52(17)	108(15)	1.0(0.3)	1.0(0.5)	185(74)	17(7)	0.3(.3)	0.7(0.5)	0.3(0.2)	0.9(0.5)	0.3(0.2)	0.7(0.3)	0.6(0.2)
N	Y	L	10	12(1)	57(11)	97(10)	0.4(0.2)	0.9(0.5)	209(51)	10(7)	0.0(0)	6.7(2.3)	1.4(1.4)	4.1(2.4)	4.7(2.5)	4.6(2.2)	1.2(0.4)
Y	N	H	10	14(1)	76(6)	96(16)	0.2(0.1)	0.8(0.4)	1214(325)	13(9)	0.0(0)	1.7(0.8)	0.6(0.3)	5.8(2.9)	5.7(2.5)	4.0(1.6)	0.6(0.3)
Y	N	L	11	10(1)	71(6)	81(8)	0.2(0.1)	1.3(0.8)	400(110)	8(6)	0.0(0)	1.3(0.8)	0.0(0.0)	3.3(1.1)	4.5(2.2)	3.7(2.3)	1.0(0.4)
Y	Y	H	8	14(1)	80(19)	115(16)	0.6(0.3)	0.8(0.3)	343(173)	19(7)	0.4(.4)	0.9(0.4)	0.5(0.4)	2.9(1.3)	4.1(3.0)	5.1(3.0)	1.9(1.1)
Y	Y	L	12	9(1)	16(8)	62(10)	0.4(0.2)	0.1(0.1)	33(18)	22(7)	0.1(.1)	0.2(0.2)	0.1(0.1)	0.9(0.5)	0.5(0.2)	0.6(0.3)	1.2(0.5)

Note: N = number of shrubs per treatment combination. F = fertilizer, P = top predator, and L = light. N = no, Y = yes, H = high, L = low.

1–6 Living Resources.

[1] FBD = food body density per 0.5-cm-diameter circle on adaxial side of sheathing leaf base (petiole chambers).

[2] Ht = stem height.

[3] SBs = Stem borers – Coleoptera: Curculionidae: *Ambates* sp. plus other coleopteran borers.

[4] FBC = Food body consumers – Diptera: Cecidiomyiidae and Coleoptera: Coccinellidae.

[5] Ants = Ants other than *Ph. bicornis* Hymenoptera: Formicidae (*Ph. campanae*, *Ph. susannae*, *Ph. specularis*, *Ph. ruida*, *Hypoponera sp.*, and *Neostruma myllorhapha*).

[6] TPs = Top predators – Diptera: Syrphidae: *Microdon* sp. and Coleoptera: Cleridae: *Tarsobaenus letourneauae*.

7–12 Detrital Resources.

[7] Dec = Decomposers (Iohmanniid mites, Entomobrya, Isotoma, other collembolans, Polydesmida, other diplopods, isopods).

[8] FF = Fungus flies – Diptera: Cecidiomyiidae, Sciaridae, Mycetophilidae, Chironomidae.

[9] Enc = Annelida: Oligochaeta: Enchytraeidae.

[10] Nem = Nematoda in the genera: *Eucephalobus*, *Rhabditidae*, *Diploscapter*, *Plectus* (bacterial feeders) and *Epidorylaimus*, *Eudorylaimus*, *Aporcelaimium*, *Thonus* (fungal feeders).

[11] Arthropoda: Acarina: Damaeoidea (= Belboidea), Polyaspidoidea, Tectocepheidae, Haplozetidae, Ascidae, Phthiracaroidea, Euphtheracaroidea, Carabodoidea, Stigmaidae, Archipteriidae, Trombidiidae, other oribatids.

[12] Acarina: Galumnoidea, Mesostigmata (= Gamasida), Ascidae, Uropodina, Hypoaspididae.

from herbivore damage, but from encroaching vines and surrounding vegetation that may act as fuel for fires. Sapling mortality increases substantially when ant colonies are excluded (Janzen 1966, 1967, 1969). Janzen's initial fieldwork prompted subsequent descriptions of ant–plant associations around the world, and experimental tests for a fitness advantage to myrmecophytes through biotic antiherbivore defense.

Risch et al. (1977) published the first empirical study of *Piper* ant-plants. They suspected a mutual fitness advantage for ants and plants in the *Ph. bicornis–P. cenocladum* association, but were not convinced that these small ants, which exhibited much more subtle recruitment and attack behaviors than do ants such as those found in Bull's Horn *Acacia*, were capable of defending their host plant against herbivores. With evidence that the ants stored frass and corpses in the interior base of *P. cenocladum* stems, Risch et al. (1977) suggested that the ants provision the plant with needed nutrients, much as other species of ants provide nutrients to epiphytes in Papua New Guinea (Huxley 1978) and Sarawak (Janzen 1974).

The primary task, then, was to demonstrate experimentally that resident ants provide a fitness advantage to *Piper* plants, such that this association could be classified as a mutualism. Using empirical and experimental tests, I examined two mechanisms by which ants could benefit the understory shrubs: antiherbivore defense and nutrient provisioning. These mechanisms are not mutually exclusive, and field observations provided support for both the antiherbivore hypothesis and the nutrient procurement hypothesis. The following sections report the evidence for a mutualistic association between *Piper* shrubs and *Pheidole* ants, and show that although ants transfer plant nutrients to *Piper* shrubs and defend *Piper* foliage against herbivores, both functions are conditionally beneficial and may complement each other. Other novel means by which ants enhance plant fitness are also described.

2.8.1. *Evidence for Nutrient Procurement by* Pheidole bicornis *Plant-Ants*

Piper ant-plants provide nutrients in the form of lipids and proteins to their resident ants (Risch and Rickson 1981), which are now known to be the major food item for developing colonies (Fischer et al. 2002). A net increase in plant nutrients afforded by their ant-inhabitants, then, would comprise either digestion and excretion of plant nutrients from food bodies (in effect a form of processing) or inputs derived from foraging by ants on other food resources. I investigated the second possibility by tracking ants foraging on *Piper* shrubs, then capturing them for microscopic analyses of their crop contents (Letourneau 1998). Full of fungal and algal spores, minute mycelial and algal fragments, and other particles such as lepidopteran wing scales, ant crop contents eventually end up in frass deposits in the base of the hollowed stem. However, calculating a rough total input of nitrogen and phosphorous based on the average observed frequency of ant foraging bouts, length of bouts, net gain in crop contents, and estimated dry mass and nutrient concentration of spores, I could account for only an extremely small plant fertilizer potential. The estimated 0.1 mg of phylloplane particles per month would yield <5 µg of N and 1 µg of P per shrub. Fischer et al. (2003) used ^{15}N-labeled glycine to demonstrate that *Piper* plants indeed take up nutrients deposited by *Pheidole* workers. Their studies with *P. fimbriulatum* showed that ant provisioning accounted for less than 1% of the shrub's shoot-tissue N. Although these

TABLE 2.2
Comparison of Herbivore Damage (Average Percent Leaf Blade Removed) on Young and Mature *Piper* Ant-Plant Leaves on Shrubs with and without Resident *Pheidole bicornis* Colonies

Forest	*Pheidole* Colonies	Leaf Age	N Leaves	N Shrubs	% Foliage Damage (Mean ± SE)[1]
Corcovado	Present	All	62	6	7.9 ± 0.6a
	Absent	All	47	4	31.9 ± 3.3b
La Selva	Present	Young	11	11	8.0 ± 7.3a
	Excluded	Young	10	10	47.0 ± 12.3b
	Present	All	233	42	12.5 ± 1.7a
	Excluded	All	188	30	27.3 ± 2.6b
Loma Linda	Present	All	234	26	16.5 ± 1.0a
	Excluded	All	255	25	27.0 ± 1.3b

Note: Excluded = Majority of ants in colony killed with dilute pesticide solution (Diazinon ™); absent = found without ants.
[1] Means followed by different letters are significantly different (Corcovado test in 1979 for all leaves on plants >50 cm tall, see Letourneau 1983; La Selva test in 1985 for young leaves only, 1 month: Kruskall-Wallis, $p < 0.05$; La Selva test in 1996–1997 for all leaves, 18 months: ANOVA, $p < 0.05$; Loma Linda, test for all leaves, 22 months, 1986–1987: Repeated measures ANOVA, $p < 0.05$).

nutrients, as well as other ant products such as CO_2, are supplied to the plant, such small amounts of nutrient provisioning are unlikely to improve plant fitness.

2.8.2. *Evidence for Defense against Folivores by* Pheidole bicornis *Plant-Ants*

Empirical evidence of antiherbivore defense was easily demonstrated. First, ants forage more commonly on young, tender leaves of *Piper* ant-plants than on mature or senescent leaves (Letourneau 1983). Second, established *Piper* ant-plants that had lost ant colonies had higher amounts of damage from herbivores than did those with active colonies (Letourneau 1983). Third, forests with higher levels of plant-ant occupancy also had lower standing levels of herbivory, on average, than did forests with low occupancy rates by *Ph. bicornis* (Letourneau and Dyer 1998a). Significantly greater levels of leaf damage were also apparent when ants were excluded in manipulative experiments (Table 2.2). Indeed, the mechanism of ant protection in *Piper* ant-plants was subtle, but effective. The ant's custodial (rather than "aggressive") behavior was responsible. Insect egg baits placed on the foliage were usually found within 60 min. The eggs were either taken into the plant (presumably as food) or dropped off the edge of the leaf onto the forest floor (Letourneau 1983). Janzen (1972) discovered that small *Pachysima* ants (Pseudomyrmicinae) could protect *Barteria* (Passifloraceae) in Nigerian rain forests in a similar manner, as did a number of other ant species in associations investigated subsequently (e.g., McKey 1984, Vansoncelos 1991). Clearly, the absence, removal, or reduction of *Pheidole bicornis* ant colonies in *Piper* ant-plants causes significant increases in folivory (Table 2.2, Letourneau 1983, Dyer and Letourneau 1999b). Gastreich (1999) also found an increase in herbivore damage to leaves when ants avoided leaves occupied by *Dipoena banksii* spiders (see Chapter 6).

2.8.3. Evidence for Higher Fitness in Establishing Fragments via Defense against Folivores

Traditionally, evidence of antiherbivore defense in ant-plant studies is used to justify a fitness advantage to the plant (Huxley and Cutler 1991). Actually demonstrating the impact of folivory on *Piper* ant-plant fitness, however, was not straightforward in the *Piper* system. Just as the defensive behavior of *Piper* plant-ants is subtle compared with the painful stings of *Pseudomyrmex* ants living in *Acacia* trees and *Azteca* ants in *Cecropia* trees, so the fitness advantage for plants with ants has been more difficult to uncover in *Piper* ant-plants. Years of attempts to measure a direct suppression of plant fitness due to leaf tissue removal were thwarted by the fact that loss of leaf tissue simply didn't seem to matter very much to established plants. The application of artificial herbivory on shrubs with ants, via the removal of leaf blade tissue at rates of 25–50% of the total leaf area of the shrub over several years, did not cause a significant decrease in either vegetative growth or flowering rate (Table 2.3). Ant defense of plants against folivory, however, emerged as a major determinant of the survival rate of ramets derived from established shrubs (Dyer and Letourneau 1999b). In 1997, 88 three-leaf "fragments" were clipped from the apex of individual shrubs. Percent herbivory for each leaf on each fragment was determined, before placing fragments in the soil, using a thermoplastic overlay grid and measuring the area of the entire leaf and of missing tissue. Fragments with relatively high herbivory ($>15\%$ on one or more leaves, $n = 46$) had a significantly lower survival rate ($\chi^2 = 18.1$, df $= 1$, $P = 0.001$) after 12 weeks than did fragments with lower levels of herbivory ($<15\%$ on all three leaves, $n = 42$). Establishment and survival of fragments with lower herbivory (71%) was nearly 3 times as likely as for fragments with high levels of leaf damage (26%). Folivory on shoot tips or branches that are destined to become newly establishing plant fragments, then, significantly reduces survival rates compared with fragments without leaf damage. In lowland wet forests, where vegetative reproduction is common, antifolivore defense pays off in terms of increasing the success rate for establishment of new shrubs from broken stems or branches (Dyer and Letourneau 1999b).

TABLE 2.3
No Detectable Difference in Growth Rate (Average Production of New Leaves per Plant) or Flower Production on *Piper* Ant-Shrubs with and without Artificial Herbivory Applied Every 6–8 Months to the Younger Leaves (50% of Half the Leaves), to the Older Leaves (50% of Half the Leaves), or to All the Leaves (50 or 33% Removed on Each Leaf)

Forest Site	% Total Leaf Area Removed (Target Leaves)	Treatment Frequency (Months)	Plants (N)	Total of New Leaves/Plant (Mean ± SE)	Time Period (Months)	Number of New Inflorescences/Plant (Mean ± SE)
La Selva	Control	0	20	12.3 ± 0.8	48	None
	25 (older)	6–8	20	13.9 ± 0.9	48	None
	25 (younger)	6–8	20	12.3 ± 1.0	48	None
	50 (all)	6–8	20	11.3 ± 0.7	48	None
Loma Linda	Control	0	25	10.9 ± 0.8	24	5.72 ± 0.9
	33 (all)	6	24	11.5 ± 0.6	24	6.54 ± 0.8

TABLE 2.4
Comparison of Stem Borer Damage, Stem Death, and Disease Incidence for Mature *Piper sagittifolium* Shrubs at Loma Linda with Ant Colonies Present and Excluded

Ants	Trees (N)	Incidence Weevil Damage/ Tree (Mean ± SE)[1]	No. Dead Stems/ Tree (Mean ± SE)[1]	% Diseased Inflorescences (Mean ± SE)[1]
Control	25	0.04 ± 0.0a	0.13 ± 0.1a	5.0 ± 2.2a
Excluded	25	0.46 ± 0.1b	0.58 ± 0.1b	52.2 ± 13.9b

[1] Means followed by different letters are significantly different (GLM ANOVA, source treatment, error plant nested in treatment, $p < 0.05$).

2.8.4. Evidence for Additional Plant Fitness Advantages Afforded by Ph. Bicornis *Plant-Ants*

Additional evidence for mutualism in the *Piper–Pheidole* association came from ant-exclusion experiments on established *P. sagittifolium*, *P. fimbriulatum*, and *P. obliquum* shrubs. At Loma Linda, these shrubs reproduce primarily by seed. Although folivory was not shown to be a critical factor, ants were found to serve other critical protective functions for their host plants. Seed set was enhanced by *Ph. bicornis* ant mutualists through (1) increased survival of shoots to reproductive maturity through protection from boring beetle larvae, and (2) increased protection of the inflorescences from disease (Table 2.4, Letourneau 1998). The main shoot of 68% of the *Piper* shrubs without ants produced no inflorescences compared with 24% of the control shrubs with active ant colonies. During a 12-month experiment, 25 shrubs with ants produced over 5,800 seeds (87 inflorescences) compared to ca. 100 seeds (37 inflorescences) on ant-exclusion shrubs subject to increased stem death from stem-boring weevils and fungal infections of inflorescences (Letourneau 1998). Thus, the *Piper–Pheidole* association was a useful experimental system that verified a dramatic gain in plant fitness with ant occupancy, while challenging simple conclusions about the role of antifolivore defense. To assess plant investments in ant defense and chemical defense, we focused on *P. cenocladum* shrubs in the lowland wet forest at La Selva, where antifolivore functions are critical because vegetative reproduction through fragmentation is much more common than sexual reproduction.

2.9. TRITROPHIC INTERACTIONS AND ANTIHERBIVORE DEFENSE

The mutualistic *Piper–Pheidole* system has been useful for studying tritrophic interactions involving a plant, its herbivores, and a generalist predator. That is, plant–herbivore interactions cannot be fully understood without inclusion of the predatory role of ants, and predator–prey interactions are strongly determined by the food and refuge provided by plants. By manipulating the presence of ant mutualists, the *Piper–Pheidole* association can be used to clarify how plants allocate material and energy for antiherbivore defense. *Piper* ant-plants have facultative food body production such that these lipid- and protein-rich cells occur in large quantities in petiole chambers only when *Ph. bicornis* ants are present in the plant. The conservation of food bodies when plant-ants are not present suggests that food body production is a costly process, and allows the ideal comparison—that of the same or similar individuals with and without this expenditure. My initial interest in using *Piper* ant-plants (compared with related species without ants or individuals, with ants excluded) to test for dual antiherbivore defense mechanisms, costs of defense, and trade-offs of material and energy was spurred in discussions with David Seigler at the 1986 Gordon Conferences. Over a decade later, at another Gordon Conference (1997), I was able to report the initial results of work by Lee Dyer and Craig Dodson using the tritrophic *Piper–Pheidole* system to explore dual defenses and possible trade-offs between chemical and biotic defenses against herbivores. These initial results will be summarized here, and further developed with additional data and different perspectives by Dyer, Richards, and Dodson (Chapter 7).

Could *Piper* ant-plants have the capability of phytochemical as well as ant defense? If so, what is the nature of these types of defense, what are their relative costs, and how do they interact and affect herbivores? Exploration of possible secondary compounds in *Piper* ant-plant foliage along with a literature review on phytochemically active compounds of Piperaceae, suggested that N-based amides are responsible for chemically based, antiherbivore defense in the genus (reviewed in Chapter 7). With a focus on amides in *P. cenocladum*, we posed several initial hypotheses about plant resource allocation trade-offs for defense: (1) resources should not be allocated maximally to both biotic (ants) and chemical (amides) defenses, (2) the cost of biotic defense should be lower or the effectiveness of biotic defenses should be greater than that of chemical defenses against *P. cenocladum* herbivores, and (3) if dual defenses occur, they should be complementary rather than redundant. Each of these will be discussed in turn.

First, do plants that provide food rewards for predators invest relatively less energy or materials for phytochemical defense against herbivores? Janzen (1973) found that *Cecropia* trees in locations lacking mutualist *Azteca* ants produced no Mullerian bodies, in stark contrast to the ubiquitous food bodies on *Cecropia* trees with resident colonies of stinging *Azteca*. Although plant chemistry was not measured in those plants, Rehr *et al.* (1973) found that ant-Acacia individuals without ant residents had higher levels of cyanogenic glycosides than did individuals with ants. Koptur's (1985) surveys of *Inga* over an elevational gradient extending higher than the range of ant associates, also demonstrated a concomitant increase in the concentration of secondary compounds responsible for plant protection from herbivores where ants were absent. These and more recent studies (Heil *et al.* 2000) are consistent with the hypothesis that, if resources are targeted to maximize fitness (reproductive success), then there should be no redundancy in antiherbivore defense.

To test the first hypothesis, then, that *Piper* ant-plants with functioning biotic defenses should have less-developed chemically based protection, shrubs with active ant colonies were compared with shrubs from which ants had been excluded (Dodson *et al.* 2000). According to general surveys of the compounds found in the foliage of *P. cenocladum*, the likely constituents associated with antiherbivore defense are *Piper* amides. Although present in measurable quantities in all shrubs and fragments tested, we found that *P. cenocladum* shrubs housing *Ph. bicornis* colonies had as much as one third lower concentrations of amides than did unoccupied plants (Dyer *et al.* 2001). Thus, both food bodies and amides are present in low quantities, but can be produced in large quantities, facultatively. The maximum output of chemical defenses does not occur in ant-occupied individuals, and resources can be allocated to compensate, to some degree, for a loss of biotic defense. In most ant-shrubs in most forests, established plants are ant-defended and presumably invest submaximally in the production of amides.

The second hypothesis, that biotic defenses against *P. cenocladum* herbivores should either cost less or be more effective than chemical defenses, is based simply on two observations: that most individuals in most forests house active colonies and that ant defense is absent in most congeneric shrubs. This hypothesis is more difficult to test, because of difficulties in estimating costs to the plant. Costs involve at least the relative availability of the raw materials needed to produce ant nutrient food bodies or amides as herbivore toxins, the metabolic pathways used to synthesize food or toxins, and the relative amounts needed of each defense. Food bodies produced by *Piper* ant-plants are ~5% cell wall and other insoluble constituents, and contain between 40 and 50% lipids by dry weight, 20% protein, and approximately 2% soluble carbohydrate (Fischer *et al.* 2002). Total N content is 4–6% of the dry weight. Plant-available N rather than C may limit food body production by *P. cenocladum*. For plants grown in a shadehouse under different levels of light and fertilizer, food body density was significantly increased by fertilizer ($F_{3,23} = 5.0$, $P = 0.008$) but not by light ($F_{1,23} = 0.0$, $P = 1.0$), and there was no significant interaction ($F_{3,23} = 0.2$, $P = 0.9$) (Dyer and Letourneau 1999b).

If the dry weight of food bodies versus amides is compared, the relative costs of the energy and materials needed for developing and maintaining biotic defense is less than for chemical defense in *Piper* ant-plants. Although both kinds of currency have an active turnover rate, the standing quantity of amides is likely to be much greater than the dry mass of food bodies. Fischer *et al.* (2002) estimated that the dry mass of a 1.5-m *Piper* ant-plant is 32 g for aboveground plant parts, on average, and will contain 115 mg ants, and 3 mg of food bodies. *Piper* amides would have to make up only 0.01% of the dry mass of the plant, to have a total dry mass as low as 3 mg. Given a concentration in *Piper* ant-plant foliage between 0.5 and 2% (see Dyer *et al.* 2001, 2003), such a low level of amides would require that the stem and petioles contain no amides and represent 80–95% of the dry mass of the shoot. However, in the shadehouse, leaf tissue biomass accounted for more than half the dry weight of the aboveground shoot (Dyer and Letourneau 1999b).

To assess the cost of defense accurately, dry weight of product is a necessary, but not sufficient, measure. What are the relative costs of producing, transporting, and storing amides versus housing ants? How costly are the respective biochemical pathways used in synthesizing lipids/proteins versus amides? Does one pathway vary minimally from those used for primary metabolite production and one differ greatly in the "machinery" and processing needed? How does one measure the cost of adaptions such as stem domatia (Mattheck *et al.* 1994, Brouat and McKey 2000)? It is probable that ant defense becomes relatively more costly for the plant than that from any estimate based on dry weight of food bodies alone, but relative assessments of such costs are difficult. It is much like the proverbial apples and oranges to compare the relative cost of, say, pith excavation and a hollow stem with the cost of synthesizing cenocladamide.

How effective is ant defense against folivores versus chemical defense? By comparing leaf tissue loss to herbivores on plants with ants (biotic defenses intact) and without ants (higher levels of amides), we found that amides (and any other compounds that may be involved) were less effective in reducing cumulative, overall herbivore damage on established plants (Table 2.5). That is, at the times and locations tested, the percentage of leaf area removed by herbivores was significantly greater when ants were absent, even though

TABLE 2.5
Comparison of Percent Leaf Area Lost to Herbivores on *Piper cenocladum* Fragments Transplanted in Understory Plots on Ultisols and Grown for Approximately 1.5 Years with Ant Colonies Present, Ant Colonies Excluded with Insecticides, and Predatory Beetles Added After a Single Insecticide Treatment

Symbionts in Established Fragments	N	% Herbivory/Leaf (Mean ± SE)	Area Remaining/Leaf (Mean ± SE) (cm^2)
Pheidole ants[1]	23 plants	14.4 ± 2.7	63.7 ± 14.2
Ants excluded[1]	17 plants	25.8 ± 3.7	50.5 ± 14.4
Pheidole ants[2]	6 plot means	9.7 ± 2.5	46.9 ± 10.6
Tarsobaenus beetle[2]	6 plot means	25.5 ± 5.0	24.9 ± 4.5

[1] Experiment described in Dyer and Letourneau (1999a); data from unfertilized plants.
[2] Experiment described in Letourneau and Dyer (1998b); data from ultisols.

amides were presumably produced at high levels in the foliage of ant-exclusion plants. Different types of herbivore damage seem to be prominent, however, when ants are removed. Whereas specialist lepidopterans were less deterred by high concentrations of amides, orthopterans and leafcutter ants rarely fed on individuals with no ants (and high amide levels) (Dyer *et al.* 2001).

Differences in the relative availability and efficacy of biotic and chemical defenses become relevant when examining the third hypothesis, that dual defenses should be complementary rather than redundant. For example, seedlings of *Piper* ant-plants can grow very slowly, taking a year or more to become large enough to house an ant colony. During this vulnerable period of development, phytochemical compounds may be a critical source of defense. Also, some patches or populations of ant-plants have low ant occupancy (Letourneau and Dyer 1998a), so that biotic defense is compromised. We also know that the *Piper–Pheidole* mutualism is not based solely on antifolivore defense (see Section 2.8), but instead on ants' playing multiple roles in enhancing plant fitness. True redundancy of biotic and abiotic defenses would require that secondary compounds actively defend against flower smut and stem-boring insects and ants deter feeding by adult beetles and orthopterans. Although we have some evidence that *Piper* ant-plant amides suppress fungal growth, and that other *Piper* spp. contain neolignins with antifungal activity (reviewed in Chapter 7), such defenses did not compensate overall for the absence of ant mutualists in field experiments (Tables 2.2 and 2.5). Stem damage to plants by the weevil larvae *Ambates melanops* Champion and *A. scutiger* Champion was also severe when ants were absent (and amides were likely to be at high concentration, at least in the foliage) (Table 2.4). In contrast to the effective disruption of herbivores as eggs or soft-bodied larvae, field observations suggest that large, sclerotized insect herbivores are either invisible or impervious to the custodial activities of the small *Ph. bicornis* plant-ants. In a field comparison using a mixed sample of ant-plants (*P. sagittifolium*, *P. fimbriulatum*, and *P. obliquum*) randomly designated as ant-exclusion plants or ant-occupied controls, ants were more effective at reducing folivory by geometrid larvae than by adult flea beetles (Letourneau 1998). Dyer *et al.* (2001) provide other examples of differential effects of ants on herbivores. The evidence, then, that phytochemical-based protection cannot fully compensate for ant defenses (Table 2.5), and that ants are absent from small plants and ineffective against a whole suite of herbivores, suggests that the two forms of defense (biotic and chemical) are complementary rather than redundant.

2.10. TROPHIC CASCADES

Research on mutualism, antiherbivore defense, and multitrophic interactions, in general, grew from theoretical and experimental roots in the 1960s, and expanded through the next two decades. Likewise, the early "seeds" of trophic cascades theory were "sown" in the 1960s with papers by Hairston *et al.* (1960) and Paine (1966). However, experimental tests of trophic cascades gained prominence later, at the end of the 1980s in aquatic systems (Brett and Goldman 1996). Alternating effects of top predators on the biomass or abundance of successively lower trophic levels in aquatic systems led to two questions: (1) are trophic cascades most pronounced, or indeed, restricted to aquatic (or relatively species-poor) communities (Strong 1992), and (2) are top–down forces stronger determinants of community structure than are bottom–up, resource-driven effects (Power 1992)? The former question was addressed in the 1990s as more investigations in terrestrial and species-rich communities were undertaken. Letourneau and Dyer (1998b) suggested that relatively greater logistical difficulties and a paucity of manipulative studies in terrestrial systems could create the impression that trophic cascades were more prominent in aquatic systems than in terrestrial communities. Conflicting conclusions derived from meta-analyses of trophic cascades (incidence and/or strength) in terrestrial versus aquatic systems by Schmitz *et al.* (2000), Halaj and Wise (2001), and Shurin *et al.* (2002) leave us with inconclusive results and some controversy. Certainly trophic cascades are prominent in aquatic systems. However, trophic cascades also have been shown to occur in terrestrial systems. The strength of terrestrial cascades varies and overlaps the range of interaction strengths demonstrated in aquatic communities. The second dichotomy, comparable to the issues of resource allocation to chemical versus biotic defense, but at the scale of communities and ecosystems, required concomitant manipulations of both predators and resources. Early encouragement from Donald Strong and Mary Power prompted me to try some experiments using the *Piper–Pheidole* mesocosm to assess effects of top–down and bottom–up forces in this complex, terrestrial community.

My field experiments with Lee Dyer in the 1990s (the "Cadenas" project at La Selva) were focused on the following theoretical models of herbivore regulation at the scale of communities: (1) top–down trophic cascades in which consumers regulate their food, (2) the thermodynamics model in which herbivores are regulated by plant biomass, and (3) the green desert model in which herbivores are regulated through plant nutritional quality and/or secondary compounds. Hairston *et al.*'s (1960) proposition concerning top–down mechanisms for herbivore regulation ("Why the world is green?") (Polis 1994, Persson 1999, Pace *et al.* 1999), supposed that predators and plants are resource-limited whereas herbivores are consumer-limited. Fretwell (1977) and Oksanen *et al.* (1981) pointed out that systems with four trophic levels should contain resource-limited even-numbered trophic levels and consumer-limited odd-numbered trophic levels.

The green desert and thermodynamic models are both based on bottom–up forces of plant quality as mediators of herbivore regulation. Following the basic laws of thermodynamics, that energy is lost as it is transferred up the trophic chain, the biomass of herbivores, and primary and secondary carnivores, attenuates and is dependent on total primary productivity (Lindeman 1942, Slobodkin 1960). The green desert hypothesis, an alternative bottom–up hypothesis, focuses on resource limitation as the factor determining community structure (Murdoch 1966, White 1978, Menge 1992, Moen et al. 1993). That is, plant nutrients and/or plant chemistry regulates herbivores by restricting the utilization of most plant parts: either they cannot digest the most common plant macromolecules (e.g., cellulose; Abe and Higashi 1991) or they are averse to the toxic secondary metabolites (e.g., Murdoch 1966, White 1978).

2.10.1. Experimental Test I: Do Trophic Cascades Operate on the Four Trophic Level System Associated with P. cenocladum Ant-Plants?

Traditionally, ant–plant associations have been used as study systems to test for mutualism, but those studies also represent some of the earliest empirical work on three–trophic-level interactions (Letourneau 1991). Recent studies of ant–plant systems have revealed patterns and complexities beyond those associated directly with mutualism, making them generally useful as experimental systems (McKey 1988, Davidson and Fisher 1991). Our initial experiments used the *P. cenocladum* mesocosm to test the relative roles of top–down (e.g., predation or herbivory) and bottom–up forces (e.g., level of primary productivity or availability of light or nutrients) on net plant productivity (see Hunter and Price 1992, Power 1992, Strong 1992, Carpenter and Kitchell 1993, Carpenter et al. 1990, Spiller and Schoener 1990, Williams and Ruckelshaus 1993, Polis and Winemiller 1996). Our experiments with *Piper* ant-plants in species-rich, tropical wet forest contrasted with the simpler terrestrial and aquatic habitats of variable complexity in which trophic cascades had been shown to occur.

To address the three models of top–down and bottom–up effects, we used factorial experiments to manipulate top-predator abundance under different conditions of plant resource availability. This design enabled us to consider variation in top–down forces as a function of bottom–up heterogeneity (*sensu* Oksanen et al. 1981 and Hunter and Price 1992). In 1994, three hundred sixty *P. cenocladum* cuttings in 36 plots were assigned to a factorial design: predator treatment (three levels), light condition (two levels), and soil quality (two levels). We monitored indicators of ant colony size (percent petioles occupied/tree), herbivore loads (leaf area loss), and tree biomass (total leaf area/tree) over a period of months.

Our results showed that when the top predator beetle was added experimentally to the three–trophic-level ant–plant system, the average abundance of *Pheidole bicornis* ants was reduced fivefold, average herbivory to *Piper cenocladum* leaves was increased nearly threefold, and total leaf area was reduced by nearly half (Letourneau and Dyer 1998b). The direct effect of predatory beetles on ants was more pronounced and more rapid than were indirect effects accumulating to the second and first trophic levels. Although light and soil quality tended to influence herbivory levels, neither potential leaf area accumulation by trees nor actual leaf area (left after herbivory) showed a direct response to the availability of soil nutrients or light. Thus, our initial experiments showed that the effects of top–down trophic cascades were present and strong in this species-rich mesocosm. The results also supported the hypotheses that indirect effects mediated by intervening trophic levels would have weaker impacts than direct effects on the respective target populations and that indirect effects would be delayed in comparison to direct effects (Lawton 1989). The direct effect of predatory beetles on ants occurred within a few months. Indirect effects accumulated to the second trophic level within a year; and significant effects of top predators on the first trophic level required over a year.

In contrast to the strong effects of top predators added to *Piper cenocladum* plots, a resource-driven, bottom–up "cascade" was not evident. Rain forest soils of generally higher nutrient content (primary forest plots) did not result in significantly greater potential or actual leaf area (remaining after herbivory) than did poorer soils (secondary forest plots). Although light levels in the forest affected the amount of herbivory the plant received, neither potential leaf area accumulation nor actual leaf area were significantly influenced by light. Although we located our plots on two soil types to create two levels for the nutrient treatments, the forests associated with these two soil types had different plant community composition, and possibly other relevant differences (Krebs *et al.* 1995). In a separate experiment, conducted simultaneously on the poor soils, we manipulated soil fertilizer level to test directly the effects of enhanced plant resources on plants. Higher trophic levels in this model community experiment also showed strong top–down cascades rather than bottom–up cascades (Dyer and Letourneau 1999b). The discovery of three amides produced by *P. cenocladum* and the truncation of effects past the direct, positive effects of nutrients on plants supported the green desert version of bottom–up regulation rather than the thermodynamic hypothesis of trophic cascades.

2.10.2. *Experimental Test II: Can Top–Down and Bottom–Up Forces Affect Animal Diversity in the Endophytic Community of* P. Cenocladum *Ant-Plants?*

A dramatic increase in terrestrial studies in the 1990s (Persson 1999) shifted the debate from the primacy of top–down forces versus bottom–up forces to a series of hypotheses about how these forces may *interact* to structure communities (Oksanen 1991, Leibold 1996, Schmitz 1992), *vary* over space, time, and taxa (Power 2000), *act differentially* on components of complex food webs (Persson 1999, Polis and Strong 1996), and *maintain* heterogeneity and biodiversity (Hunter and Price 1992, Terborgh 1992) in complex ecosystems. Using trophic groups (trophic levels within component food webs based on living or detrital resources) we measured direct effects of both top predators and plant resources on biodiversity of endophytic animals in *P. cenocladum* ant-shrubs (Dyer and Letourneau 2003).

In 1995, eighty established *P. cenocladum* shrubs on infertile ultisols in a Costa Rican wet forest were assigned three treatments in a factorial design: high versus low light environment, fertilizer additions versus no fertilizer, and introduction of top predators versus no top predator. Hollow stems and petiole chambers of all shrubs were initially inhabited by *Pheidole bicornis* ant colonies. After 15 months, we dissected 77 surviving *P. cenocladum* shrubs and collected the entire endophytic fauna (Arthropoda, Annelida, and Nematoda) to determine the number of invertebrate species and their relative abundance for each shrub. More than 50 species of animals were among the 43,188 individuals found living within

TABLE 2.6
Multivariate Analysis of Variance and Profile Analysis for the Effects of Top–Down (Top Predator Treatment) and Bottom–Up (Fertilizer × Light) Forces on Community-Level Parameters—Richness, Abundance, and Diversity (H)—of the Endophytic Invertebrates Inhabiting *Piper cenocladum*. The Profile Analysis Statistically Compares the Response of Abundance vs. Richness

Factor	Wilks' Lambda	F Value, df $= 1, 69$	P Value
MANOVA			
Fertizer (Fert)	0.97	2.21	0.1418
Predator (Pred)	0.73	25.49	**<0.0001**
Fert × Pred	0.99	0.03	0.8678
Light	0.96	3.03	0.0861
Fert × Light	0.81	16.30	**0.0001**
Pred × Light	0.99	0.05	0.8219
Pred × Fert × Light	0.96	2.75	0.1018
		F Value, df $= 2, 68$	P Value
PROFILE			
Fertizer (Fert)		0.98	0.3254
Predator (Pred)		38.11	**<0.0001**
Light		1.29	0.2598
Fert × Light		3.06	0.0845

the petiole chambers and stems of the harvested shrubs. *Pheidole bicornis* ants accounted for as few as 0 and as many as 2,842 individuals per shrub. Other ants numbered over 1,000 individuals, usually inhabiting plants or portions of plants not occupied by *Ph. bicornis*. The remaining 969 individuals were annelids, nematodes, and other arthropods, including crustaceans, collembolans, dipterans, coleopterans, and at least 17 mite species (Table 2.1).

Biodiversity, as described by species richness (S), abundance, and diversity (H') was significantly affected by both top-predator addition and by the bottom–up interaction of fertilizer and light (Table 2.6). Shrubs with top-predator additions had 3 times the animal diversity, on average, of shrubs without additions of the predatory clerid beetle ($H' = 0.48 \pm 0.04$ versus $H' = 0.16 \pm 0.07$). The addition of top predators increased diversity of secondary and primary consumers supported by living plant tissue, whereas balanced plant resources (light and nutrients) increased the diversity of primary through tertiary consumers in the detrital resources food web. Thus, the endophytic community was structured by both top–down and bottom–up forces, such that predators *and* plant resources were determinants of biodiversity in this terrestrial system.

Ant species richness was higher when shrubs contained top predators ($\chi = 0.28 \pm 0.08$ SE other-ant-species (without top predator additions) versus $\chi = 0.57 \pm 0.11$ SE other-ant-species (with top predators added): $\chi^2 = 14.05$, $P = 0.0002$) because of the colonization of ants other than the dominant species *Ph. bicornis*. The average species richness of stem borers per shrub was also greater in shrubs treated with top predators ($\chi = 0.22 \pm 0.07$ SE) than with no top predator additions ($\chi = 0.08 \pm 0.04$ SE). No significant effect of fertilizer, light, or significant interaction among those factors was detected, even when the food web was reanalyzed placing all consumers of food bodies, including omnivorous ants, on the second trophic level (ANOVA, $F_{7,69} = 1.55$, $P = 0.1643$).

In contrast to the top–down effects on the living-plant-tissue–based food resource web, the diversity of organisms on all three trophic levels of the detrital-based community was affected by an interaction between light availability and fertilizer (Table 2.6). Diversity (H') was higher under conditions of balanced plant resources compared with unbalanced light and fertilizer conditions (Dyer and Letourneau 2003). In plants growing under conditions of skewed carbon, nitrogen ratios simply supported a lower biodiversity of organisms associated with dead tissue, frass, and the bacteria and fungi associated with those resources. The mean augmentation of animal diversity for shrubs with high light and high nutrients or low light and low nutrients was, for the second, third, and fourth trophic levels, respectively, 4.3, 1.7, and 1.9 times that in shrubs with unbalanced resources (low light and high fertilizer or high light and no fertilizer). These results suggest an attenuated series of indirect, bottom–up effects of light and fertilizer plant resources over four trophic levels in the detrital component food web.

2.10.3. Experimental Test III: Can Indirect Effects of Top Predators Extend to Other Plants in the Understory Community of Piper Ant-Plants?

In natural stands of *P. cenocladum*, we compared amounts of folivory on plants that harbored clerid beetle ant-predators with plants that did not contain these beetle larvae. Beetle-containing shrubs were located by examining all individuals found within 40 m of all the trails at La Selva where *P. cenocladum* occurs (approximately 30 km of trails). To test if herbivore damage was greater on shrubs containing top predators than on shrubs without beetles otherwise exposed to the same conditions, we categorized damage (high = at least one leaf with >50% leaf area removed, or low = all leaves with less than 50% damage) on *P. cenocladum* shrubs containing beetles ($n = 145$), on their nearest-neighbor conspecific without beetles ($n = 145$), and a plant in the next nearest patch without a beetle ($n = 145$). The distance from each shrub with a beetle larva to its nearest neighbor (with ants and without a beetle) was measured. To test the additional hypothesis that herbivore damage is reduced as distance increases from the beetle-containing plant, we used logistic regression, with herbivory levels as the dependent variable and distance from the beetle-containing plant as a predictor variable.

We found that *P. cenocladum* shrubs found containing *Tarsobaenus* beetles had the highest frequency of severe herbivore damage (61% of 145 shrubs), followed by their nearest neighbors (41% of 145 shrubs had high herbivory) and then plants from patches without beetles (21% of 145 shrubs with high herbivory) ($\chi^2 = 12.78$, df $= 2$, $P = 0.0017$). In addition, distance from a beetle-containing shrub was a significant predictor of herbivory levels on locally occurring *P. cenocladum* ($\chi^2 = 11.18$, df $= 1$, $P = 0.0008$), with *P. cenocladum* shrubs near beetle-containing individuals suffering higher levels of herbivory than those further away. Thus, the effects of beetles shown in our initial experiments were not restricted to the manipulated host plant. Indirect effects also cascaded to nearby individuals of the same species (Dyer and Letourneau 1999a).

To test for an association between herbivory rates on *Piper* ant-plants and survival rates of understory plants of other species, we monitored all understory plants between 20 cm and 2 m in height in sixteen 2-m × 4-m survey plots (8 with *P. cenocladum* fragments to which top predator beetles were added, and 8 in sites at least 75 m from beetle additions). Individual plants in the understory plots were tagged in 1995, identified, and followed for nearly 2 years. Presence or absence of each individual and the maximum level of herbivory (the proportion of leaf lost on the most damaged leaf of each plant was estimated visually) were recorded before beetles were applied to treatment plants, and at 8, 12, 16, 20, and 24 months after beetles were added to plants surrounding half the survey plots. Survivorship was monitored for all plants tagged in the initial sample. To test if herbivore damage can determine survivorship of established understory plants, the frequency of plants that lived for 2 years versus those that died during the experiment was compared for plants with high initial herbivory (at least one leaf with >50% leaf tissue lost) and low initial herbivory (0–5%) using Fisher's Exact test. Finally, we compared herbivore damage levels (frequency of high and low damage) after 1 year on understory plants in survey plots within the *P. cenocladum* beetle-addition plots with damage on understory plants in survey plots away from beetle-addition plots, also using Fisher's Exact test (Dyer and Letourneau 1999a).

If the survival or reproductive success of *P. cenocladum* in the forest understory affects other plant species through competitive or positive interactions, then effects of top predators potentially extend to the plant community through plant–plant interactions. Insect–plant interactions within the understory are also possible: given the high diversity of the wet forest understory (113 species of plants in 47 families were identified in understory survey plots), other species may also serve as hosts to the herbivores released by the presence of top predators. Indeed, most of the lepidopteran larvae that eat *P. cenocladum* are generalists on plants in the Piperaceae (geometrids, limacodids, and hesperiids) as well as plants in several other families (apatelodids). Thus, the effect of beetles could potentially cascade to the local plant community via species that are alternative hosts of the generalist herbivores. Our survey data supported this notion, with 53% of understory plants exhibiting high herbivory in plots with beetle additions versus 25% of understory plants in areas with no top predators added. Overall, we did not find a significant effect of initially high herbivory rates on the survivorship of understory plants (53% of 87 plants with high herbivory died compared with 47% of 76 plants with low herbivory; Fisher's Exact test, $P > 0.05$). However, herbivore damage may be a strong determinant of survival for seedlings or young plants. When only plants less than 60 cm tall were used in the analysis, 60% of plants ($n = 40$) with high herbivore damage died compared with 40% of those with low herbivory ($n = 41$) (Fisher's Exact test, $P = 0.07$).

In conclusion, top–down trophic cascades occurred in a species-rich, terrestrial community with four trophic levels, despite predictions that such cascades cannot occur in such complex systems. A specialist top predator caused cascading effects through three trophic levels, resulting in fewer predators, greater herbivory, and smaller plants with less leaf area. These effects cascade to other plants of the same species in the understory vegetation, and possibly to different species of plants that share the same herbivores. Such a four–trophic-level cascade can ultimately alter community variables such as plant species richness (Letourneau et al., 2004).

2.11. CONCLUSIONS

Over almost three decades the *Piper–Pheidole* interaction and associated community has served as a useful, manipulatable system for mesocosm experiments on a range of critical questions in ecology and evolution. In this chapter, I have highlighted contributions made with field and lab tests within a historical context of emerging themes in the science of ecological interactions among plants and animals. Certainly, component communities such as the one living on, and in, *Piper* ant-shrubs are commonplace in tropical rain forests—subtly existing with little external bravado, but slowly unfolding with painstaking scientific examination to expose intricate interrelationships among myriad species. This community exhibits complex mutualistic interactions between plants and ants (Fig. 2.3); parasitic, competitive, and predaceous actions of a clerid beetle (Fig. 2.5); specialist predation by spiders, neuropterans, and dipterans; herbivory in every imaginable form on all plant parts; and hosts a species-rich endophytic community of detritivores (Fig. 2.6). It has demonstrated several new ways in which ants protect plants, a capacity for facultative protection through ants, chemistry, and both types of defenses, and broadly cascading effects through four functional trophic levels with interaction strengths as strong as those found in aquatic systems.

Although many questions have been answered for the *Piper–Pheidole* system, future research directions comprise an impressive array of possibilities. This mesocosm may make possible an examination of actual costs of defenses and a benefit assessment of trade-offs because of its ability to facultatively change its levels of investment in different avenues of defense. The issue of dissolution of mutualisms is an obvious study for this system, because the clerid beetle disrupts ants, and in some (rare) locations, is more common than ant colonies. And, what is the role of parasitism in maintaining the mutualism as a stable interaction? Theoretically, an important regulating factor can be a third species that interacts with a tightly coevolved pair of mutualists. In this case, the clerid beetle may act as an "aprovechado" (*sensu* Boucher 1985) that takes advantage of the mutualistic products and services, but the role of beetles as parasites versus predators has yet to be explored. The discovery of Dyer *et al.* (2000) of another beetle–*Piper* interaction calls into question which relationship came first? Ants and plants? Or beetles and plants? More extensive exploration of *Piper*–arthropod relationships, phylogenetic relationships among plants, beetles, and ants, as well as analyses of gland formation in *Piper* spp. will help to clarify the evolutionary history of these interactions. In essence, does the *Piper–Pheidole* system represent a shift from exploitation to mutualism, or from mutualism to exploitation? Or is it simply a case where mutualism can arise out of a parasitism between two other species? Finally, what are the dynamics of trophic cascades over time? For example, do bottom–up forces build a system to a plateau (extensive patches or populations of plants supporting multiple trophic levels), with top–down forces then becoming dominant until the trends are reversed and plants become sparse? This model of resilience in complex communities has not been demonstrated, but may simply be a matter of long-term research and tracking of such dynamics at different scales. We are in the process of investigating these and other questions, including the construction of a model to assess the population dynamics of *Piper* ant-shrubs and their major herbivores. Though subject to system-specific peculiarities, long-term studies of specific-component communities (such as Whitham's or Price's long-term studies of willows and arthropods, Barbosa's work on tritrophic communities on eastern trees, Langenheim's *Hymanea courbaril* studies, Strong's above- and belowground lupine community, or Schoener and Spiller's tropical island mesocosms) as model systems have served to deepen our understanding of the intricacies of multitrophic, ecological interactions, generating new lines of inquiry, and advancing the field in novel directions.

2.12. ACKNOWLEDGMENTS

I thank Lee Dyer for sharing my enthusiasm for the *Piper* ant-plant mesocosm, and for his brilliant research contributions on this promising system. The Organization for Tropical Studies, National Park System of Costa Rica, and Darryl Cole allowed the use of their facilities and gave valuable logistical help. *Piper* ant-plant research was supported by NSF (DEB-9318543 and DEB-0074806), the Academic Senate and Divisional Research funds of the University of California, University Research Expeditions Program (UREP), Tulane University faculty research support, and Earthwatch grants. G. Vega C., A. Barberena, F. Arias G., D. Arias G., R. Gomez M., L. Gomez M., L. Chavez M., R. Krach, R. DiGaudio, T. Haff, H. Kloeppl, J. Sorensen, S. Klass, A. Shelton, C. Squassoni, A. Lewis, K. Harvey, W. Williams, and volunteers from UREP and Earthwatch provided field and lab assistance. Identifications for endophytic animals were provided by M. Paniagua G. (mites), J. Longino (ants) (ALAS) and Soil FoodWebs, Inc. (arthropods, annelids, and nematodes), and we thank V. Behan-Pelletier, R. Colwell, A. Moldenke, R. Norton, and H. Schatz for consultations on the feeding biology of the organisms in the detrital food web.

REFERENCES

Abe, T., and Higashi, M. (1991). Cellulose centered perspective on terrestrial community structure. *Oikos* 60:127–133.

Addicott, J. F. (1974). Predation and prey community structure: An experimental study of the effect of mosquito larvae on the protozoan communities of pitcher plants. *Ecology* 55:475–492.

Begon, M., Harper, J. L., and C. R. Townsend (1986). *Ecology: Individuals, Populations and Commnities*. Sinauer Associates, Sunderland, MA.

Beyers, R. J., and Odum, H. T. (1993). *Ecological Microcosms*. Springer-Verlag, New York, p. 557.

Boucher, D. H. (ed.). (1985). *The Biology of Mutualism*. Oxford University Press, New York.

Brett, M. T., and Goldman, C. R. (1996). A meta-analysis of the freshwater trophic cascade. *Proceedings of the National Academy of Sciences* 93:7723–7726.

Brouat, C., and McKey, D. (2000). Origin of caulinary ant domatia and the timing of their onset in plant ontogeny: Evolution of a key trait in horizontally transmitted ant-plant symbioses. *Biological Journal of the Linnean Society* 71:8012–819.

Burger, W. (1971). Flora Costaricensis. *Fieldiana Botany* 35:1–227.

Carpenter, S. R., and Kitchell, J. F. (1993). *The Trophic Cascade in Lakes*. Cambridge University Press, New York.

Carpenter, S. C., Frost, T. M., Kitchell, J. F., Kratz, T. K., Schindler, D. W., Shearer, J., Sprules, W. G., Vanni, M. J., and Zimmerman, A. P. (1990). Patterns of primary production and herbivory in 25 North American lake ecosystems. In: Cole, J., Findlay, S., and Lovett, G. (eds.), *Comparative Analyses of Ecosystems: Patterns, Mechanisms, and Theories*. Springer-Verlag, New York, pp. 67–96.

Davidson, D. W., and Fisher, B. L. (1991). Symbiosis of ants with *Cecropia* as a function of the light regime. In: Huxley, C. R., and Cutler, D. K. (eds.), *Ant–Plant Interactions*. Oxford University Press, Oxford, pp. 289–309.

Dodson, C., Dyer, L. A., Searcy, J., Wright, Z., and Letourneau, D. K. (2000). Cenocladamide, a dihydropyridone alkaloid from *Piper cenocladum*. *Phytochemistry (Oxford)* 53(1):51–54.

Dyer, L. A., and Letourneau, D. K. (1999a). Trophic cascades in a complex, terrestrial community. *Proceedings of the National Academy of Sciences* 96:5072–5076.

Dyer, L. A., and Letourneau, D. K. (1999b). Relative strengths of top–down and bottom–up forces in a tropical forest community. *Oecologia* 119:265–274.

Dyer, L. A., and Letourneau, D. (2003). Top–down and bottom–up diversity cascades in detrital vs. living food webs. *Ecology Letters* 6:60–68.

Dyer, L. A., Letourneau, D. K., Williams, W., and Dodson, C. (2000). A commensalism between *Piper marginatum* Jacq. (Piperaceae) and a coccinellid beetle at Barro Colorado Island, Panama. *Journal of Tropical Ecology* 15:841–846.

Dyer, L. A., Dodson, C. D., Beihoffer, J., and Letourneau, D. K. (2001). Trade-offs in antiherbivore defenses in *Piper cenocladum*: Ant mutualists versus plant secondary metabolites. *Journal of Chemical Ecology* 27:581–592.

Dyer, L. A., Dodson, C. D., Stireman, J. O., Tobler, M. A., Smilanich, A. M., Fincher, R. M., and Letourneau, D. K. (2003). Synergistic effects of three Piper amides on generalist and specialist herbivores. *Journal of Chemical Ecology* 29: 2499–2514.

Erlich, P. R., and J. Roughgarden (1987). *Ecology*. Macmillan, New York.

Fischer, R. C., Richter, A., Wanek, W., and Mayer, V. (2002). Plants feed ants: Food bodies of myrmecophytic *Piper* and their significance for the interaction with *Pheidole bicornis* ants. *Oecologia* 133:186–192.

Fischer, R. C., Richter, A., Wanek, W., Richter, A., and Mayer, V. (2003). Do ants feed plants? A 15N labelling study of nitrogen fluxes from ants to plants in the mutualism of *Pheidole* and *Piper*. *Journal of Ecology* 91:126–134.

Fretwell, S. D. (1977). The regulation of plant communities by food chains exploiting them. *Perspectives in Biology and Medicine* 20:169–185.

Gastreich, K. R. (1999). Trait-mediated indirect effects of a theridiid spider on an ant–plant mutualism. *Ecology* 80:1066–1070.

Greig, N. (1993). Predispersal seed predation on five *Piper* species in a tropical rainforest. *Oecologia* 93:412–420.

Hairston, N. G., Smith, F. E., and Slobodkin, L. B. (1960). Community structure, population control, and competition. *American Naturalist* 94:421–424.

Halaj, J., and Wise, D. H. (2001). Terrestrial trophic cascades: How much do they trickle? *American Naturalist* 157:262–281.

Heil, M., Staehelin, C., and McKey, D. (2000). Low chitinase activity in *Acacia* myrmecophytes: A potential trade-off between biotic and chemical defences? *Naturwissenschaften* 87:555–558.

Hunter, M. D., and Price, P. W. (1992). Playing chutes and ladders: Heterogeneity and the relative roles of bottom–up and top–down forces in natural communities. *Ecology* 73:724–732.

Huxley, C. R. (1978). Ant-plants *Myrmecodia* and *Hydnophytum* (Rubiaceae), and relationships between their morphology, ant occupants, physiology and ecology. *New Phytologist* 80:231–245.

Huxley, C. R., and Cutler, D. K. (eds.). (1991). *Ant–Plant Interactions*. Oxford University Press, Oxford.

Janzen, D. H. (1966). Coevolution of mutualism between ants and acacias in Central America. *Evolution* 20:249–275.

Janzen, D. H. (1967). Interaction of the Bull's-Horn acacia (*Acacia cornigera*) with an ant inhabitant (*Pseudomyrmex ferruginea*) in eastern Mexico. *Kansas University Science Bulletin* 47:315–558.

Janzen, D. H. (1969). Birds and ant × Acacia interaction in Central America, with notes on birds and other myrmecophytes. *Condor* 71:240–256.

Janzen, D. H. (1972). Protection of *Barteria* (Passifloraceae) by *Pachysima* ants (Pseudomyrmecinae) in a Nigerian rain forest. *Ecology* 53:885–892.

Janzen, D. H. (1973). Dissolution of mutualism between *Cecropia* and its *Azteca* ants. *Biotropica* 5:15–28.

Janzen, D. H. (1974). Epiphytic myrmecophytes in Sarawak Indonesia: Mutualism through the feeding of plants by ants. *Biotropica* 6:237–259.

Koptur, S. (1985). Alternative defenses against herbivores in Inga (Fabaceae: Mimosoideae) over an elevational gradient. *Ecology* 66:1639–1650.

Krebs, C. J. (1972). *Ecology: The Experimental Analysis of Distribution and Abundance*. Harper & Row, New York.

Krebs, C. J. (1978). *Ecology: The Experimental Analysis of Distribution and Abundance*, 2nd ed. Harper & Row, New York.

Krebs, C. J. (1994). *Ecology: The Experimental Analysis of Distribution and Abundance*, 4th Ed. Harper & Row, New York.

Krebs, C. J., Boutin, S., Boonstra, R., Sinclair, A. R. E., Smith, J. N. M., Dale, M. R. T., Martin, K., and Turkington, R. (1995). Impact of food and predation on the snowshoe hare cycle. *Science* 269:1112–1115.

Lawton, J. H. (1989). Food webs. In: Cherrett, L. M. (ed.), *Ecological Concepts*. Blackwell, Oxford, pp. 43–78.

Leibold, M. A. (1996). A graphical model of keystone predators in food webs: Trophic regulation of abundance, incidence, and diversity patterns in communities. *American Naturalist* 147:784–812.

Letourneau, D. K. (1983). Passive aggression: An alternative hypothesis for the *Piper–Pheidole* association. *Oecologia (Berlin)* 60:122–126.

Letourneau, D. K. (1988). Conceptual framework of three-trophic-level interactions. In: Barbosa, P., and Letourneau, D. K. (eds.), *Novel Aspect of Insect–Plant Interactions*. Wily and Sons, Inc., New York, pp. 1–9.

Letourneau, D. K. (1990). Code of ant–plant mutualism broken by parasite. *Science* 248:215–217.

Letourneau, D. K. (1991). Parasitism of ant–plant mutualisms and the novel case of *Piper*. In: Huxley, C. R., and Cutler, D. K. (eds.), *Ant–Plant Interactions*. Oxford University Press, Oxford, pp. 390–396.

Letourneau, D. K. (1998). Ants, stem-borers, and fungal pathogens: Experimental tests of a fitness advantage in *Piper* ant-plants. *Ecology* 79:593–603.

Letourneau, D. K., and Dyer, L. A. (1998a). Density patterns of *Piper* ant-plants and associated arthropods: Top predator trophic cascades in a terrestrial system? *Biotropica* 30:162–169.
Letourneau, D. K., and Dyer, L. A. (1998b). Experimental test in lowland tropical forest shows top–down effects through four trophic levels. *Ecology* 79:1678–1687.
Letourneau, D. K., Dyer, L. A., and Vega, C. G. (2004). Indirect effects of top predator on rain forest understory plant community. *Ecology* (in press).
Lindeman, R. L. (1942). The trophic–dynamic aspect of ecology. *Ecology* 23:399–418.
Maguire, B., Jr. (1971). Phytotelmata: Biota and community structure determination in plant-held waters. *Annual Review of Ecological Systems* 2:439–464.
Marquis, R. J. (1991). Herbivore fauna of *Piper* (Piperaceae) in a Costa Rican wet forest: Diversity, specificity, and impact. In: Price, P. W., Lewinsohn, T. M., Fernandes, G. W., and Benson, W. W. (eds.), *Plant–Animal Interactions: Evolutionary Ecology in Tropical and Temperate Regions*. John Wiley & Sons, Inc., New York, pp. 179–208.
Mattheck, C., Bethge, K., and West, P. W. (1994). Breakage of hollow tree stems. *Trees* 9:47–50.
McKey, D. (1984). Interaction of the ant-plant *Leonardoxa africana* (Caesalpiniaceae) with its obligate inhabitants in a rainforest in Cameroon. *Biotropica* 16:81–99.
McKey, D. (1988). Promising new directions in the study of ant–plant mutualisms. In: Greuter, W., and Zimmer, B. (eds.), *Proceedings of the XIV International Botanical Congress*. Konigstein/Taunus, Koeltz, pp. 335–355.
McNaughton, S. J., and Wolf, L. L. (1979). *General Ecology*, 2nd ed. Holt, Rinehart and Winston, New York.
Menge, B. A. (1992). Community regulation: Under what conditions are bottom–up factors important on rocky shores. *Ecology* 73:755–765.
Moen, J., Garfjell, H., Oksanen, L., Ericson, L., and Ekerholm, P. (1993). Grazing by food-limited microtine rodents on a productive experimental plant community: Does the "green desert" exist? *Oikos* 68:401–413.
Murdoch, W. W. (1966). Community structure, population control, and competition. *American Naturalist* 100:219–226.
Odum, E. P. (1971). *Fundamentals of Ecology*, 3rd Ed. Saunders, Philadelphia, PA.
Odum, E. P. (1983). *Basic Ecology*. Saunders College Publishing, Philadelphia, PA.
Oksanen, L. (1991). Trophic levels and trophic dynamics: A consensus emerging? *Trends in Ecology and Evolution* 6:58–60.
Oksanen, L., Fretwell, S., Arruda, J., and Niemela, P. (1981). Exploitation ecosystems in gradients of primary productivity. *American Naturalist* 118:240–261.
Pace, M. L., Cole, J. J., Carpenter, S. R., and Kitchell, J. F. (1999). Trophic cascades revealed in diverse ecosystems. *Tree* 14:483–487.
Paine, R. T. (1966). Food web complexity and species diversity. *American Naturalist* 100: 65–76.
Persson, L. (1999). Trophic cascades: Abiding heterogeneity and the trophic level concept at the end of the road. *Oikos* 85:385–397.
Polis, G. A. (1994). Food webs, trophic cascades and community structure. *Australian Journal of Ecology* 19:121–136.
Polis, G. A., and Strong, D. R. (1996). Food web complexity and community dynamics. *American Naturalist* 147:813–846.
Polis, G. A., and Winemiller, K. O. (eds.). (1996). *Food Webs: Integration of Patterns and Dynamics*. Chapman & Hall, New York.
Power, M. E. (1992). Top–down and bottom–up forces in food webs: Do plants have primacy? *Ecology* 73: 733–746.
Power, M. E. (2000). What enables trophic cascades? Commentary on Polis et al. *Trends in Ecology and Evolution* 15:443–444.
Price, P. W. (1975). *Insect Ecology*. Wiley, New York.
Price, P. W. (1984). *Insect Ecology*, 2nd Ed. Wiley, New York.
Price, P. W. (1997). *Insect Ecology*, 3rd ed. John Wiley & Sons, New York.
Price, P. W., Bouton, C. E., Gross, P., McPheron, B. A., Thompson, H. N., and Weis, A. E. (1980). Interactions among three trophic levels: Influence of plants on interactions between insect herbivores and natural enemies. *Annual Review of Ecological Systems* 11:1141–1165.
Rehr, S. S., Feeny P. P., and Janzen, D. H. (1973). Chemical defense in Central American nonant *Acacias*. *Journal of Animal Ecology* 42:405–416.
Ricklefs, R. E. (1973). *Ecology*. Chiron Press, Newton, MA.
Ricklefs, R. E. (1979). *Ecology*, 2nd ed. Chiron Press, Portland, OR.
Risch, S., McClure, M., Vandermeer, J., and Waltz, S. (1977). Mutualism between three species of tropical *Piper* (Piperaceae) and their ant inhabitants. *American Midlands Naturalist* 98:433–444.
Risch, S. J., and Rickson, F. R. (1981). Mutualism in which ants must be present before plants produce food bodies. *Nature* 291:149–150.

Schmitz, O. J. (1992). Exploitation in model food chains with mechanistic consumer–resource dynamics. *Theoretical Population Biology* 41:161–181.

Schmitz, O. J., Hamback, P. A., and Beckerman, A. P. (2000). Trophic cascades in terrestrial systems: A review of the effects of carnivore removals on plants. *American Naturalist* 155:141–153.

Shurin, J. B., Borer, E. T., Seabloom, E. W., Kurt, A., Blanchette, C. A., Broitman, B., Cooper, S. C., and Halpern, B. S. (2002). A cross-ecosystem comparison of the strength of trophic cascades. *Ecology Letters* 5:785–791.

Slobodkin, L. B. (1960). Ecological energy relationships at the population level. *American Naturalist* 94:213–236.

Smith, R. L. (1966). *Ecology and Field Biology*. Harper & Row, New York.

Smith, R. L. (1992). *Elements of Ecology*, 3rd Ed. Harper-Collins, New York.

Speight, M. R., Hunter, M. D., and Watt, A. D. (1999). *Ecology of Insects: Concepts and Applications*. Blackwell Science, Oxford, UK.

Spiller, D. A., and Schoener, T. W. (1990). A terrestrial field experiment showing the impact of eliminating top predators in foliage damage. *Nature* 347:469–472.

Spurr, S. H., and Barnes, B. V. (1973). *Forest Ecology*, 2nd Ed. Ronald Press Co., New York, NY.

Stiling, P. D. (1992). *Introductory Ecology*. Prentice-Hall, Upper Saddle River, NJ.

Stiling, P. D. (2002). *Ecology: Theories and Applications*. Prentice-Hall, Upper Saddle River, NJ.

Strong, D. R. (1992). Are trophic cascades all wet? Differentiation and donor control in speciose ecosystems. *Ecology* 73:747–754.

Terborgh, J. (1992). Maintenance of diversity in tropical forests. *Biotropica* 24:283–292.

Vasconcelos, H. L. (1991). Mutualism between *Maieta guianensis* Augl. a myrmecophytic melastome, and one of its ant inhabitants: Ant protection against insect herbivores. *Oecologia* 87:295–298.

White, T. C. R. (1978). The importance of a relative shortage of food in animal ecology. *Oecologia* 33:71–86.

Williams, S. L., and Ruckelshaus, M. H. (1993). Effects of nitrogen availability and herbivory on eelgrass (*Zostera marina*) and epiphytes. *Ecology* 74(3):904–918.

Wilson, E. O. (2003). *Pheidole in the New World: A Dominant, Hyperdiverse Ant Genus*. Harvard University Press, Cambridge, MA.

3
Pollination Ecology and Resource Partitioning in Neotropical *Pipers*

Rodolfo Antônio de Figueiredo[1] *and Marlies Sazima*[2]
[1] *Centro Universitário Central Paulista, 13563-470, São Carlos, São Paulo, Brazil*

[2] *Universidade Estadual de Campinas, Campinas, São Paulo, Brazil*

3.1. INTRODUCTION

Interspecific and intraspecific competition between plants is mediated by factors such as resources, pollinators, dispersers, herbivores, or microbial symbionts (Goldberg 1990). Many biotic communities are competitively structured, where interspecific competition has been an important force in determining species composition, relative abundance, and resource use patterns (Turkington and Mehrhoff 1990). Thus, the species assemblages of communities are a result of physical factors, opportunities for colonization, disturbance, historical biotic conditions, predators, pathogens, parasites, and mutualists (Clay 1990, Louda *et al.* 1990, Hunter *et al.* 1992). Darwin (1859) proposed that evolution via natural selection is a consequence of these community-level interactions, and his studies on pollination systems helped him strengthen his theory (e.g., Darwin 1876).

Plants compete among one another for abiotic and biotic resources essential to their growth and reproduction, including water, light, minerals, pollinators, and seed dispersers. Competition for biotic resources can be the proximate cause of the patterns of coexistence of different plant species (Giller 1984). Competition and coexistence could lead to resource partitioning, which in turn reduces niche overlap and the magnitude of competition between species (Silvertown and Lovett Doust 1993). Competing species with a high degree of overlap in resource use may, over evolutionary time, diverge in their morphological, behavioral, or physiological character states, resulting in character displacement and stability (Lawlor and Maynard Smith 1976, Giller 1984, Bazzaz 1990).

Studies on patterns of resource partitioning in coexisting, potentially competing taxa provide information on the dissimilarity of coexisting species. The supporting data on resource competition come from studies on sympatric, congeneric species (Schoener 1975).

Such studies usually focus on animal groups, since plants are intractable because of their sedentary nature, circumscribed interactions, plasticity, and essentially similar resource requirements (Silander and Pacala 1990). Nevertheless, different species of the same plant family, which are genetically similar and show similar features of flower morphology, such as members of the genus *Piper*, could be suitable for speculative studies of character displacement. The hypothesis of such studies is that these species originated in allopatry, and that features that are similar in an allopatric state differ in sympatry (Giller 1984).

There are approximately 283 *Piper* species in Brazil (Yuncker 1972, 1973), and several species grow sympatrically in both disturbed and undisturbed locations. These species constitute an interesting model on which to test ecological hypotheses. Because of their abundance in clearings and in the understory (Ingram and Nadkarni 1993, Kotchetkoff-Henriques and Joly 1994), their key mutualism with frugivorous bats (Herbst 1986, Findley 1993, Bizerril and Raw 1997; see Chapter 4), their role as a food resource for birds (Palmeirim *et al.* 1989, Loiselle 1990), and their importance as medicinal sources for traditional human societies (Leitão-Filho 1995a, Milliken and Albert 1996), pipers are important ecosystem components.

To date, little ecological information exists on the pollination of Neotropical *Piper* species (Semple 1974, Fleming 1985, de Figueiredo and Sazima 2000), and few published studies have focused on their coexistence in a given area (e.g., Fleming 1985). In our study on pipers growing in sympatry in a Brazilian forest, we hypothesized that they would show a complex pollination system, in which wind and animals would be pollen carriers, and that they would differ to some extent in at least one of the niche axes analyzed. The objectives of this chapter are to present this study and current information about the pollination of pipers, to speculate about resource partitioning among sympatric Neotropical *Piper* species in a Brazilian forest, and to discuss both aspects in light of evolutionary and conservation ecology.

3.2. POLLINATION AND RESOURCE PARTITIONING IN *Piper*

The modern economic importance of the pepper family (Piperaceae) is related to such chemical components as amides and safrole, which are used in medicine and industry (e.g., Abe *et al.* 2001, Ang-Lee *et al.* 2001; Chapter 7). The Paleotropical species *Piper nigrum* produces peppercorns that have long been among the most important spices globally. In addition, some current studies reveal pipers as one of the fundamental components of tropical forests (Galindo-Gonzalez *et al.* 2000, Hartemink 2001). In spite of their importance, only a few studies have been carried out on the floral biology of the pepper family.

Our study of sympatric Brazilian pipers addresses the hypothesis that the coexistence of congeners results in character displacement and resource partitioning. This theme was addressed by Fleming (1985), who demonstrated resource partitioning between sympatric pipers in Costa Rica. The plant characters examined in Brazil included growth habits, habitat utilization, vegetative reproduction, reproductive phenology, flower visitors, and pollination of each species. Differences in one or more of these characters could result in reduced competition with congeners, allowing coexistence of multiple species in a single area.

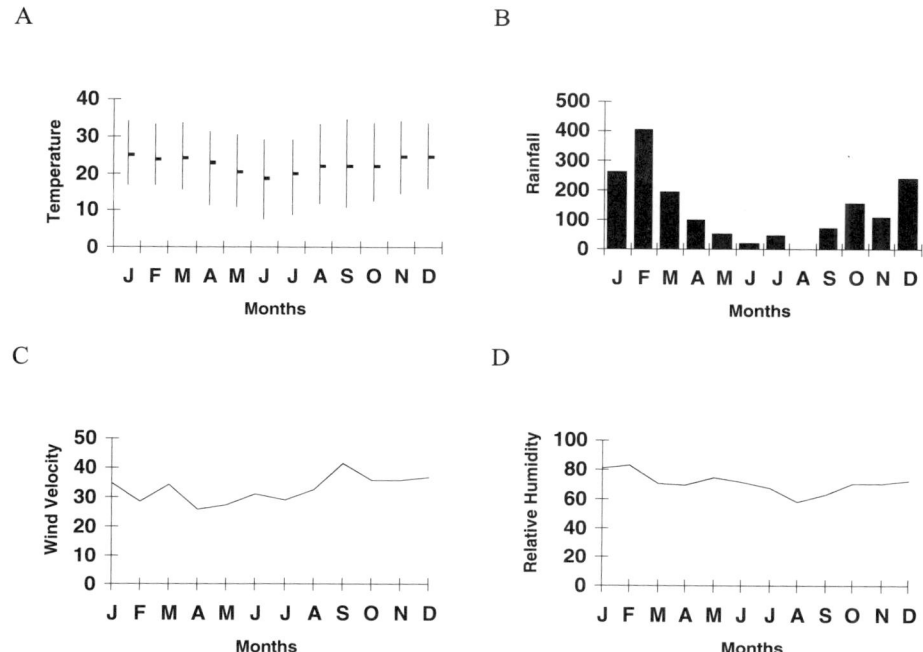

FIGURE 3.1. Climate variables recorded at Santa Genebra Reserve, southeastern Brazil, in 1995: (A) Maximum, minimum, and average temperature (°C); (B) Rainfall (mm); (C) Wind velocity (km/h); and (D) Relative humidity (%). Data from Feagri/Unicamp.

3.2.1. Study Site and Species of the Brazilian Study

The data presented here were collected between January and December 1995 in the Santa Genebra Reserve (22°49'S, 47°07'W, 670 m alt.), a 251.77-ha area of semideciduous forest surrounded by crop fields and human dwellings. This area consists of two forest subtypes: the semideciduous forest *sensu stricto*, which corresponds to 85% of the reserve, and the hygrophilous or swampy forest, which covers ca. 15% of the total area (Leitão-Filho 1995b). Five percent of the semideciduous forest burned in 1981; this area is still undergoing a natural regeneration process (Leitão-Filho 1995b). Figure 3.1 shows the climate conditions during the study period in the Santa Genebra Reserve. Seasonality in this area is pronounced, with a rainy season between October and March and a dry season between April and September (de A. Mello *et al.* 1994).

Data were collected on marked plants along transects on the edge and through the interior of both the semideciduous and swampy forests (Goldsmith and Harrison 1976). Twelve *Piper* species were recorded (Table 3.1), and identified with the help of the literature (Yuncker 1972, 1973, 1975). Because some authors have suggested that *Ottonia* is actually a subclade of *Piper* (Tebbs 1989a, 1993; Chapter 10), species within this subclade were included within the total number of pipers used in this study.

TABLE 3.1
Habits and Habitats of the *Piper* Species Studied in Southeastern Brazil

Species	Density[1]	Relative Frequency	Habitat[2]	Habit	Fidelity[3]
Piper aduncum	4.40	39.12	1,2,a,b	Shrub	Indifferent
P. amalago	1.76	15.68	1,2,a,b	Shrub	Indifferent
P. arboreum	0.34	03.04	1,2,a,b	Tree	Selective
P. crassinervium	0.51	04.50	1,2,a,b	Shrub	Selective
P. gaudichaudianum	0.28	02.51	2,b	Shrub	Exclusive
P. glabratum	0.58	05.18	2,a,b	Shrub	Selective
P. macedoi	0.11	01.00	2,a,b	Shrub	Selective
P. mikanianum	0.14	01.22	1,a,b	Herb	Selective
P. mollicomum	0.45	04.03	1,2,a,b	Shrub	Selective
P. regnelli	0.63	05.63	2,a,b	Shrub	Selective
P. martiana	0.16	01.45	2,b	Shrub	Exclusive
P. propinqua	0.15	01.37	1,a,b	Shrub	Selective

[1] Individuals/hectare.
[2] Habitat: 1 – semideciduous forest; 2 – swampy forest; a – edge; b – inside.
[3] For establishment of the degrees of fidelity was considered the number of individuals recorded in each habitat.

3.2.2. *Habit and Habitat Utilization*

As several studies have already shown (Gomez-Pompa 1971, Chazdon and Field 1987, Chazdon *et al.* 1988, Tebbs 1989a,b, Williams *et al.* 1989, Sanchez-Coronado *et al.* 1990), pipers have very different habits and habitats. Habitat diversification could diminish competition among pipers by resulting in the use of different pollination and dispersal agents and by inducing some differences in nutrient requirements. Fleming (1985) suggested that pipers restricted to the interior of humid forests develop self-compatibility as a character displacement to compensate for the diminished number of visitors. Some studies on other plant species have also suggested the evolution of self-fertilization because of selective pressures related to a lower number of pollinators (Levin 1972, Solbrig and Rollins 1977, Inoue *et al.* 1995).

As in Central American forests, most Brazilian pipers are shrubs, with high densities of each species found in a given habitat. The study area in Brazil was characterized by four different habitats:

a) Edge of the semideciduous forest: a dry site with direct sunlight for several hours a day
b) Inside the semideciduous forest: a dry, shadowed site, beginning 10 m from the forest edge
c) Edge of the swampy forest: a constantly wet site, receiving direct sunlight
d) Inside the swampy forest: a wet, shadowed site, beginning 10 m from the forest edge

We surveyed 3,200 m^2 on the edge and 910 m^2 in the interior of the semideciduous forest, as well as 800 m^2 on the edge and 1,120 m^2 in the interior of the swampy forest. Transects inside the forest were randomly selected on a map and surveyed for pipers. Within 1-m × 1-m square plots we calculated density (number of individuals of each species in a plot), relative frequency (number of occurrences of a given species as compared to the number of occurrences of all species), and habitat fidelity (degree of restriction of a given species to a specific type of habitat, which may be exclusive, selective, preferential, or indifferent) of each *Piper* species (Braun-Blanquet 1936, Goldsmith and Harrison 1976).

Because of their growth habits, the herbaceous *Piper mikanianum* and the arboreous *P. arboreum* were studied separately from the other pipers, which are shrubs (Table 3.1). Regarding habitat, *P. mikanianum* solely grows both at the edges and inside of the semideciduous forest subtype, whereas *Piper arboreum* was found in all habitat types, although most of its individuals occurred in the semideciduous forest (Table 3.1). Five species, *P. gaudichaudianum, P. glabratum, P. macedoi, P. regnelli,* and *P. martiana,* are habitat-selective and only grow in the swampy areas (Table 3.1). They form a guild separate from *P. mikanianum* and *P. propinqua,* which occur exclusively in the semideciduous habitat, and from the other pipers, which could be found in both the semideciduous and swampy forests (Table 3.1). *Piper aduncum* and *P. amalago* were shown as indifferent to habitat, because their individuals are equally distributed among them (Table 3.1).

3.2.3. Vegetative Reproduction

Asexual reproduction allows the clonal growth of the plant, which is an advantage both for occupying spaces where soil resources are in patches and for invading and displacing competitors. Silvertown and Lovett Doust (1993) hypothesized that clonal growth would be found where selection favors early reproduction, such as in disturbed habitats.

Gartner (1989) first demonstrated that sympatric pipers reproduce vegetatively. She showed that pipers are morphologically well adapted to survive in a habitat with a high incidence of breakage due to falling debris. In addition, pipers rapidly regrow after damage, which helps them persist in changing sites. In another study at the same Costa Rican site, Greig (1993a) divided the pipers observed into two groups: the light-demanding, pioneer species, whose main mode of reproduction is seed production, and the shade-tolerant, understory species, which are able to reproduce vegetatively.

We verified incidences of vegetative reproduction of the Brazilian *Piper* species through an analysis of subterranean plant parts. The absence of a pivotal root and/or the presence of traces of rhizomes were indicative of a vegetative origin. Vegetative reproduction is also known to occur through sprouting on broken branches, and when roots emerge from fragmented or prostrate twigs (cf. Gartner 1989, Greig 1993a). We found four modes of vegetative propagation in pipers: branch repositioning, prostration of branches on the soil, fragmentation of branches, and fragmentation of rhizomes. No vegetative reproduction was registered for *P. amalago, P. crassinervium,* or *P. mollicomum.*

Branch repositioning occurs when branches arise from the base of the piper after injuries to their main branches. These repositioned branches are slimmer than the original ones, present a greater length between nodes, and have no secondary ramifications. They grow faster and taller than the original branches. This kind of vegetative propagation is found in *P. aduncum, P. arboreum, P. gaudichaudianum, P. glabratum, P. macedoi, P. mikanianum, P. regnelli, P. martiana,* and *P. propinqua.*

Branches prostrated on soil produce roots and secondary branches at the nodes. The original branch can rot away, resulting in independent individuals. This type of vegetative reproduction occurs in *P. arboreum*, *P. mikanianum*, *P. martiana*, and *P. propinqua*. Fragments of branches on soil may develop secondary branches and roots at the nodes, and establish as independent individuals. This type of vegetative reproduction is found in *P. glabratum*, *P. mikanianum*, *P. martiana*, and *P. propinqua*. Finally, rhizomes and rooted stolons may give rise to new individuals after their fragmentation. This type of asexual reproduction occurs in *P. mikanianum*, *P. regnelli*, *P. martiana*, and *P. propinqua*.

We identified three guilds for our Brazilian pipers based upon characteristics related to vegetative reproduction. The first one comprises pipers that show no vegetative reproduction: *P. amalago*, *P. crassinervium*, and *P. mollicomum*. The first two of these species are light-demanding whereas the latter is a shade-tolerant one. The second guild is represented by *P. mikanianum*, *P. martiana*, and *P. propinqua*, all of which are shade-tolerant species. These three species show all four aforementioned modes of asexual reproduction. The third guild is composed of shade-tolerant and light-demanding pipers that show only one or two vegetative reproduction modes.

Vegetative propagation is an important mode of reproduction in herbs and shrubs growing in both temperate (Bierzychudek 1982) and tropical forests (Gartner 1989, Kinsman 1990). Greig (1993a) found that repositioning branches and rooting after branch prostration were common mechanisms of vegetative reproduction among the Central American pipers.

3.2.4. Reproductive Phenology

Flowering and fruiting phenologies affect plant success and integrate selective pressures from both abiotic and biotic factors (van Schaik *et al*. 1993). Gentry (1974) ascribed one of the mechanisms for maintenance of high species diversity in tropical plant communities to variations in flowering phenology. Variability in phenological patterns decreases the effects of competition among sympatric plants (Borchert 1983, Newstrom *et al*. 1994). Webber and Gottsberger (1999) demonstrated phenology displacement in six species within the same genus in Amazonia; this phenology displacement allowed for a better partitioning of pollinators. Wright and Calderon (1995), however, showed that the flowering times were similar among sympatric congeners in Panama.

The phenological displacement hypothesis was tested on the Brazilian sympatric pipers. The few studies about the piper phenology that were previously carried out in Central and South America (Opler *et al*. 1980, Fleming 1985, Marquis 1988, Marinho-Filho 1991, Wright 1991) found that pipers flower mainly in the dry season. The phenology of each Brazilian *Piper* species was studied at the population level, on five randomly chosen adult individuals, between January and December 1995. The phenophases were defined as follows:

1) Inflorescence formation: individuals showing inflorescences with flower buds
2) Flowering: individuals showing inflorescences with anthesed flowers
3) Fruiting: individuals showing infructescences

Flowering strategies at the specific level were defined (modified from Morellato 1991, Newstrom *et al*. 1994) as follows:

1) Continual: at least one individual of a given species incessantly flowers for the entire year (admitting only one single interval of 2 months without flowers in anthesis)
2) Episodic: individuals flower in several intervals of the year, each one separated by months without flowers
3) Seasonal: individuals flower once a year, for at least 4 months.

Continual flowering occurred in *P. aduncum*, *P. amalago*, *P. gaudichaudianum*, *P. glabratum*, and *P. macedoi*, which presented flowers throughout the year. Episodic flowering was found in *P. arboreum* and *P. crassinervium*, which flower simultaneously. Seasonal flowering was observed in *P. mikanianum*, *P. mollicomum*, *P. regnelli*, *P. martiana*, and *P. propinqua*, which flower at the end of the dry season (Fig. 3.2(A)).

Pipers found in different habitats differed in their phenological patterns. *Piper* species from the semideciduous forest presented a seasonal pattern, whereas those indifferent to habitat had a continual flowering pattern. Among the pipers with continual flowering, *P. amalago*, *P. glabratum*, and *P. macedoi* are insect pollinated; *P. gaudichaudianum* is wind pollinated; and *P. aduncum* is both wind and insect pollinated. The episodic *P. arboreum* is

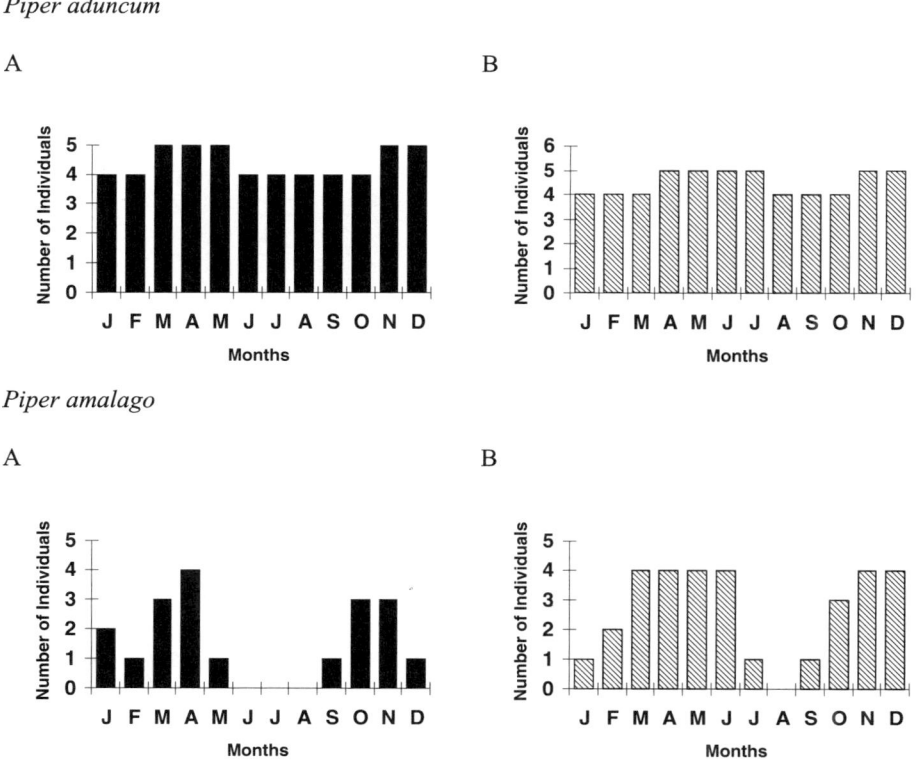

FIGURE 3.2. Flowering and fruiting phenology of *Piper* species in 1995: (A) Number of flowering individuals; (B) Number of fruiting individuals.

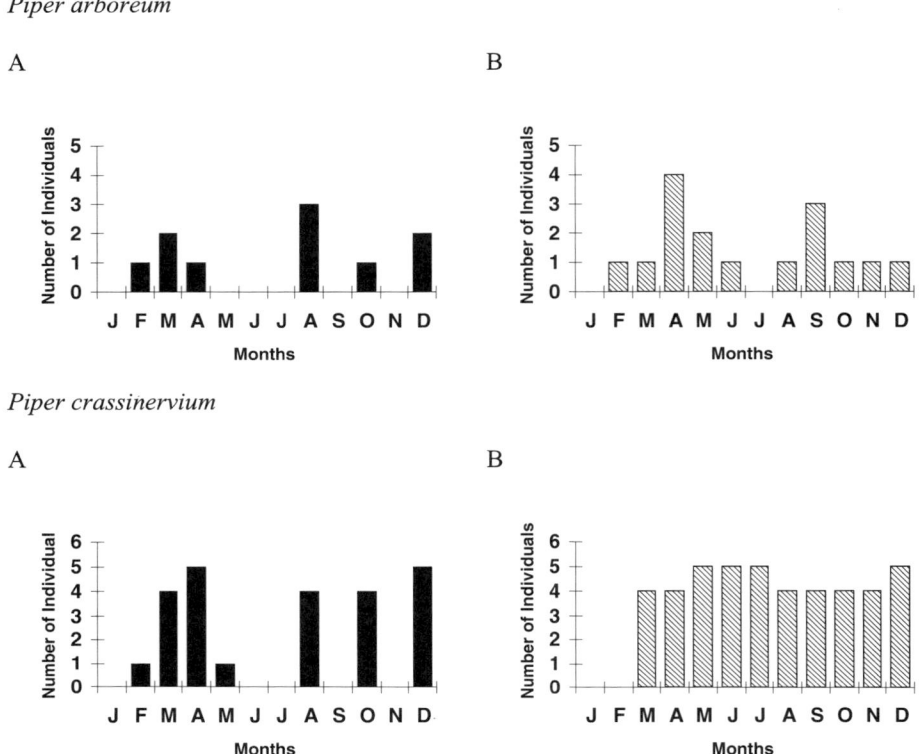

FIGURE 3.2. (cont.)

wind pollinated whereas the episodic P. *crassinervium* is insect pollinated. Three of the seasonal species (*P. mikanianum*, *P. martiana*, and *P. propinqua*) are wind pollinated, whereas the other two species (*P. mollicomum* and *P. regnelli*) are both insect and wind pollinated (de Figueiredo and Sazima 2000).

Infructescences were present throughout the year in populations of *P. aduncum*, *P. amalago*, *P. arboreum*, *P. crassinervium*, *P. gaudichaudianum*, *P. glabratum*, and *P. macedoi* (Fig. 3.2(B)). The fruiting period was long (8–10 months) in *P. regnelli*, *P. martiana*, and *P. propinqua* and short (3–4 months) in *P. mikanianum* and *P. mollicomum* (Fig. 3.2(B)). The overlap in fruiting periods is more pronounced than the overlap in flowering periods. This may indicate that pipers share seed dispersers in the studied area, although no information is available on that matter. Future research is still needed to confirm if the same animal group eats and disperses different *Piper* seeds in the Santa Genebra Reserve.

The recorded fruiting seasons for *P. aduncum*, *P. gaudichaudianum*, and *P. glabratum* were very different from those found by Marinho-Filho (1991) in the Serra do Japi Reserve (45 km away from the Santa Genebra Reserve). In addition, the fruiting seasons of *P. arboreum* in Brasília, Brazil (Bizerril and Raw 1997), and *P. amalago* in Costa Rica (Fleming 1981) differ from that of the Santa Genebra study. Such differences suggest that *Piper* species can modify their phenological responses according to their distribution. The

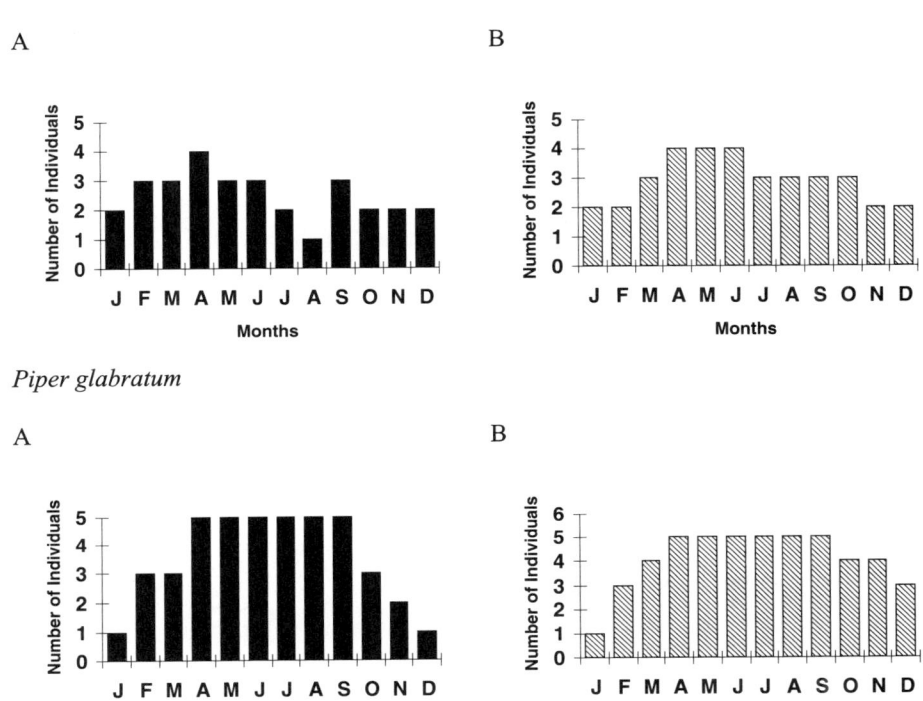

FIGURE 3.2. (cont.)

continual fruiting and the alteration of phenological responses may be important to gap colonizers, as such gaps arise randomly in time and space (Greig 1993b).

3.2.5. Pollination and Visitors

Pollinators, in their role as gamete vectors, can influence reproductive isolation and speciation in plants, particularly when congeners partition animal visitors (Weis and Campbell 1992). For example, *Ficus* plants have a very close association with their pollinators, where each species of *Ficus* is pollinated by only one or a few wasp species (e.g., Ramirez 1970, Wiebes 1979, de Figueiredo and Sazima 1997). Plants visited by few pollinator species with sharply contrasting preferences for floral form are likely to be under disruptive selection (Galen *et al.* 1987). On the other hand, partitioning of different individual parts of a given plant species by many different pollinator species should not generate reproductive isolation or plant speciation (Weis and Campbell 1992).

The flowers of pipers are tiny and aggregated in inflorescences called spikes. Most species are hermaphroditic and bear flowers with no perianth, 2 to 10 anthers, 1 to 5 stigmas, and airborne pollen. These morphological features comply with the anemophilous syndrome

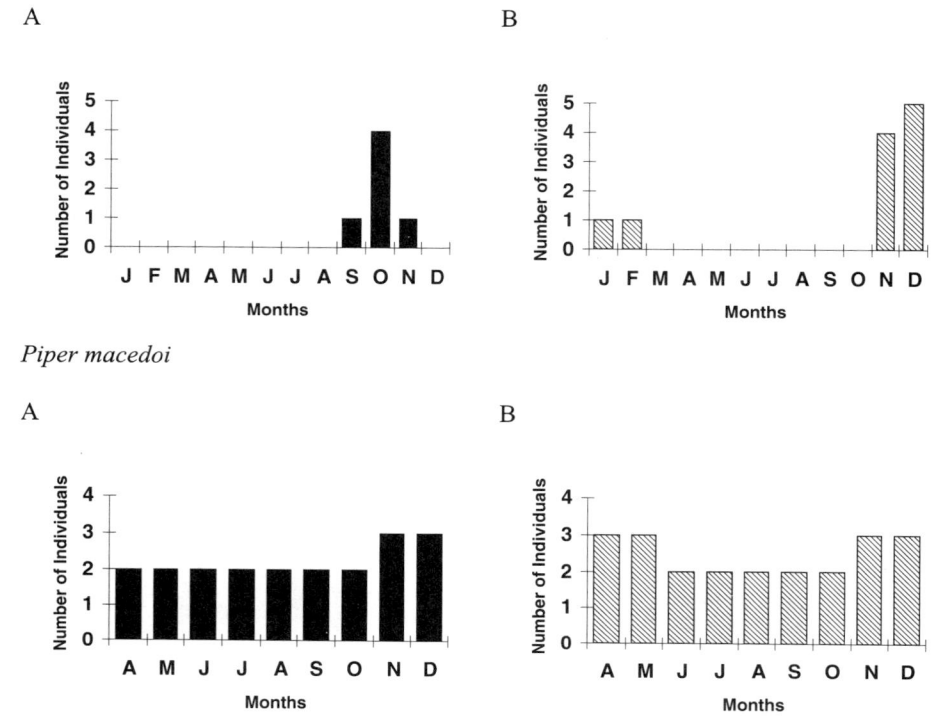

FIGURE 3.2. (cont.)

(*sensu* Faegri and van der Pijl 1979). Indeed, early researchers (Barber 1906, Anandan 1924) recorded abiotic modes of pollen transmission as the most important pollination mechanism in pipers, and until recently it was widely assumed that most *Piper* species were pollinated by wind and/or water. These initial studies were carried out on black pepper (*P. nigrum*), a Paleotropical species of great economic importance.

In India, Anandan (1924) found that *P. nigrum* vines protected from rain failed to set fruit even when visited by bees. Although he did not mention the bee species or their visiting behavior, he was the only author to refer to insects as visitors of *P. nigrum* flowers. Menon (1949) did not record insects on the flowers of black pepper and concluded that only rain was responsible for pollen flow. In his examination of *P. nigrum* fruiting, Gentry (1955a,b) concluded that, because of its small dimensions, the pollen of this species could be vectored by wind. He also observed apomixis in individuals bearing unisexual flowers.

Martin and Gregory (1962) studied the pollination of black pepper cultivars in Puerto Rico as well as four native Neotropical species (*P. amalago*, *P. scabrum*, *P. citrifolium*, and *P. blattarum*). They found self-pollination in black pepper, with pollen flow mediated by rain and wind, with small insects (Collembola) functioning as potential pollinators. Martin and Gregory (1962) suggested that the pollination of three of the Neotropical species took place through wind action; in the case of *P. blattarum*, which bears glutinous pollen similar

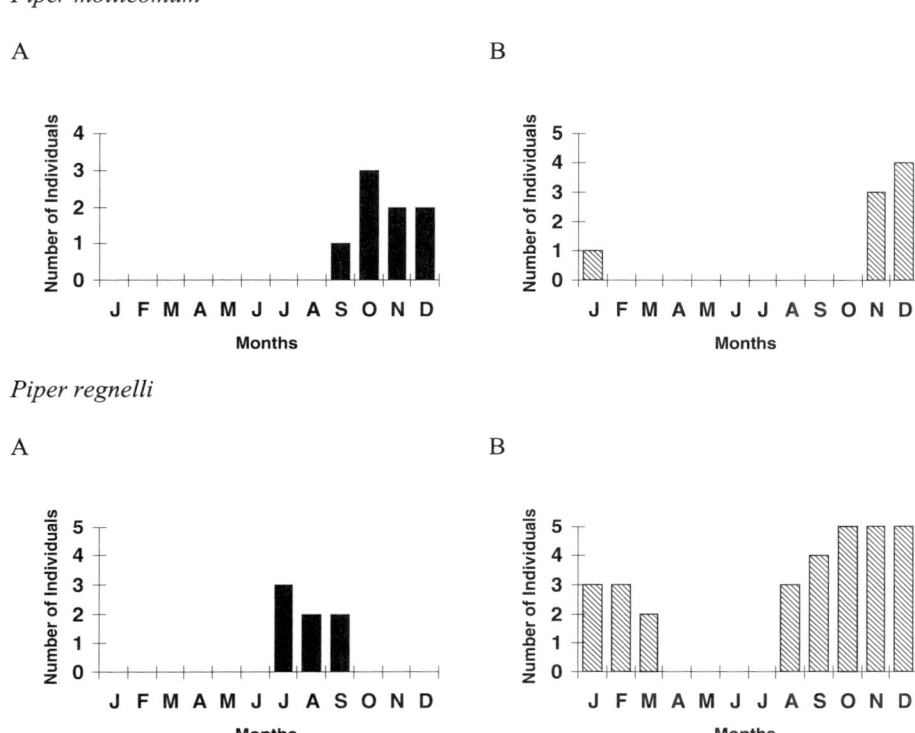

FIGURE 3.2. (cont.)

to that of *P. nigrum*, pollination was reported to occur through the forces of gravity, wind, and rain water. They described the self-pollination and self-fertility of the black pepper as an adaptation to cultivation. Another consideration is that Martin and Gregory (1962) found cultivars of black pepper with airborne pollen, which differed from the glutinous pollen described in the previous studies. Sasikumar *et al.* (1992) reexamined *P. nigrum* reproduction and showed that rain water is not essential for pollination, and even recorded self-compatibility and apomixis in this species.

Further investigations helped produce evidence that Neotropical pipers and Australian pipers are pollinated by insects. Semple (1974) studied the Costa-Rican *P. aduncum*, *P. auritum*, *P. friedrichsthalii*, *P. villiramulum*, and *P. peltata* and considered small Neotropical bees (Hymenoptera: Apidae and Halictidae) as their main pollinators, along with small beetle species (Coleoptera), ruling out wind and water pollination for these Neotropical species because of the glutinous nature of their pollen. Later, Fleming (1985) carried out a study on five piper species in Costa Rica: *P. amalago*, *P. jacquemontianum*, *P. marginatum*, *P. pseudofuligeum*, and *P. tuberculatum*. He examined the niche overlap among these species and reported that several bees (mainly Apidae) and flies (Diptera, mainly Syrphidae) visited the flowers. Dealing with the phenological characteristics of a Neotropical piper, *P. arieianum*, Marquis (1988) pointed out that its inflorescences were visited by bee and fly

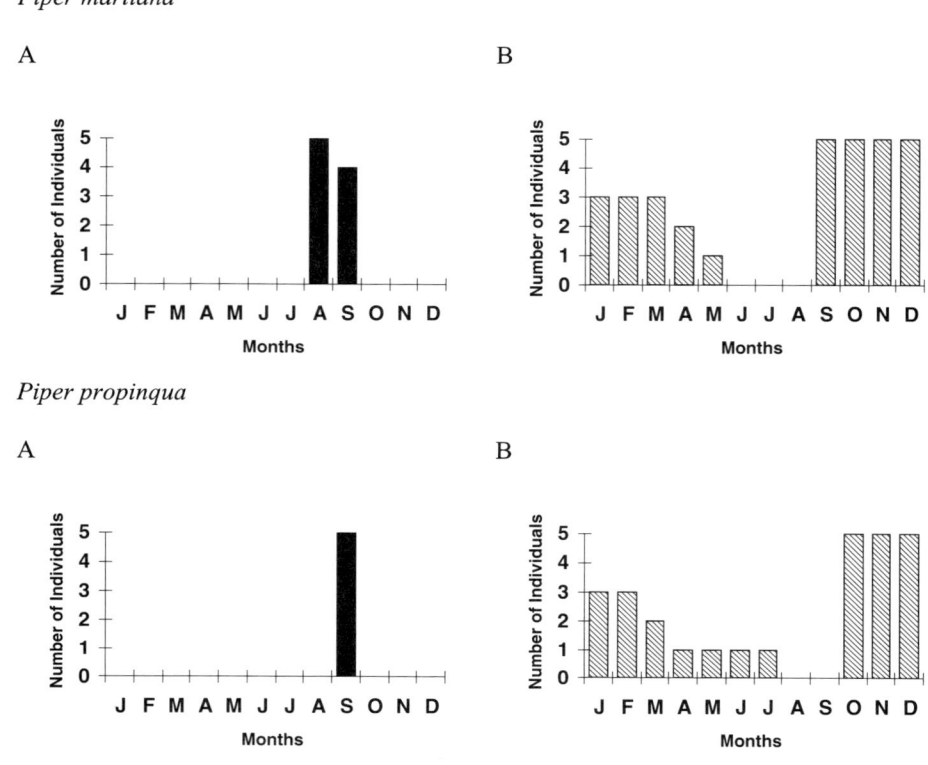

FIGURE 3.2. (cont.)

species and needed outcrossed pollen for maximum fruit set. The first and only study on Australian pipers was published by Ollerton (1996), who did not observe any other insects visiting the flowers of *P. novae-hollandiae* and *P. rothiana* than the plant-feeding gall midges (Diptera: Cecidomyiidae), which, he suggested, may act as pollinators of these species.

Because the evolution of floral diversity seems to be based upon specialized relationships with pollinators, and given that the floral characteristics of pipers are very similar, we expected Brazilian pipers to share different types of generalist visitor insects. In addition, we tested the hypothesis that Brazilian pipers could show both biotic and abiotic pollination systems.

Anemophily was tested by isolating 30 immature inflorescences in cloth bags that prevented insect visits, but allowed the entrance of airborne pollen (Kearns and Inouye 1993). We identified insect visitors to the inflorescences by observing 10–20 inflorescences, with a total of 60 h of observations per *Piper* species. Each visitor species was collected and observed under a microscope in order to verify the presence of pollen grains adhered to its body.

For each *Piper* species, richness (S), diversity (Shannon Index: H') and evenness $[E = H'/\ln(S)]$ of the flower visitors (Magurran 1988) were calculated. Grouping analysis was utilized to calculate the similarity of the visitor fauna between *Piper* species, employing

FIGURE 3.3. The hoverfly *Salpingogaster nigra* visiting an inflorescence of *Piper amalago* in southeastern Brazil.

a divisive method of cluster analysis (Ludwig and Reynolds 1988, Jongman *et al.* 2001). The most similar pipers were grouped according to visitor guild. The Euclidean distance was used as a measurement of similarity among pipers in relation to the composition and abundance of visitors. Links between pipers were elaborated through group averages, which calculate the average of all the distances between pipers of a given formed group and those of other groups. The delimitation of the major groups of pipers was based on a subjective interpretation of the data.

Seven of the Brazilian pipers studied showed anemophily; three showed entomophily; and four showed ambophily (de Figueiredo and Sazima 2000). Flies (mainly hoverflies, Syrphidae) and bees (mainly Apidae) visited flowers of all of the *Piper* species, and some pipers also received visits from butterflies and bugs. The 83 flower visitor species are presented in Table 3.2, together with the *Piper* species they visited.

Hoverflies were the dominant floral visitors in terms of both species richness and abundance (see Fig. 3.3). *Ornidia obesa*, a large hoverfly, was the primary visitor of *P. amalago*, *P. crassinervium*, and *P. regnelli*. *Ocyptamus* spp. were the main visitors to the inflorescences of *P. gaudichaudianum*, *P. glabratum*, *P. macedoi*, *P. mollicomum*, *P. martiana*, and *P. propinqua*; these hoverflies have small, thin bodies and were also reported by Fleming (1985) to visit the inflorescences of Costa Rican pipers. *Piper arboreum* and *P. mikanianum* were visited mainly by *Copestylum* spp., slightly smaller hoverflies than *O. obesa*. The main visitor of *P. aduncum* was *Salpingogaster nigra*, a hoverfly with a thin body, though larger than *Ocyptamus* spp. Thus, although the Brazilian pipers shared the same visitor groups, they showed a distinct pattern of resource partitioning of their main pollinators.

TABLE 3.2
Insects (N = Number of Records) on Inflorescences of the Pipers Studied in Southeastern Brazil

Visitor Taxon	Visitor Species	N	*Piper* Species
DIPTERA			
Syrphidae			
Syrphini	*Salpingogaster nigra* Schiner	338	adu, ama, arb, cra, mac, mik, mol
	Salpingogaster minor Austen	15	cra, mik, mol
	Salpingogaster sp.1	02	mol
	Salpingogaster sp.2	05	mol, mar
	Salpingogaster sp.3	02	adu
	Syrphus phaeostigma (Wiedemann)	05	reg
	Ocyptamus arx (Fluke)	17	adu, cra, mik, mol, reg, mar, pro
	Ocyptamus clarapex (Wiedemann)	169	gau, gla, mac, mik, mol, reg, pro
	Ocyptamus zenia (Curran)	19	adu, mol, pro
	Ocyptamus sp.1	65	mac, cra, gau, gla, mac, mik, mol, mar, pro
	Ocyptamus sp.2	132	arb, gau, gla, mac, mol, pro
	Ocyptamus sp.3	37	adu, mac, cra, mol, reg, mar
	Ocyptamus sp.4	25	ama, gla, reg, pro
	Ocyptamus sp.5	24	ama, mol, reg
	Ocyptamus sp.6	03	gau, mol
	Ocyptamus sp.7	16	gau, mar
	Ocyptamus sp.8	16	adu, gla, mol
	Allograpta neotropica Curran	03	cra, reg
	Leucopodella sp.	07	reg
	Trichopsomyia sp.1	69	adu, ama, cra, reg
	Trichopsomyia sp.2	24	ama, cra, reg
	Trichopsomyia sp.3	01	mac
Volucellini	*Copestylum lanei* (Curran)	14	arb, mac, reg, pro
	Copestylum tripunctatum (Hull)	99	arb, gla, mik, mol
	Copestylum mus (Williston)	51	arb, cra
	Copestylum belinda Hull	11	ama, arb, mik
	Copestylum vagum (Wiedemann)	04	cra
	Copestylum virtuosa (Hull)	38	cra, mik, reg
	Copestylum pallens (Wiedemann)	25	reg
	Copestylum sp.1	59	ama, arb, gla, mik, mol
	Copestylum sp.2	50	arb, gla
	Copestylum sp.3	01	mik
	Ornidia obesa (Fabricius)	1422	adu, ama, cra, mik, reg
Eristalini	*Palpada furcata* (Wiedemann)	195	mik, reg
	Palpada obsoleta (Wiedemann)	04	reg
	Palpada sp.1	173	mol, reg
	Palpada sp.2	170	mik, reg
	Palpada sp.3	21	reg
	Orthonevra neotropica (Shannon)	29	cra, mol, reg
Toxomerini	*Toxomerus* sp.	37	gau, gla, mol, reg
Melanostomatini	*Xanthandrus bucephalus* (Wiedemann)	02	mac
Muscidae	*Limnophora* sp.1	02	cra
	Limnophora sp.2	21	reg
Sarcophagidae	sp.	01	adu
Lauxaniidae	*Setulina* sp.	07	cra, mik, reg, pro
Calliphoridae	*Chrysomya putoria*	04	cra, reg
Family unknown	sp.1	04	cra, mol, reg
	sp.2	33	gla, reg
	sp.3	18	gla, reg

TABLE 3.2
(*cont.*)

Visitor Taxon	Visitor Species	N	*Piper* Species
HYMENOPTERA			
Halictidae	*Augochloropsis cuprella* (Smith)	24	pro
	Augochloropsis hebecens (Smith)	01	reg
	Augochloropsis cf. *brasiliana* (Cockerell)	10	gla, mik, mol, pro
	Augochloropsis sp.	02	ama, mar
	Augochlora caerulior Cockerell	05	adu, gla
	Augochlora sp.	16	ama, mac, mik, mol
	Notophus sp.	28	cra, mol, reg
	Neocorynura sp.	04	cra, gau, mac
	Dialictus (Chloralictus) sp.1	03	gau
	Dialictus (Chloralictus) sp.2	13	gla, mik, mol, reg
Megachilidae	*Megachile (Ptilasarus) bertonii* (Schrottky)	14	arb, reg
	sp.1	26	adu, ama, arb, gla
Andrenide	*Oxaea flavescens*	02	reg
Anthophoridae	*Exomalopsis analis* Spinola	11	gau, reg
	sp.1	01	cra
	sp.2	01	cra
Apidae	*Trigona spinipes* (F.)	276	reg
	Apis mellifera (L.)	335	cra, mik, reg, pro
	Tetragonisca angustula (Latreille)	93	adu, ama, cra, gau, mac, mik, pro
	Nannotrigona testaceicornis (Lepeletier)	41	reg, pro
	sp.	01	gla
Colletidae	*Colletes petropolitanus* Dalla Torres	20	arb, cra, gla
Vespidae	*Stelopolybia* sp.	03	reg, pro
LEPIDOPTERA			
Nymphalidae	*Actinote* sp.	02	cra, reg
	sp.	02	cra
Dioptidae	*Josia adiante* Walker	03	cra, reg
Hesperiidae	sp.	01	adu
Família indet.	sp.	02	gla, pro
COLEOPTERA			
Chrysomelidae			
Alticinae	sp.1	05	cra, reg
	sp.2	02	arb, gla
Scarabaeidae			
Rutelinae	sp.	06	adu, gau
HEMIPTERA			
Pentatomidae	*Sibaria armata* Stal	23	adu, arb, pro
Alydidae	*Megalotomus* sp.	03	arb, gla
Miridae	*Tropidosteptes* sp.	01	gau

Note: adu – *Piper aduncum*; ama – *P. amalago*; abr – *P. arboreum*; cra – *P. crassinervium*; gau – *P. gaudichaudianum*; gla – *P. glabratum*; mac – *P. macedoi*; mik – *P. mikanianum*; mol – *P. mollicomum*; reg – *P. regnelli*; mar – *P. martiana*; pro – *P. propinqua*.

Syrphids are known to be important pollinators of other plants (Robertson 1928, Waldbauer 1984, Arruda and Sazima 1996). In some species, hoverflies and small bees (mainly Halictidae) showed complementary action as pollinators (Ornduff 1971, 1975, Schlessman 1985), and facilitated cross-pollinating in anemophilous plants (Leereveld *et al.*

TABLE 3.3
Number (Σ), richness (S), evenness (E), and Diversity (H') of Visitor Insects on the Pipers Studied in Southeastern Brazil

Species	Σ	S	E	H'
Piper aduncum	264	13	0.36	0.92
P. amalago	297	14	0.65	1.71
P. arboreum	201	13	0.69	1.77
P. crassinervium	709	25	0.45	1.49
P. gaudichaudianum	21	09	0.94	2.07
P. glabratum	147	18	0.70	2.02
P. macedoi	85	11	0.60	1.43
P. mikanianum	112	21	0.76	2.30
P. mollicomum	111	23	0.81	2.61
P. regnelli	2106	37	0.55	2.01
P. martiana	38	06	0.86	1.54
P. propinqua	188	15	0.71	1.92
Averages	317.77	15.4	0.63	1.60

1976, Stelleman and Meeuse 1976). In Brazilian pipers, the syrphids primarily consumed pollen, and only less frequently visited inflorescences with receptive stigmas. This is a characteristic of hoverflies (Robertson 1924, Gilbert 1981, Bierzychudek 1987), which in addition to their long-distance flight (Banks 1951, Lorence 1985) and daily movements inside the forest due to their changing preferences for microhabitats (Maier and Waldbauer 1979), makes them both predators and mutualists of pipers. The position of the plant–insect relationship along this continuum varies according both to fly and *Piper* species.

Brazilian pipers also showed differences in the number of visiting individuals of each insect species. Table 3.3 summarizes the diversity measurements of visiting insects on piper inflorescences. A grouping analysis allowed the separation of the pipers into major groups, according to the similarity of their visitors. The first group is composed of *P. regnelli* and *P. crassinervium*. These species were, by far, the most attractive to insects, showing high values in the number, richness, and diversity of visiting insects (Table 3.3). Nevertheless, *P. crassinervium* differs from *P. regnelli* by showing a lower diversity of visitors (Table 3.3). A second group is composed only of *P. propinqua*, whose inflorescences attracted the greatest diversity and abundance of bees. The other pipers form the third group, with slight differences between species in visiting insects. The wide geographic distribution of pipers and their variety of habits and habitats may explain the promiscuity of these species, since regional variations in the diversity and abundance of visiting insects could reduce the selection of floral traits that favor specialized pollinators (Pettersson 1991, Herrera 1995).

This Brazilian study recorded 49 species of Diptera (41 in Syrphidae) and 23 species of Hymenoptera visiting piper inflorescences. In Costa Rica, Semple (1974) recorded only 7 species of Hymenoptera, and Fleming (1985) reported 12 species of Syrphidae and 17 species of Hymenoptera. The Meso-American pipers differ from those studied in Brazil since they were visited mainly by *Trigona* bees, considered to be their main pollinators (Semple 1974, Fleming 1985). *Piper amalago*, the only piper studied both in Brazil and Costa Rica, is mainly pollinated by the hoverfly *Salpingogaster nigra* in Brazil and by *Trigona* bees in Costa Rica.

Although they suffer from seasonal fluctuations, hoverfly populations are more stable than other insect populations (Bankowska 1989, 1995a, Owen and Gilbert 1989). Smaller hoverfly species, such as *Ocyptamus* spp. reported in the present study, are probably less stable throughout the year than larger species (Owen and Gilbert 1989). Pipers that had *Ocyptamus* spp. as their main pollinators were also pollinated by wind (de Figueiredo and Sazima 2000), which suggests that anemophily could compensate for a possible seasonal deficiency of insect pollinators. Cox (1991) suggested that species that use both insects and wind as their pollen dispersers can have a wide distribution and phenological flexibility, as well as a great ability to colonize and survive in places with low insect diversity.

The Brazilian pipers shared visitor insects, which could place them as effective mutualists with the hoverflies (cf. Waser and Real 1979). There was a discrete partitioning of pollinator species among the pipers, resembling what other studies found (Fleming 1985; Haslett 1989). These slight differences in visitor guilds may be influenced by the preferences of hoverflies for certain plant species. Such preferences have been explained by differing efficiencies of pollen on the maturation of eggs and oviposition (Maier 1978; Schemske *et al.* 1978; Gilbert 1985), as well as the level of leaf herbivory shown by the plants (Strauss *et al.* 1996).

3.3. CONCLUSIONS: POLLINATION AND RESOURCE PARTITIONING OF *Pipers* IN LIGHT OF EVOLUTIONARY AND CONSERVATIVE ECOLOGY

This Brazilian study tested the hypothesis that growth habit, habitat, vegetative reproduction, flowering and fruiting periods, and pollination agents are important factors allowing sympatric *Piper* species to diminish interspecific competition. It has been suggested that interspecific competition for limited resources determines divergences in the type, timing, or rate of resource use (Lawlor and Maynard Smith 1976). We separated Brazillian pipers growing in the same area into guilds on the basis of resource use (Fig. 3.4).

Only three species, *P. aduncum*, *P. arboreum*, and *P. mikanianum*, seem not to compete with the other pipers (Fig. 3.4) for habit, habitat, phenology, and main pollination agent. However, these species could be competing, not through direct interference or indirect exploitation of shared resources, but through mechanisms of apparent competition caused by shared herbivores, mutualists, and commensalists (Connell 1990).

Piper amalago, *P. glabratum*, *P. macedoi*, *P. regnelli*, and *O. martiana* are the main species competing with each other (Fig. 3.4). With the exception of *P. amalago*, these pipers are mature forest shade species. The pipers that presented the highest number of competitors (4.8 species in average) grew in the hygrophilous forest. On the other hand, the semideciduous forest showed the lowest level of competition among pipers (1.14 species in average). Since habitat segregation is a major means of coexistence between competing species (Giller 1984), pipers inhabiting a more stable environment show a higher degree of niche overlap. Despite this competitive condition, pipers could coexist in the area, maybe through equivalent competitive abilities (Aarssen 1983).

In our Brazilian pipers, pollinators do not appear to have a major role in structuring guilds of competing species. This could be explained by the importance of anemophily in

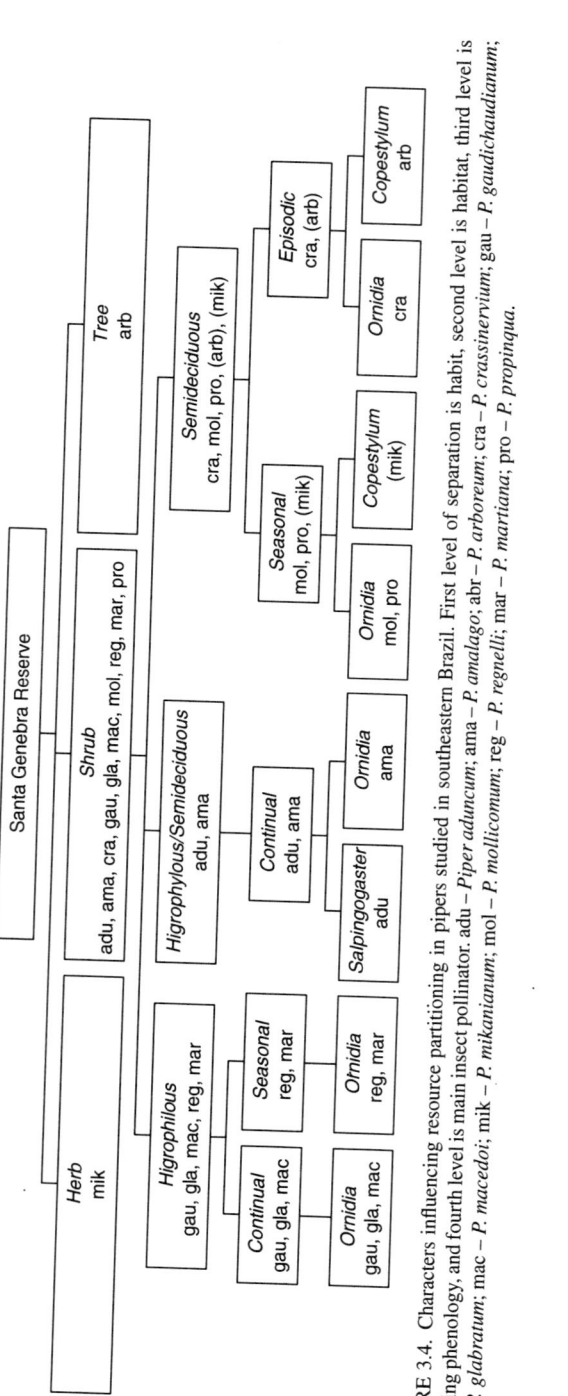

FIGURE 3.4. Characters influencing resource partitioning in pipers studied in southeastern Brazil. First level of separation is habit, second level is habitat, third level is flowering phenology, and fourth level is main insect pollinator. adu – *Piper aduncum*; ama – *P. amalago*; abr – *P. arboreum*; cra – *P. crassinervium*; gau – *P. gaudichaudianum*; gla – *P. glabratum*; mac – *P. macedoi*; mik – *P. mikanianum*; mol – *P. mollicomum*; reg – *P. regnelli*; mar – *P. martiana*; pro – *P. propinqua*.

several species and by the generalist (and even predatory) behavior of hoverflies. Two pipers (*P. regnelli* and *P. crassinervium*), however, seem to be pollen-limited and appear to be competing for pollination services. Their inflorescences emit a strong citric scent and they were the only two species where Thomisid spiders were seen preying upon visitors (de Figueiredo and Sazima 2000). *P. propinqua* was the only species whose inflorescences received more bee than hoverfly visits. This species could then be experiencing disruptive selection at the Brazilian study site. Still, studies on this species in different communities should be carried out to explore this possibility.

Differences in habitat and phenology seem to be important for the coexistence of multiple *Piper* species. Nonetheless, pipers that share the same habitat may be competing and thus may invest in different kinds of asexual reproduction. The suggestion of evolutionary lines based on ecological interpretation is problematic, since floral construction is not necessarily shaped by ecology (Wilson and Thomson 1995). But, as Grant (1949) proposed, tenuous variations in flower morphology and pollinator behavior could drive speciation. For instance, *P. nigrum* cultivated under natural pollination conditions gives rise to both vegetative (Ravindran and Sasikumar 1993) and chromosomal (Nair *et al.* 1993) variations.

Piperaceae is considered to be one of the most primitive families of angiosperms, derived from the herbaceous protoangiosperms with simple, minute flowers (Heywood and Fleming 1986, Taylor and Hickey 1990, 1992). Pipers share characteristics with two other primitive families, Chloranthaceae and Winteraceae. Chloranthaceae have floral morphologies similar to those of Piperaceae and are also pollinated by wind, flies, and other nonspecialized insects (Walker and Walker 1984, Crane *et al.* 1986, 1989, 1995, Endress 1987, Proctor *et al.* 1996). Winteraceae also show several reproductive characteristics similar to those of Piperaceae, such as vegetative reproduction and pollination by Diptera (including hoverflies) and Hymenoptera (Ehrendorfer *et al.* 1979, Gottsberger *et al.* 1980, Sampson 1980, Thien 1980, Godley and Smith 1981, Gottsberger 1988). These three families probably show ancient pollination systems, strengthening the hypothesis that primitive angiosperms used insects as well as wind as pollen vectors (Crepet 1979, Dilcher 1979).

Pipers show characteristics that comply with the anemophilous syndrome, but the development of very proximal flowers in the inflorescences could also be associated with the activity of hoverflies and small bees, which take pollen from inflorescence surfaces (Burger 1972, Eyde and Morgan 1973, Maier and Waldbauer 1979). Most Neotropical pipers were shown to be entomophilous or ambophilous (Semple 1974, Fleming 1985, de Figueiredo and Sazima 2000). Considering the hypothesis that entomophily is the primitive condition (Faegri and van der Pijl 1979), wind-pollinated pipers could provide evidence that anemophily is a derived condition because of the presence of nectaries and osmophores in their flowers (de Figueiredo and Sazima 2000).

The Brazilian study offers some suggestions for the management of this dominant plant group in Neotropical forests. Pipers are likely to have an influence on the maintenance of insect diversity, especially that of hoverflies, in the study area. Insects are very sensitive to forest fragmentation (Didham *et al.* 1996, Kerr *et al.* 1996, Turner 1996), and areas that suffered environmental impact disturbance showed low richness and abundance of hoverflies (Bankowska 1980, 1989, 1994, 1995b). Pipers provide food for these insects, which are important pollen vectors of several other plant species (Proctor *et al.* 1996). In addition, frugivorous vertebrates of the study area, a forest fragment, may use piper infructescences as a food supply throughout the year.

The Brazilian pipers studied showed significant variation in growth habits, habitats, vegetative reproduction, phenology, and pollinator agents. This information was used to separate most of them into groups of competing species. The results of this study, based on a natural history approach, could strengthen the hypothesis that congeners could partition resources, allowing their coexistence in a given area.

3.4. GUIDELINES FOR FUTURE RESEARCH ON THE POLLINATION OF *Pipers*

The floral biology of pipers, at a general level, is still poorly known. The studies carried out to date describe the reproduction of less than 5% of the total number of known *Piper* species. Although Piperaceae occur in every Neo- and Paleotropical region, only one study has been carried out in South America, one in Australia, and two in Central America. In India, several studies have been performed, but only on *P. nigrum*. The pollination ecology of Piperaceae species in ecosystems with high plant diversity, such as the Brazilian Atlantic Forest and Amazon Rainforest, has not yet been studied. The hypothesis presented in this chapter must be tested in other ecosystems. Because the studies presented here characterized the pollinator agents of pipers without any experimental evidence, they can be considered working hypotheses, and the unequivocal confirmation of pollen flow through biotic agents remains to be accomplished.

Piper nigrum, black pepper, is an important commercial plant, and additional research on pollination and fruit set is needed for a number of regions where it is cultivated. Brazil, for example, is the fourth major peppercorn exporter, and a study on the floral biology of its cultivars could potentially improve the fruit set in this country. Comparing native species in different regions is important in understanding the floral modifications of piper species with broad ranges and the possible similarities or differences that may exist in pollination ecology. *Piper aduncum*, for example, has been introduced in Asia and Africa and it could be interesting to compare the floral biology of these populations with that of the populations studied in America. Finally, the studies carried out until now have not reported on gamete flow in pipers; thus the population structure and evolution of these species are virtually unknown.

3.5. ACKNOWLEDGMENTS

We are grateful to Lee Dyer and two anonymous reviewers for their extensive comments, which improved the earlier version of this chapter. Research reported in this chapter was supported by a fellowship from Conselho Nacional de Desenvolvimento Científico e Tecnológico—CNPq (process number 140055/94-0) to R. A. de Figueiredo and (process number 300993/79-0) to M. Sazima.

REFERENCES

Aarssen, L. (1983). Ecological combining ability and competitive combining ability in plants: Towards a general evolutionary theory of coexistence in systems of competition. *American Naturalist* 122:707–731.

Abe, Y., Takikawa, H., and Mori, K. (2001). Synthesis of gibbilimbols A-D, cytotoxic and antibacterial alkenulphenols isolated from *Piper gibbilimbum*. *Bioscience, Biotechnology, and Biochemistry* 65:732–735.
Anandan, N. (1924). Observations on the habits of the pepper vine with special reference to the reproductive phase. *Madras Department of Agriculture Yearbook* (1924), pp. 49–69.
Ang-Lee, M. K., Moss, J., and Yuan, C. S. (2001). Herbal medicines and perioperative care. *JAMA* 286:208–216.
Bankowska, R. (1980). Fly communities of the family Syrphidae in natural and anthropogenic habitats of Poland. *Memorabilia Zoologica* 33:3–93.
Bankowska, R. (1989). Hover flies (Diptera, Syrphidae) of moist meadows on the Mazovian lowland. *Memorabilia Zoologica* 43:329–347.
Bankowska, R. (1994). Diversification of Syrphidae (Diptera) fauna in the canopy of Polish pine forests in relation to forest stand age and forest health zones. *Fragmenta Faunistica* 36:469–484.
Bankowska, R. (1995a). Fauna Syrphidae (Diptera) Puszczy Bialowieskiej. *Fragmenta Faunistica* 37:451–483.
Bankowska, R. (1995b). Dipterans (Diptera) of pine canopies of the Berezinsky Biosphere Reserve in Byelorussia. *Fragmenta Faunistica* 38:181–185.
Banks, C. J. (1951). Syrphidae as pests of cucumbers. *Entomologist's Monthly Magazine* 86:239–240.
Barber, C. A. (1906). *The Varieties of Cultivated Pepper*. Bull. No. 56, Madras Department of Agriculture, India.
Bazzaz, F. A. (1990). Plant–plant interactions in successional environments. In: Grace, J. B., and Tilman, D. (eds.), *Perspectives on Plant Competition*. Academic Press, New York, pp. 239–263.
Bierzychudek, P. (1982). Life histories and demography of shade-tolerant temperate forest herbs: A review. *New Phytologist* 90:757–776.
Bierzychudek, P. (1987). Pollinators increase the cost of sex by avoiding female flowers. *Ecology* 68:444–447.
Bizerril, M. X. A., and Raw, A. (1997). Feeding specialization of two species of bats and the fruit quality of *Piper arboreum* in a Central Brazilian gallery forest. *Revista de Biologia Tropical* 45:913–918.
Borchert, R. (1983). Phenology and control of flowering in tropical trees. *Biotropica* 15:81–89.
Braun-Blanquet, J. (1936). *Plant Sociology*. McGraw-Hill Book Co., London.
Burger, W. C. (1972). Evolutionary trends in the Central American species of *Piper* (Piperaceae). *Brittonia* 24:356–362.
Chazdon, R. L., and Field, C. B. (1987). Determinants of photosynthetic capacity in six rainforest *Piper* species. *Oecologia* 73:222–230.
Chazdon, R. L., Williams, K., and Field, C. B. (1988). Interactions between crown structure and light environment in five rain forest *Piper* species. *American Journal of Botany* 75:1459–1471.
Clay, K. (1990). The impact of parasitic and mutualistic fungi on competitive interactions among plants. In: Grace, J. B., & Tilman, D. (eds.), *Perspectives on Plant Competition*. Academic Press, New York, pp. 391–412.
Connell, J. H. (1990). Apparent versus "real" competition in plants. In: Grace, J. B., and Tilman, D. (eds.), *Perspectives on Plant Competition*. Academic Press, New York, pp. 9–26.
Cox, P. A. (1991). Abiotic pollination: An evolutionary escape from animal-pollinated angiosperms. *Philosophical Transactions of the Royal Society of London B*333:217–224.
Crane, P. R., Friis, E. M., and Pedersen, K. R. (1986). Lower Cretaceous angiosperm flowers: Fossil evidence on early radiation of dicotyledons. *Science* 232:852–854.
Crane, P. R., Friis, E. M., and Pedersen, K. R. (1989). Reproductive structure and function in Cretaceous Chloranthaceae. *Plant Systematics and Evolution* 165:211–226.
Crane, P. R., Friis, E. M., and Pedersen, K. R. (1995). The origin and early diversification of Angiosperms. *Nature* 374:27–33.
Crepet, W. L. (1979). Some aspects of the pollination biology of Middle Eocene angiosperms. *Review of Palaeobotany and Palynology* 27:213–238.
Darwin, C. (1859). *On the Origin of Species by Natural Selection*. John Murray, London.
Darwin, C. (1876). *On the Effects of Cross and Self Fertilization in the Vegetable Kingdom*. John Murray, London.
de Arruda, V. L. V., and Sazima, M. (1996). Flores visitadas por sirfídeos (Diptera: Syrphidae) em mata semidecídua de Campinas, SP. *Revista Brasileira de Botanica* 19:109–117.
de Figueiredo, R. A., and Sazima, M. (1997). Phenology and pollination ecology of three Brazilian fig species (Moraceae). *Botanica Acta* 110:73–78.
de Figueiredo, R. A., and Sazima, M. (2000). Pollination biology of Piperaceae species in southeastern Brazil. *Annals of Botany* 85:455–460.
de A. Mello, M. H., Pedro Junior, M. J., Ortolani, A. A., and Alfonsi, R. R. (1994). Chuva e temperatura: cem anos de observações em Campinas. *Boletim Técnico do Instituto Agronômico* 154:1–48.
Didham, R. K., Ghazoul, J., Stork, N. E., and Davis, A. J. (1996). Insects in fragmented forests: A functional approach. *Trends in Ecology and Evolution* 11:255–260.
Dilcher, D. L. (1979). Early angiosperm reproduction: An introductory report. *Review of Palaeobotany and Palynology* 27:291–328.
Ehrendorfer, F., Silberbauer-Gottsberger, I., and Gottsberger, G. (1979). Variation on the population, racial and species level in the primitive relic angiosperm genus *Drimys* (Winteraceae) in South America. *Plant Systematics and Evolution* 132:53–83.

Endress, P. K. (1987). The early evolution of the angiosperm flower. *Trends in Ecology and Evolution* 2:300–304.
Eyde, R. H., and Morgan, J. T. (1973). Floral structure and evolution in Lopezieae (Onagraceae). *American Journal of Botany* 60:771–787.
Faegri, K., and van der Pijl, L. (1979). *The Principles of Pollination Ecology*. Pregamon Press, Oxford.
Findley, J. S. (1993). *Bats: A Community Perspective*. Cambridge University Press, Cambridge.
Fleming, T. H. (1981). Fecundity, fruiting pattern, and seed dispersal of *Piper amalago* (Piperaceae), a bat-dispersed tropical shrub. *Oecologia* 51:42–46.
Fleming, T. H. (1985). Coexistence of five sympatric *Piper* (Piperaceae) species in a tropical dry forest. *Ecology* 66:688–700.
Galindo-Gonzalez, J., Guevara, S., and Sosa, V. J. (2000). Bat- and bird-generated seed rains at isolated trees in pastures in a tropical rainforest. *Conservation Biology* 14:1693–1703.
Galen, C., Zimmer, K. A., and Newport, M. E. A. (1987). Pollination in floral scent morphs of *Polemonium viscosum*: A mechanism for disruptive selection on flower size. *Evolution* 41:599–606.
Gartner, B. L. (1989). Breakage and regrowth of *Piper* species in rain forest understorey. *Biotropica* 21:303–307.
Gentry, A. H. (1974). Coevolutionary patterns in Central American Bignoniaceae. *Annals of the Missouri Botanical Garden* 61:728–759.
Gentry, H. S. (1955a). Introducing black pepper into America. *Economic Botany* 9:256–268.
Gentry, H. S. (1955b). Apomixis in black pepper and jojoba? *Journal of Heredity* 46:8.
Gilbert, F. S. (1981). Foraging ecology of hoverflies: Morphology of the mouthparts in relation to feeding on nectar and pollen in some commom urban species. *Ecological Entomology* 6:245–262.
Gilbert, F. S. (1985). Ecomorphological relationships in hoverflies (Diptera, Syrphidae). *Proceedings of the Royal Society of London B* 224:91–105.
Giller, P. S. (1984). *Community Structure and the Niche*. Chapman & Hall Ltd., London.
Godley, E. J., and Smith, D. H. (1981). Breeding systems in New Zealand plants. 5. *Pseudowintera colorata* (Winteraceae). *New Zealand Journal of Botany* 19:151–156.
Goldberg, D. E. (1990). Components of resource competition in plant communities. In: Grace, J. B., and Tilman, D. (eds.), *Perspectives on Plant Competition*. Academic Press, New York, pp. 27–49.
Goldsmith, F. B., and Harrison, C. M. (1976). Description and analysis of vegetation. In: Chapman, S. B. (ed.), *Methods in Plant Ecology*. Blackwell Scientific Publications, Oxford, pp. 85–155.
Gomez-Pompa, A. (1971). Possible papel de la vegetacion secundaria en la evolución de la flora tropical. *Biotropica* 3:125–135.
Gottsberger, G. (1988). The reproductive biology of primitive angiosperms. *Taxon* 37:630–643.
Gottsberger, G., Silberbauer-Gottsberger, I., and Ehrendorfer, F. (1980). Reproductive biology in the primitive relic angiosperm *Dimys brasiliensis* (Winteraceae). *Plant Systematics and Evolution* 135:11–39.
Grant, V. (1949). Pollination systems as isolating mechanisms in angiosperms. *Evolution* 3:82–97.
Greig, N. (1993a). Regeneration mode in Neotropical *Piper*: Habitat and species comparisons. *Ecology* 74:2125–2135.
Greig, N. (1993b). Predispersal seed predation on five *Piper* species in tropical rainforest. *Oecologia* 93:412–420.
Hartemink, A. E. (2001). Biomass and nutrient accumulation of *Piper aduncum* and *Imperata cylindrica* fallows in the humid lowlands of Papua New Guinea. *Forest Ecology and Management* 144:19–32.
Haslett, J. R. (1989). Interpreting patterns of resource utilization: Randomness and selectivity in pollen feeding by adult hoverflies. *Oecologia* 78:433–442.
Herbst, L. H. (1986). The role of nitrogen from fruit pulp in the nutrition of the frugivorous bat *Carollia perspicillata*. *Biotropica* 18:39–44.
Herrera, C. M. (1995). Floral traits and plant adaptation to insect pollinators: A devil's advocate approach. In: Lloyd, D. G., and Barret, S. C. H. (eds.), *Floral Biology: Studies on Floral Evolution in Animal-Pollinated Plants*. Chapman & Hall, New York, pp. 65–87.
Heywood, J. S., and Fleming, T. H. (1986). Patterns of allozyme variation in three Costa Rican species of *Piper*. *Biotropica* 18:208–213.
Hunter, M. D., Ohgushi, T., and Price, P. W. (eds.). (1992). *Effects of Resource Distribution on Animal–Plant Interactions*. Academic Press, New York.
Ingram, S. W., and Nadkarni, N. M. (1993). Composition and distribution of epiphytic organic matter in a Neotropical cloud forest, Costa Rica. *Biotropica* 25:370–383.
Inoue, K., Maki, M., and Masuda, M. (1995). Evolution of *Campanula* flowers in relation to insect pollinators on Islands. In: Lloyd, D. G., and Barret, S. C. H. (eds.), *Floral Biology: Studies on Floral Evolution in Animal-Pollinated Plants*. Chapman & Hall, New York, pp. 377–400.
Jongman, R. H. G., Ter Braak, C. J. F., and Van Tongeren, O. F. R. (2001). *Data Analysis in Community and Landscape Ecology*. Cambridge University Press, Cambridge.
Kearns, C. A., and Inouye, D. W. (1993). *Techniques for Pollination Biologists*. University Press of Colorado, Niwot.

Kerr, W. E., Carvalho, G. A., and Nascimento, V. A. (1996). *Abelha uruçu: biologia, manejo e conservação*. Fundação Acangaú, Belo Horizonte.
Kinsman, S. (1990). Regeneration by fragmentation in tropical montane forest shrubs. *American Journal of Botany* 77:1626–1633.
Kotchetkoff-Henriques, O., and Joly, C. A. (1994). Estudo florístico e fitossociológico de uma floresta semidecídua na Serra do Itaqueri, Itirapina, estado de São Paulo, Brasil. *Revista Brasilliria de Biologia* 54:477–487.
Lawlor, L. R., and Maynard Smith, J. (1976). The coevolution and stability of competing species. *The American Naturalist* 110:79–99.
Leereveld, H., Meeuse, A. D. J., and Stelleman, P. (1976). Anthecological relations between reputedly anemophilous flowers and syrphid flies. II. *Plantago media* L. *Acta Botanica Neerlandica* 25:205–211.
Leitão-Filho, H. F. (1995a). Plantas úteis da mata de Santa Genebra. In: Morellato, L. P. C., and Leitão-Filho, H. F. (eds.), *Ecologia e preservação de uma floresta tropical urbana: Reserva de Santa Genebra*. Editora da Unicamp, Campinas, pp. 121–129.
Leitão-Filho, H. F. (1995b). A vegetação da Reserva de Santa Genebra. In: Morellato, L. P. C., and Leitão-Filho, H. F. (eds.), *Ecologia e preservação de uma floresta tropical urbana: Reserva de Santa Genebra*. Editora da Unicamp, Campinas, pp. 19–36.
Levin, D. A. (1972). Competition for pollinator service: A stimulus for the evolution of autogamy. *Evolution* 26:668–669.
Loiselle, B. S. (1990). Seeds in droppings of tropical fruit-eating birds: Importance of considering seed composition. *Oecologia* 82:494–500.
Lorence, D. H. (1985). A monograph of the Monimiaceae (Laurales) in the Malagasy region (southwest Indian Ocean). *Annals of the Missouri Botanical Garden* 72:1–165.
Louda, S. M., Keeler, K. H., and Holt, R. D. (1990). Herbivore influence on plant performance and competitive interactions. In: Grace, J. B., and Tilman, D. (eds.), *Perspectives on Plant Competition*. Academic Press, New York, pp. 413–444.
Ludwig, J. A., and Reynolds, J. F. (1988). *Statistical Ecology: A Primer on Methods and Computing*. John Wiley & Sons, New York.
Maier, C. T. (1978). The immature stages and biology of *Mallota posticata* (Fabr.) (Diptera, Syrphidae). *Proceedings of the Entomological Society of Washington* 80:424–440.
Maier, C. T., and Waldbauer, G. P. (1979). Diurnal activity patterns of flower flies (Diptera: Syrphidae) in an Illinois sand area. *Annals of the Entomological Society of America* 72:237–245.
Magurran, A. E. (1988). *Ecological Diversity and Its Measurements*. Croom Helm Ltd., London.
Marinho-Filho, J. S. (1991). The coexistence of two frugivorous bat species and the phenology of their food plants in Brazil. *Journal of Tropical Ecology* 7:59–67.
Marquis, R. J. (1988). Phenological variation in the Neotropical understory shrub *Piper arieianum*: Causes and consequences. *Ecology* 69:1552–1565.
Martin, F. S., and Gregory, L. E. (1962). Mode of pollination and factors affecting fruit set in *Piper nigrum* L. in Puerto Rico. *Crop Science* 2:295–299.
Menon, K. K. (1949). The survey of pollu and root diseases of pepper. *Indian Journal of Agricultural Science* 19:89–136.
Milliken, W., and Albert, B. (1996). The use of medicinal plants by the Yanomani indians of Brazil. *Economic Botany* 50:10–25.
Morellato, L. P. C. (1991). *Estudo da fenologia de árvores, arbustos e lianas de uma floresta semidecídua no sudeste do Brasil*. Ph.D. Thesis, Universidade Estadual de Campinas, Campinas.
Nair, R. R., Sasikumar, B., and Radindran, P. N. (1993). Polyploidy in a cultivar of black pepper (*Piper nigrum* L.) and its open pollinated progenies. *Cytologia* 58:27–31.
Newstrom, L. E., Frankie, G. W., and Baker, H. G. (1994). A new classification for plant phenology based on flowering patterns in lowland tropical rainforest trees at La Selva, Costa Rica. *Biotropica* 26:141–159.
Ollerton, J. (1996). Interactions between gall midges (Diptera: Cecidomyiidae) and inflorescences of *Piper novae-hollandiae* (Piperaceae) in Australia. *The Entomologist* 115:181–184.
Opler, P. A., Frankie, G. W., and Baker, H. G. (1980). Comparative phenological studies of shrubs and treelets in wet and dry forests in the lowlands of Costa Rica. *Journal of Ecology* 68:167–186.
Ornduff, R. (1971). The reproductive system of *Jepsonia heterandra*. *Evolution* 25:300–311.
Ornduff, R. (1975). Complementary roles of halictids and syrphids in the pollination of *Jepsonia heterandra* (Saxifragaceae). *Evolution* 29:371–373.
Owen, J., and Gilbert, F. S. (1989). On the abundance of hoverflies (Syrphidae). *Oikos* 55:183–193.
Palmeirim, J. M., Gorchov, D. S., and Stoleson, S. (1989). Trophic structure of a Neotropical frugivore community: Is there competition between birds and bats? *Oecologia* 79:403–411.
Pettersson, M. W. (1991). Pollination by a guild of fluctuating moth populations: Option for unspecialization in *Silene vulgaris*. *Journal of Ecology* 79:591–604.

Proctor, M., Yeo, P., and Lack, A. (1996). *The Natural History of Pollination*. Harper Collins Publishers, London.
Ramírez, W. (1970). Host specificity of fig wasps (Agaonidae). *Evolution* 24:680–691.
Ravindran, P. N., and Sasikumar, B. (1993). Variability in open pollinated seedlings of black pepper (*Piper nigrum* L.). *Journal of Spices & Aromatic Crops* 2:60–65.
Robertson, C. (1924). Flowers visits of insects. II. *Psyche* 31:93–111.
Robertson, C. (1928). Flowers and insects. XXV. *Ecology* 9:505–526.
Sampson, F. B. (1980). Natural hybridism in *Pseudowintera* (Winteraceae). *New Zealand Journal of Botany* 18:43–51.
Sanchez-Coronado, M. E., Rincon, E., and Vazquez-Yanes, C. (1990). Growth responses of three contrasting *Piper* spp. growing under different light conditions. *Canadian Journal of Botany* 68:1182–1186.
Sasikumar, B, George, J. K., and Ravindran, P. N. (1992). Breeding behaviour of black pepper. *Indian Journal of Genetics* 52:17–21.
Schemske, D. W., Willson, M. F., Melampy, M. N., Miller, L. J., Verner, L., Schemske, K. M., and Best, L. B. (1978). Flowering ecology of some spring woodland herbs. *Ecology* 59:351–366.
Schlessman, M. A. (1985). Floral biology of American ginseng (*Panax quinquefolium*). *Bulletin of the Torrey Botanical Club* 112:129–133.
Schoener, T. W. (1975). Resource partitioning in ecological communities. In: Hazen, W. E. (ed.), *Readings in Population and Community Ecology*. W. B. Saunders Co., Philadelphia, pp. 370–390.
Semple, K. S. (1974). Pollination in Piperaceae. *Annals of the Missouri Botanical Garden* 61:868–871.
Silander, J. A., Jr., Pacala, S. W. (1990). The application of plant population dynamic models to understanding plant competition. In: Grace, J. B., and Tilman, D. (eds.), *Perspectives on Plant Competition*. Academic Press, New York, pp. 67–91.
Silvertown, J. W., and Lovett Doust, J. (1993). *Introduction to Plant Population Biology*. Blackwell Scientific Publications, Oxford.
Solbrig, O. T., and Rollins, R. C. (1977). The evolution of autogamy in species of the mustard genus *Leavenworthia*. *Evolution* 31:265–281.
Stelleman, P., and Meeuse, A. D. J. (1976). Anthecological relations between reputedly anemophilous flowers and syrphid flies I. The possible role of syrphid flies as pollinators of *Plantago*. *Tijdschrift Voor Entomologie* 119:15–31.
Strauss, S. Y., Conner, J. K., and Rush, S. L. (1996). Foliar herbivory affects floral characters and plant attractiveness to pollinators: implications for male and female plant fitness. *American Naturalist* 147:1098–1107.
Taylor, D. W., and Hickey, L. J. (1990). An Aptian plant with attached leaves and flowers: Implications for angiosperm origin. *Science* 247:702–704.
Taylor, D. W., and Hickey, L. J. (1992). Phylogenetic evidence for the herbaceous origin of angiosperms. *Plant Systematics and Evolution* 180:137–156.
Tebbs, M. C. (1989a). Revision of *Piper* (Piperaceae) in the New World. 1. Review of characters and taxonomy of *Piper* section *Macrostachys*. *Bulletin of the British Museum of Natural History (Botany)* 19:117–158.
Tebbs, M. C. (1989b). The climbing species of New World *Piper* (Piperaceae). *Willdenowia* 19:175–189.
Tebbs, M. C. (1993). Piperaceae. In: Kubitzki, K., Rohwer, J. G., and Bittrich, V. (eds.), *The Families and Genera of Vascular Plants*. Springer-Verlag, Berlin, pp. 516–520.
Thien, L. B. (1980). Patterns of pollination in the primitive angiosperms. *Biotropica* 12:1–13.
Turkington, R., and Mehrhoff, L. A. (1990). The role of competition in structuring pasture communities. In: Grace, J. B., and Tilman, D. (eds.), *Perspectives on Plant Competition*. Academic Press, New York, pp. 307–340.
Turner, I. M. (1996). Species loss in fragments of tropical rain forest: A review of the evidence. *Journal of Applied Ecology* 33:200–209.
van Schaik, C. P., Terborgh, J. W., and Wright, S. J. (1993). The phenology of tropical forests: Adaptive significance and consequences for primary consumers. *Annual Review of Ecological Systems* 24:353–377.
Waldbauer, G. P. (1984). Mating behavior at blossoms and the flower association of mimetic *Temnostoma* spp. (Diptera, Syrphidae) in northern Michigan. *Proceedings of the Entomological Society of Washington* 86:295–304.
Walker, J. W., and Walker, A. G. (1984). Ultrastructure of Lower Cretaceous angiosperm pollen and the origin and early evolution of flowering plants. *Annals of the Missouri Botanical Garden* 71:464–521.
Waser, N. M., and Real, L. A. (1979). Effective mutualism between sequentially flowering plant species. *Nature* 281:670–672.
Webber, A. C., and Gottsberger, G. (1999). Phenological patterns of six *Xylopia* (Annonaceae) species in central Amazonia. *Phyton* 39:293–301.
Weis, A. E., and Campbell, D. R. (1992). Plant genotype: A variable factor in insect–plant interactions. In: Hunter, M. D., Ohgushi, T., and Price, P. W. (eds.), *Effects of Resource Distribution on Animal–Plant Interactions*. Academic Press, New York, pp. 75–111.

Wiebes, J. T. (1979). Co-evolution of figs and their insect pollinators. *Annual Review of Ecological Systems* 10:1–12.

Williams, K., Field, C. B., and Mooney, H. A. (1989). Relationships among leaf construction cost, leaf longevity, and light environment in rain-forest plants of the genus *Piper*. *American Naturalist* 133:198–211.

Wilson, P., and Thomson, J. D. (1995). How do flowers diverge? In: Lloyd, D. G, and Barret, S. C. H. (eds.), *Floral Biology: Studies on Floral Evolution in Animal-Pollinated Plants*. Chapman & Hall, New York, pp. 88–111.

Wright, S. J. (1991). Seasonal drought and the phenology of understory shrubs in a tropical moist forest. *Ecology* 72:1643–1657.

Wright, S. J., and Calderon, O. (1995). Phylogenetic patterns among tropical flowering phenologies. *Journal of Ecology* 83:937–948.

Yuncker, T. G. (1972). The Piperaceae of Brazil I: *Piper*—Group I, II, III, IV. *Hoehnea* 2:19–366.

Yuncker, T. G. (1973). The Piperaceae of Brazil II: *Piper*—Group V; *Ottonia; Pothomorphe; Sarcorhachis*. *Hoehnea* 3:29–284.

Yuncker, T. G. (1975). The Piperaceae of Brazil IV: Index. *Hoehnea* 5:123–145.

4
Dispersal Ecology of Neotropical *Piper* Shrubs and Treelets

Theodore H. Fleming
Department of Biology, University of Miami, Florida

4.1. INTRODUCTION

Along with the Melastomataceae, Rubiaceae, and Solanaceae, shrubs and treelets of the Piperaceae (primarily *Piper*) are numerically dominant members of the understories of many Neotropical forests. This dominance occurs both in number of species and number of individuals. Gentry and Emmons (1987) reported that *Piper* species richness ranged from 0.3 to 6.5 species per 500-m transect along a rainfall/soil fertility gradient in Central and South America compared with ranges of 0.7–5.8 and 1.0–8.4 in the Melastomataceae and Rubiaceae, respectively. Data summarized in Gentry (1990) indicate that *Piper* species richness in certain well-studied moist or wet tropical forests ranges from 18 (Barro Colorado Island, Panama) to 60 or more species (La Selva, Costa Rica). Tropical dry forests have much lower diversity (e.g., five species at Parque Nacional Santa Rosa, Costa Rica; Fleming 1985). Forests around Manaus, Brazil, also contain few *Piper* species (Prance 1990). Overall, the pantropical genus *Piper* has been especially successful evolutionarily in the lowland Neotropics, which contains over twice as many species as the Asian tropics (700 vs. 300 species; Jaramillo and Manos 2001).

To what extent has *Piper* reproductive biology, especially its dispersal ecology, contributed to its ecological and evolutionary success? Most Neotropical pipers produce tiny self-incompatible, hermaphroditic flowers arrayed in spike-like inflorescences that are pollinated by generalized bees and flies (see Chapter 3). *Piper* pollination biology does not appear to involve specialized coevolutionary relationships with a restricted subset of insect pollinators. Successfully pollinated flowers develop into small, single-seeded fruits in infructescences (which will hereafter be called "fruits"). Number of seeds (fruits) per infructescence in a series of Costa Rican pipers ranged from just over 100 to about 3,000; seed mass ranged from 0.14 to 6.23 mg (Fleming 1985, Greig 1993a). In contrast to the pollination situation, a small number of frugivorous phyllostomid bats are the most important dispersers of *Piper* seeds. These bats appear to be specialized consumers and dispersers

TABLE 4.1
Characteristics of Bats of the Phyllostomid Subfamily Carolliinae

Species	Mass (g)	Distribution
Carollia brevicauda	20.1	Southern Mexico to eastern Brazil
C. castanea	14.7	Honduras to western Brazil
C. perspicillata	19.5	Southern Mexico to Paraguay
C. subrufa	16.2	Southern Mexico to western Costa Rica
Rhinophylla alethina	—	Colombia and Ecuador
R. fischerae	—	Colombia to central Brazil
R. pumilio	8.3	Colombia to eastern Brazil

Note: Data comes from Fleming (1991), Koopman (1993), and Simmons and Voss (1998). Central American forms of *C. brevicauda* are now recognized as *C. Sowell*: (Baker *et al*. 2002).

of these fruits. In this chapter, I review the dispersal ecology of Neotropical pipers in an attempt to answer the above question. I also speculate about the importance of *Piper*'s dispersal ecology for its speciation and evolutionary radiation.

4.2. THE *Piper* BATS

Bats of the family Phyllostomidae (New World leaf-nosed bats) are ubiquitous and species-rich in Neotropical lowlands. Depending on location, from 31 to 49 species of phyllostomid bats have been captured in Neotropical rain forests (Simmons and Voss 1998). Of these, 12 to 25 are frugivores classified into two sister clades—the Carolliinae and Stenodermatinae (Wetterer *et al*. 2000). These clades neatly reflect an ecological separation between understory (Carolliinae) and canopy (Stenodermatinae) fruit eaters. Because of their feeding specialization, carolliinine bats can be considered to be "*Piper* bats."

Subfamily Carolliinae contains two genera (*Carollia* and *Rhinophylla*) containing at least four and three species, respectively (Table 4.1). Species of *Carollia* are larger in body size and are much more widely distributed in the Neotropics than those of *Rhinophylla*, which occur only in South America. One or two species of *Carollia* occur in tropical dry or moist forests; up to three species co-occur in tropical wet forests (Fleming 1991). Species of *Carollia* appear to be much more specialized on a diet of *Piper* than are species of *Rhinophylla* and are probably more important dispersers of *Piper* seeds. *Piper* is a year-round dietary staple in species of *Carollia* (Fig. 4.1). In Central America, these bats eat fruits of 5–12 species per season, and percent *Piper* in *Carollia* diets is negatively correlated with a species' body size (Fleming 1991). In lowland Peruvian rain forests, Gorchov *et al.* (1995) reported that three species of *Carollia* eat seven species of *Piper* whereas two species of *Rhinophylla* eat no pipers. In French Guiana, *Piper* is a minor item in the diet of *Rhinophylla pumilio* (Charles-Dominique 1993, Charles-Dominique and Cockle 2001). Other common items in the diets of *Carollia* bats include fruits of *Solanum* (Solanaceae) and *Vismia* (Hypericaceae), two early successional shrub taxa that often co-occur with pipers.

In addition to major dietary differences, these two genera differ in their roosting behavior and degree of gregariousness. *Carollia* bats live in small to moderately large colonies (a few 100s to a few 1,000s) in caves, hollow trees, and man-made structures

FIGURE 4.1. *Carollia perspicillata* approaching a fruit of *Piper tuberculatum*, Guanacaste Province, Costa Rica. Photo reprinted with permission, courtesy of Merlin D. Tuttle/Bat Conservation International.

(e.g., wells, road culverts). In contrast, *Rhinophylla* bats live solitarily or in small groups (≤6) in "tents" formed from clipped palm or *Philodendron* leaves in the forest understory (Charles-Dominique 1993, Simmons and Voss 1998). Rather than creating their own tents, these bats appear to take over abandoned tents created by stenodermatine bats. At localities where both genera occur, *Carollia* bats are often at least an order of magnitude more common than *Rhinophylla* bats (Gorchov *et al.* 1995, Simmons and Voss 1998).

The foraging and fruit-harvesting behavior of *Carollia perspicillata* has been studied extensively and is known in great detail (Fleming *et al.* 1977, Heithaus and Fleming 1978, Fleming and Heithaus 1986, Bonaccorso and Gush 1987, Fleming 1988, Charles-Dominique 1991, Bizerril and Raw 1998, Thies *et al.* 1998, Thies and Kalko, unpublished manuscript). These bats are relatively sedentary and many forage within a kilometer of their day roosts. Their foraging ranges are larger during dry seasons, when fruit levels are low, than during wet seasons. Most individuals remain away from their day roost all night (e.g., from 1900 to 0500 h) and harvest fruits in one to three feeding areas often separated by a few hundred meters. Within their feeding areas, which usually overlap among individuals, bats harvest single fruits and take them to feeding roosts located 20–100 m from a fruiting plant to eat. An exception to this is the large (9.4 g) fruits of *P. arboreum*, which are harvested piecemeal by *C. perspicillata* (Bizerril and Raw 1998). Fruits are eaten quickly (in less than 3 min), and bats harvest a new fruit every 15–30 min. In one night, a bat will consume just over 100% of its body mass in fruit pulp and seeds (e.g., 40–50 *Piper* fruits). Passage time of fruit pulp and seeds is rapid (about 5 min, early in the evening when bats are most active, and 20–30 min, later in the night).

Carollia bats use a combination of echolocation and olfaction to locate ripe Piper fruits. They use echolocation to avoid obstacles in the cluttered understory of tropical forests and olfaction to discriminate between ripe and unripe fruit. Their final approach to ripe fruits is guided by echolocation information. They usually grab ripe fruits by the distal tip in flight and carry them back to their feeding roosts to eat. Fruit relocation experiments with P. amalago and P. pseudofuligineum in western Costa Rica indicate that Carollia bats are acutely aware of ripe Piper fruits in their environment. Regardless of whether they are located in expected (i.e., in Piper patches) or unexpected locations (e.g., in flyways hundreds of meters away from the nearest Piper patch), ripe fruits have nearly a 100% chance of being located and removed from experimental "shrubs." Observations on fruit removal rates from actual Piper plants confirm that first-night removal probabilities of ripe fruits are very high (typically $\geq 90\%$). Detailed analysis of the temporal pattern of fruit harvesting during a night indicates that members of C. perspicillata usually feed on Piper fruit early in the evening before switching to other kinds of fruit.

Social status is known to have a strong effect on the foraging behavior of C. perspicillata and presumably other species of Carollia. As described by Williams (1986) and Fleming (1988), this species has a harem-polygynous mating system in which a small number of adult males ($\leq 20\%$) are harem (or territorial) males that guard groups of up to about 20 adult females ("harems") in the day roost. All subadult and most adult males are "bachelors" that reside in a different part of the day roost away from the harems. Harem males defend their groups of females or harem sites from the intrusions of bachelor males all day and night. All females and most bachelor males leave the day roost to forage all night. Harem males, in contrast, often use the day roost as their feeding roost and only leave the roost to obtain a fruit. As a result, harem males are less likely to be effective seed dispersers than females or bachelor males; most of the seeds they ingest end up either in or very close to the day roost.

Charles-Dominique (1993) used radio telemetry to document the feeding rhythms of different social and reproductive classes of C. perspicillata in French Guiana. He found that the rhythms of harem males and females in late pregnancy and lactation differed quantitatively from those of bachelor males and females that are not in late pregnancy (Fig. 4.2). Harem males were more active, and late pregnant/lactating females were less active, than other bats. Despite their high levels of foraging activity, harem males centered this activity around their day roosts and foraged less widely than bachelor males and females.

Radio tracking observations of Rhinophylla pumilio in French Guiana and Ecuador indicate that, like Carollia, it is a sedentary bat (Rinehart 2002). Foraging areas or home ranges of these bats averaged 10–15 ha, and bats used tents as their feeding roosts. Small groups, sometimes containing a single adult male and several adult females (i.e., a harem), roost together and frequently move between several tents in their home range (Simmons and Voss 1998, Rinehart 2002). Rinehart (2002) has suggested that since adult male ranges do not overlap, they are likely to be territories.

Two other common phyllostomid bats, Glossophaga soricina (Glossophaginae) and Sturnira lilium (Stenodermatinae), sometimes eat Piper fruit, but neither species can be considered to be a Piper specialist (Heithaus et al. 1975, Charles-Dominique 1986, Fleming 1988). G. soricina is an omnivorous species whose diet includes nectar and pollen

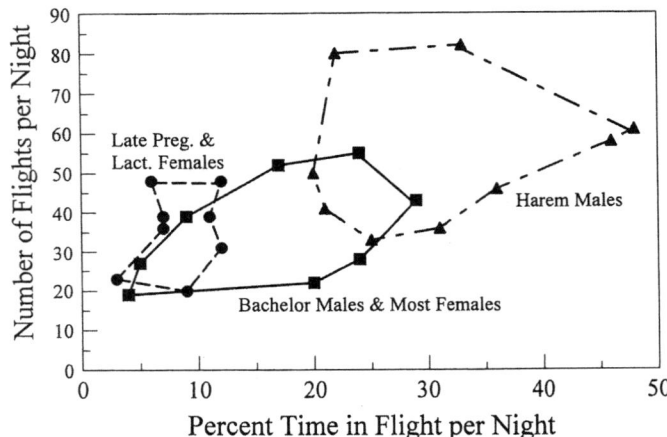

FIGURE 4.2. Foraging activity levels in three groups of adult *Carollia perspicillata* in French Guyana. Reproductively active females undergo the least, and harem males the most, flight activity each night. Data are from Charles-Dominique (1991).

(its principal specialization), fruit, and insects. *S. lilium* is an understory bat that specializes on fruits of *Solanum* (Solanaceae). At mid-montane elevations, however, *Piper* fruits become common dietary items in *S. lilium* and *S. ludovici* (Dinerstein 1986).

In summary, species of *Carollia* focus their feeding behavior on ripe *Piper* fruits whenever they are available and are the main dispersers of *Piper* seeds in the lowland Neotropics. They do this by systematically harvesting ripe fruits in one to several feeding areas each night. Most of the seeds of *Piper* (and other taxa) that they ingest are defecated relatively close to fruiting plants. Quantitative estimates of seed dispersal distances and the likelihood that defecated seeds will give rise to new seedlings will be discussed in the next section.

4.3. *Piper* FRUITING PHENOLOGY AND DISPERSAL ECOLOGY

4.3.1. Fruiting Phenology

Piper fruiting phenology has been studied in detail in tropical dry forest (Fleming 1985), tropical moist forest (Thies and Kalko, unpublished manuscript), and tropical wet forest (only one species, Marquis 1988). Opler *et al.* (1980) documented general flowering and fruiting patterns of pipers and other shrubs and treelets in tropical wet and dry forests in Costa Rica. Their data (Fig. 4.3) indicate that in wet forest, peak flowering occurs in April and peak fruiting occurs in June. In dry forest, there are no conspicuous flowering or fruiting peaks. In both habitats, flowering and fruiting occurs year round in the *Piper* flora. Year-round availability of *Piper* fruits in the understory of Neotropical forests has undoubtedly been a major factor in the evolution of dietary specialization in *Carollia* bats (Fleming 1986).

Studies of species-specific fruiting patterns tend to reveal staggered fruiting peaks and significant differences between the fruiting patterns of early successional (or large gap)

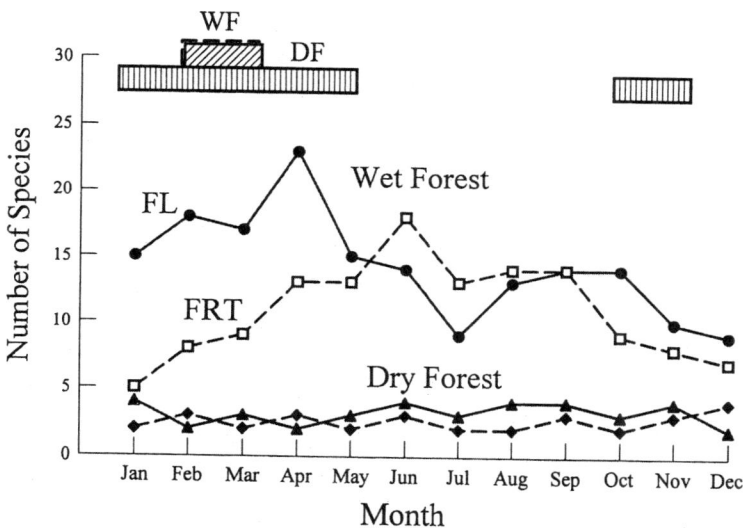

FIGURE 4.3. Flowering (FL, solid lines) and fruiting (FRT, dashed lines) phenology of *Pipers* in two Costa Rican habitats. Upper rectangles indicate the dry season in the two habitats. Data are from Opler *et al.* (1980).

species and late successional or forest species. Staggered fruiting peaks have been documented in five tropical dry forest pipers in Costa Rica (Fleming 1985), four subtropical humid forest pipers in southern Brazil (Marhino-Filho 1991), and eight tropical moist forest pipers in Panama (Thies and Kalko, unpublished manuscript). On the basis of seed size (small: 0.29–0.35 mg), number of seeds per fruit (over 1,000), and habitat distributions (i.e., in heavily disturbed sites), two of the five pipers in Costa Rican dry forest are early successional species. These species fruit in different seasons (wet season in *P. pseudofuligineum*; dry season in *P. marginatum*) (Fig. 4.4). The other three species produce fewer (100–200) and larger (0.86–1.36 mg) seeds per fruit and occur in less-disturbed, later successional habitats (dry forest, *P. amalago*; moist ravines, *P. jacquemontianum*; and riparian sites, *P. tuberculatum*). These species have fruiting peaks that differ by 1–3 months from each other; two species have two fruiting episodes per year (Fig. 4.4).

Thies and Kalko (unpublished manuscript) studied the phenology of 12 pipers in Panama, including eight forest species (*P. aequale, P. arboretum, P. carrilloanum, P. cordulatum, P. culebranum, P. dariense, P. grande,* and *P. perlasense*) and four gap species (*P. dilatatum, P. hispidum, P. marginatum,* and *P. reticulatum*), and reported that the phenological patterns of these two groups differed significantly. The forest species tended to have relatively short (1–4 months), staggered fruiting seasons in which they produced very few ripe fruits per night (means of 0.2–0.4 per plant). In contrast, gap species had longer fruiting seasons (8–11 months) with multiple peaks; fruiting periods overlapped broadly; and plants produced more fruit per night than forest species (means of 0.4–8.5 per plant). *Piper arieanum*, a common Costa Rican wet forest species, has a long fruit maturation time of 6–8 months (cf. about 2 months in dry forest *P. amalago*) and a relatively long fruiting season (3–4 months) in which very low numbers of fruits ripen per night (Marquis 1988).

As a result of different fruiting strategies, gap and forest pipers attract different groups of seed dispersers. Larger fruit crops and more ripe fruits per plant per night serve

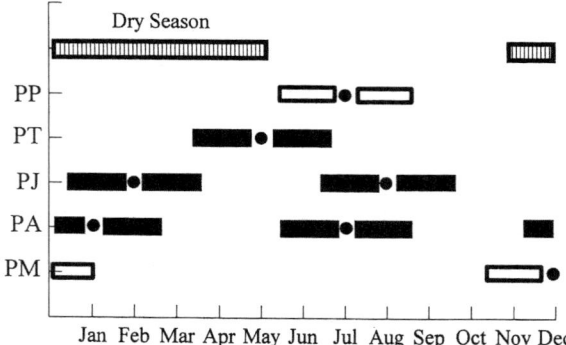

FIGURE 4.4. Fruiting phenology of five species of *Piper* in Costa Rican tropical dry forest. Data indicate peak fruiting times (circles) and extent of fruiting seasons (bars). Open bars are early successional species; closed bars are late successional species. Species are *P. amalago* (PA), *P. jacquemontianum* (PJ), *P. marginatum* (PM), *P. pseudofuligineum* (PP), and *P. tuberculatum* (PT). Data are from Fleming (1985).

to attract second growth birds as well as bats to early successional pipers. Several species of understory tanagers (Thraupinae), but not the equally common manakins (Pipridae), occasionally consume the fruit of early successional pipers (e.g., *P. auritum*, *P. hispidum*, and *P. sancti-felis* at La Selva, Costa Rica) (O'Donnell 1989, Palmeirim *et al*. 1989, Loiselle 1990). Tanagers feed by mandibulating or "mashing" fruits (Denslow and Moermond 1985, Levey 1987) and hence can more easily strip chunks of ripe fruit pulp and seeds from *Piper* infructescences than can manakins, which feed by swallowing or "gulping" entire fruits. In contrast, slow-fruiting forest pipers appear to attract only bats as fruit consumers and hence have more specialized dispersal systems than early successional species.

4.3.2. *Patterns of Seed Dispersal*

What kinds of dispersal patterns arise from seeds that are ingested by either birds or bats? How far, on average, do *Piper* seeds move from fruiting plants before being defecated into the environment? I will first describe the general results of seed rain studies before describing more detailed quantitative estimates of seed dispersal distances. Before doing this, however, I wish to point out that researchers simultaneously studying seed dispersal patterns produced by tropical birds (mainly tanagers in the case of *Piper*) and bats have concluded that these two groups produce fundamentally different patterns of seed rain. Birds are much more likely to defecate seeds from perches (e.g., in plants where they are currently feeding or elsewhere) than while they are flying (Charles-Dominique 1986, Thomas *et al*. 1988, Gorchov *et al*. 1995). Bats, on the other hand, defecate many seeds in flight in addition to the ones they defecate from perches in their feeding roosts. As a consequence, bats are more likely to deposit seeds in a greater variety of sites than birds. Also, because understory bats appear to have a greater tendency to fly through open habitats (e.g., gaps or abandoned fields) than understory birds, they are more likely to deposit seeds in early successional sites.

Patterns of seed rain of small-seeded, vertebrate-dispersed plants around and away from fruiting plants have been documented in Mexico, Costa Rica, French Guiana, and Peru

(Fleming and Heithaus 1981, Charles-Dominique 1986, Fleming 1988, Gorchov et al. 1993, Medellin and Gaona 1999, Galindo-Gonzalez et al. 2000). These studies indicate that *Piper* seed rain has both a temporal and a spatial component. In terms of time of seed deposition, *Piper* seeds are much more likely to be deposited in seed traps (or on vegetation and leaf litter and/or soil) at night than during the day, as expected if bats are the major dispersers of these seeds. In terms of the spatial patterns of seed deposition, *Piper* seeds are deposited in a wide variety of sites, including closed forest, forest gaps, abandoned fields or pastures, and under isolated fruiting trees in open pastures. In a year-long study of seed rain in two forest habitats at Santa Rosa National Park, Costa Rica, Fleming (1988) reported that *Piper* seeds were much less common than those of several species of *Ficus* (Moraceae), *Cecropia peltata* (Cecropiaceae), and *Muntingia calabura* (Eleocarpaceae). At this same site, Fleming and Heithaus (1981) reported that seeds of *P. amalago* and *P. pseudofuligineum* occurred in transects around only three of 10 fruiting trees compared with 10 of 10 trees for *C. peltata* and *M. calabura* and 7 of 10 trees for *Ficus* spp. These results suggest that despite being bat-dispersed, *Piper* seed mobility is somewhat lower than that of the other two species, whose seeds are eaten by a greater number of species (Fleming et al. 1985, Fleming and Williams 1990). Dispersal by birds, bats, and monkeys may provide wider dispersal (at least within forests) than dispersal by bats alone.

Galindo-Gonzalez et al. (2000) captured bats and measured seed rain under isolated fruiting trees in pastures in Veracruz, Mexico. They reported that fecal samples from two species of *Carollia* and *S. lilium* contained six species of *Piper*, including three early successional species (*P. auritum*, *P. hispidum*, and *P. yzabalanum*) and three early/late successional species (*P. aequale*, *P. amalago*, and *P. sanctum*). Seed rain in traps placed under these trees also contained these species. Seed traps placed in a cornfield, an abandoned field, and a cacao plantation near a tropical wet forest in Chiapas, Mexico, received seeds from three early successional *Piper* species (*P. auritum*, *P. hispidum*, and *P. nitidum*).

Quantitative estimates of *Piper* seed dispersal distances are available from tropical dry forest in western Costa Rica (Fleming 1981, 1988). These estimates come from radio tracking studies of *C. perspicillata* plus studies of its seed retention times. Results of these studies indicate that about 67% of the seeds of *P. amalago* and *P. pseudofuligineum* ingested by *C. perspicillata* are defecated in its current feeding area (i.e., <100 m from parent plants). Many of these seeds are deposited under feeding roosts. Nonetheless, some seeds (probably ≤5%) move relatively long distances (i.e., ≥1 km) when bats change feeding areas. Relatively long-distance moves occur more frequently during the dry season, when fruit densities are low and very patchy, than during the wet season. The overall deposition curves (*fide* Janzen 1970) of *Piper* seeds are thus likely to be highly leptokurtic but with a long tail (Fig. 4.5).

4.3.3. Fates of Seeds

What are the fates of *Piper* seeds once they have been defecated by birds or bats? Are they at risk from seed predators? What are their germination requirements and do they accumulate in the soil to form seed banks? *Piper* seeds, though small, are likely to suffer from postdispersal predators such as ants and rodents, at least when they occur in Petri dishes on the ground. Perry and Fleming (1980) and Fleming (1988) reported that most seeds (≥75%

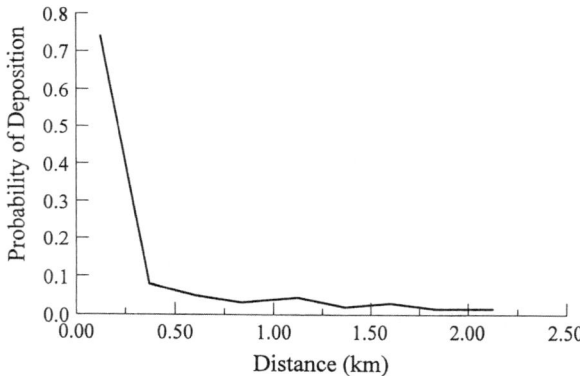

FIGURE 4.5. A generalized seed dispersal curve produced by *Carollia perspicillata* in Costa Rican tropical dry forest. The graph indicates the probability that a seed will be dispersed a given distance from the parent plant. Data are from Fleming (1988).

in most trials) of *P. amalago* were removed from dishes in 4 days of exposure in a variety of habitats at Parque Nacional Santa Rosa, Costa Rica. Ants (particularly *Pheidole* spp.) removed over twice as many seeds as two species of rodents. Despite potentially heavy predation by ants and rodents, seeds of *P. amalago* at this site occur in substantial densities (e.g., 64 seeds/m^2 in light gaps and 133 seeds/m^2 in the forest understory) in the soil. Seeds of *P. amalago* are relatively large, and it would be nice to know if the smaller seeds of early successional *Pipers* (e.g., *P. pseudofuligineum* at Santa Rosa) are at similar risk of postdispersal predation. In a study of predispersal seed predation in five species of *Piper* at La Selva, Costa Rica, Greig (1993a) found that early successional species experienced lower predation by hemipterans and weevils than late successional species. Does a similar situation hold for postdispersal predation?

In addition to predation, interspecific competition can affect the fate of *Piper* seeds. It is common to find two or more kinds of seeds in fecal samples from *Carollia* bats. During the wet season in tropical dry forests, for example, the following species can co-occur in fecal samples of *C. perspicillata*: *P. amalago, P. pseudofuligineum, P. jacquemontianum, Muntingia calabura, Cecropia peltata*, and *Chlorophora tinctoria* (Fleming 1988). Similarly, Loiselle (1990) reported finding mixed species loads of *Piper* and other species in fecal samples from tanagers in Costa Rican wet forest. Whenever seeds are defecated in intra- or interspecific clumps, they are likely to compete for resources during seedling establishment (Howe 1989). Apparently only Loiselle (1990), however, has systematically studied this competition. She found that both growth and survival were affected when she grew different combinations of two species that co-occurred in fecal samples from tanagers in the lab. When *P. auritum* was paired with *P. sancti-felicis*, for example, the former species had higher growth and survival rates than the latter. More studies of this kind are clearly needed.

Germination probabilities of *Piper* seeds are independent of bat (and bird?) gut passage but are highly dependent on deposition microhabitats. *Piper* seeds typically have very high germination percentages (80–100%) whether or not they are ingested and excreted by bats (Fleming 1988, Bizerril and Raw 1998, Galindo-Gonzalez *et al.* 2000). But unless these seeds are deposited in light gaps, or become exposed to high levels of sunlight from soil or forest disturbance, their germination probabilities are low (e.g., ≤15% in forest understory compared with 30% in light gaps in *P. amalago*; Fleming 1981). Because *Carollia* feeding roosts tend to be located under dark bowers of vegetation, these sites are especially poor places for germination of *Piper* seeds (<5% germination). Greig (1993b) found a striking difference in the abundance of seedlings of early and late successional pipers at La Selva, Costa Rica. Whereas seedlings of the former species were common in treefall gaps, seedlings of the latter species were rare in the forest understory. From these and other observations and experiments, she concluded that shade-tolerant pipers (e.g., *P. arieanum*, *P. gargaranum*, and *P. melanocladum*) are much more likely to recruit new individuals (ramets) via vegetative reproduction than via seed dispersal. Shade-intolerant species (e.g., *P. aduncum*, *P. culebranum*, and *P. sancti-felicis*), in contrast, are more likely to recruit by seed dispersal than by vegetative reproduction.

The germination characteristics of seeds produced by Mexican tropical wet forest pipers occupying different habitats have been carefully studied by Orozco-Segovia and Vazquez-Yanes (1989). Germination rates of fresh seeds and length of dormancy and photoblastic responses of seeds in the soil differed among four species. Seeds of a large gap species, *P. umbellatum*, can remain dormant for long periods in the soil and have a long-lived photoblastic response to high light conditions. In contrast, seeds of *P. auritum*, another large gap species, and *P. hispidum*, which occurs in a variety of gap and forest habitats, exhibit short soil dormancy and a short-lived photoblastic response. Both of these species probably require continuous "broadcast" dispersal by bats and birds to recruit in light gaps. Finally, seeds of a shade-tolerant forest species, *P. aequale*, exhibit long soil dormancy and a long-lived photoblastic response that allows this species to "dribble" its seeds out and wait in the soil for new gaps to form.

In summary, *Piper* seeds experience both pre- and postdispersal predation, and Greig (1993a) has suggested this predation can sometimes be severe enough to limit seedling recruitment. These seeds do not need to pass through bats to have high germination rates, but they do need exposure to high light levels for maximum germination. It is likely, therefore, that when pipers recruit by seed, they do so in forest gaps of various sizes. Shade-tolerant species can probably recruit by seed in smaller gaps than shade-intolerant species, but vegetative recruitment is also important in the former species. Seed dispersal by bats and birds appears to be most important for early successional or large gap species.

4.3.4. *Postdispersal Distribution Patterns*

Regardless of their dispersal method, species of *Piper* occur in relatively high densities in many Neotropical habitats. In this section, I describe the distribution patterns of pipers in a Costa Rican tropical wet forest at La Selva Biological Station whose *Piper* flora contains nearly 60 species (Laska 1997). I do this to address two questions: (1) How does *Piper* species diversity and density vary in space and time, and (2) are forest gaps colonized by short- or long-distance dispersal? These data were collected in a series of forty-nine 1,000-m² belt transects placed in three major habitats: primary forest on old alluvial or weathered basaltic soils (10 locations, 30 transects), secondary forest derived from abandoned pastures on old alluvial or weathered basaltic soils (3 locations, 10 transects),

TABLE 4.2
Summary of the La Selva Forest Transect Study for the Entire Data Set (A) and the Matched Primary and Secondary Sites (B)

	Habitat		
Parameter	Primary Forest ($N = 30$)	Plantations ($N = 9$)	Secondary Forest ($N = 10$)
Total *Piper* species	8.6 ± 0.3	9.0 ± 0.8	14.0 ± 0.8
Total *Piper* individuals	59.6 ± 4.3	134.2 ± 28.8	163.2 ± 22.6

	Site and Habitat			
	East Boundary		West Boundary	
Parameter	Primary ($N = 3$)	Secondary ($N = 3$)	Primary ($N = 3$)	Secondary ($N = 3$)
Total *Piper* species	8.7 ± 0.9	15.7 ± 1.5	9.0 ± 0.6	14.7 ± 1.3
Total *Piper* individuals	82.0 ± 2.6	212.0 ± 51.6	60.7 ± 6.5	170.7 ± 11.3

Note: Area of each transect was 1,000 m². Data include mean ± 1 SE. N = number of transects.

and tree plantations adjacent to primary forest on recent alluvial soil (4 locations, 9 transects) [for La Selva soil maps and data, see Vitousek and Denslow (1987) and Sollins *et al.* (1994)]. Two sites, one along the east boundary line of the original La Selva tract and another along the west boundary line, were specifically chosen to compare adjacent primary and secondary forests while controlling for soil type (also see Laska 1997). In each transect, I identified and counted the number of adult-sized individuals. Identifications were based on an unpublished key devised by Michael Grayum and Barry Hammel. Robert Marquis provided Latin binomials for several species bearing only descriptive names in the Grayum-Hammel key.

To document the species composition of pipers in gaps, my field crew and I surveyed 37 gaps of various sizes and ages in a primary forest. For each gap, we recorded its greatest length and width, gap type (i.e., tree snap, tree fall, or branch fall), and relative age (e.g., "young" = still full of plant debris and sometimes foliage; "old" = debris and foliage gone and 4–6 m tall saplings present). In each gap we identified and counted pipers of all sizes. To compare pipers in gaps with those in intact forest, we censused pipers in one or two 300-m² belt transects in forest adjacent to 20 gaps.

Results of our transect censuses are summarized in Table 4.2. We encountered a total of 39 species of *Piper* in the 49 transects (Table 4.3). Six of these were vines and the others were shrubs or small trees. The mean number of *Piper* species per transect differed significantly among habitats in the following order: second growth > plantations > primary forest (Kruskal-Wallis ANOVA, $\chi^2 = 18.33$, $P = 0.0001$). Species richness was 63% higher, on average, in second growth transects than in primary forest transects. Comparisons between primary and secondary transects at the two paired sites indicated that species richness was 72% higher in secondary forest. Total density of *Piper* plants per transect paralleled trends in species richness (Table 4.2). *Piper* density in the second growth transects was nearly 3 times higher than in the primary forest transects. Five species (*P. arieianum*,

TABLE 4.3
Summary of *Piper* Density (Number of Adults per 1,000 m^2) and Distribution Data from the La Selva Transect and Gap Study

Species	Frequency of Occurrence		Density	
	Transects ($N = 49$)	Gaps ($N = 37$)	Transects (Mean ± SD)	Gaps (Mean ± SD)
aduncum	0.04	0	1.5	—
aequale	0.04	0.05	10	39.2
arboreum	0.04	0	5.5	—
arieanum	0.9	0.86	30 ± 42.0	49.9 ± 101.0
auritum	0.1	0.11	16.8 ± 11.0	23.6 ± 28.8
biolleyi	0.12	0.14	3.2 ± 3.4	13.7 ± 13.5
biseriatum	0.29	0.16	2.7 ± 2.1	15.5 ± 13.2
carilloanum	0.04	0	9.5	—
cenocladum	0.78	0.73	7.3 ± 5.9	23.2 ± 19.6
colonense	0.29	0.16	22.6 ± 29.6	85.3 ± 176.6
concepcionis	0.14	0.08	1.6 ± 0.7	10.9 ± 9.1
decurrens	0.06	0.03	1.3 ± 0.5	26.7
dolicotrichum	0.57	0.59	4.6 ±3.7	15.9 ± 16.4
friedrichsthallii	0.06	0	4.3 ± 2.6	—
"forest biggie"	0.33	0.27	3.3 ± 3.2	12.5 ± 7.4
gargaranum	0.12	0.16	1.8 ± 1.5	5.5 ± 2.0
glabrescens	0.12	0.08	1.7 ± 1.1	25.9 ± 24.5
holdridgeianum	0.63	0.76	18.2 ± 18.7	87.9 ± 117.6
"HSF1"	0.16	0.11	3.5 ± 4.2	7.6 ± 3.8
"HSF2"	0	0.02	—	4
"HSF4"	0.02	0.03	6	0.4
imperiale	0.14	0.11	2.1 ± 1.6	8.6 ± 4.4
"lemon-lime"	0.08	0	1.8 ± 1.3	—
melanocladum	0.51	0.43	3.2 ± 2.6	7.7 ± 4.5
multiplinervum	0.31	0.51	7.1 ± 7.8	13.0 ± 15.9
nudifolium	0.04	0	10	—
phytolaccaefolium	0.1	0	57.2 ± 89.3	—
"Phillipe's pubescent"	0.12	0.08	2.8 ± 1.2	2.5 ± 0.9
riparense	0.06	0.14	1	8.7 ± 9.1
reticulatum	0.18	0.05	5.9 ± 4.2	2.6
peracuminatum	0.1	0	4.2 ±3.9	—
pseudobumbratum	0.39	0.22	5.1 ± 5.2	9.2 ± 4.8
sancti-felicis	0.47	0.27	17.6 ± 19.5	13.7 ± 7.5
silvivagum	0.06	0	1.3 ± 0.5	—
"stachyum"	0.29	0.14	1.5 ± 0.5	4.9 ± 1.5
"swamp glabrous"	0.06	0	1.3 ± 0.5	—
tonduzii	0.73	0.76	3.5 ± 2.8	19.0 ± 14.4
urostachyum	0.86	0.73	7.0 ± 13.3	25.7 ± 22.5
virgulatorum	0.14	0	11.4 ± 8.9	—

P. cenocladum, P. holdridgeianum, P. gargaranum, and *P. urostachyum*) occurred in ≥60% of the transects and can be considered to be "common"; 28 species (76%) occurred in ≤20% of the transects and can considered to be "uncommon" or "rare" (Table 4.3).

I used information in Hartshorn (1983) and from Daviu Clark (personal communication) to determine the approximate time since last clearing of the second growth and plantation sites we surveyed. These times ranged from 10 to 30 years. Over this time span, *Piper* diversity increased from 7 to 8 species per transect at 10 years to a peak of 16–18

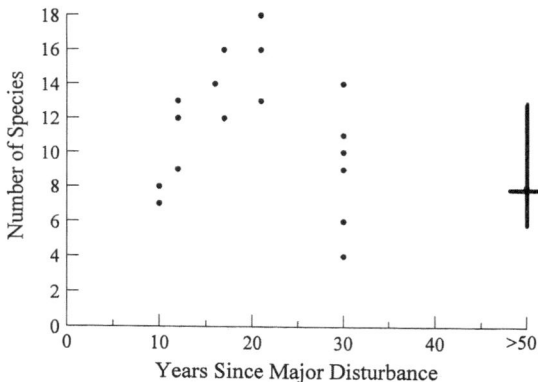

FIGURE 4.6. Number of *Piper* species in 1,000-m² belt transects in a variety of successional habitats in tropical wet forest at La Selva, Costa Rica. Each point represents one transect. Data for primary forest (≫50 years old) indicate median and range of values.

species at 15–20 years before falling to a median value of 9 species at 30 years (Fig. 4.6). Median species richness in intact primary forest was 8 species per transect. Although we did not survey cleared sites less than 10 years old, my casual observations in the systematically cleared "successional strips" (Hartshorn 1983) and in other recently cleared land around La Selva indicated that only a few species (e.g., *P. auritum, P. biseriatum, P. aduncum,* and *P. sancti-felicis*) occur in new, large clearings. Thus, *Piper* diversity increased with time since clearing, reached a peak at about 20 years postclearing, and declined as forest succession continued.

We encountered a total of 27 species of *Piper* in the 37 gaps. Each of these species occurred in one or more of our 49 transects (Table 4.3). There was a strong positive correlation between the frequency of occurrence of the 27 species in gaps and their frequency in transects (arcsine-transformed data, Pearson's $r = 0.94$, $F_{1,25} = 206.8$, $P < 0.0001$). The number of *Piper* species per gap was positively correlated with gap area (Pearson's $r = 0.57$, $F_{2,35} = 16.48$, $P = 0.0003$) but did not differ between "young" and "old" gaps (Mann–Whitney U test, $P > 0.32$). On the basis of Sørensen's (1948) similarity index, which is based on presence–absence data, similarity between the gaps and adjacent intact forest was high and averaged 0.61 ± 0.03 (SE) (out of a maximum value of 1.0) in the 20 gap–transect comparisons.

From these results, I reach the following conclusions. First, the rich *Piper* flora of this lowland wet forest contains a few (ca. five) common and broadly distributed species and many uncommon species, a pattern that characterizes most biotic communities. Second, *Piper* species richness and density varies substantially among habitats where the second growth forest has higher point diversities (i.e., the number of species co-occurring in a small area) than in other habitats. Third, *Piper* diversity has a temporal or successional component. It is initially low on newly cleared land and increases as forest succession proceeds. Point diversity peaks in relatively young forests and is low in mature forests. Fourth, colonization of gaps in primary forests appears to involve local, rather than long-distance (i.e., from distant disturbances or second growth), dispersal. Composition of the *Piper* flora in the

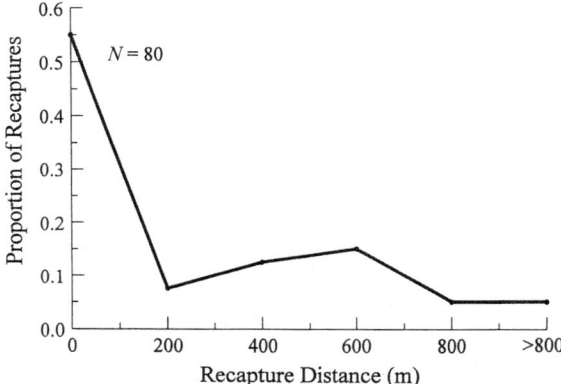

FIGURE 4.7. Distribution of recapture distances for three species of *Carollia* (data combined) in tropical wet forest at La Selva Biological Station, Costa Rica.

gaps we censused largely reflected composition of the flora in adjacent intact forests. The relative importance of fresh seed rain versus germination from the soil seed bank (or from plant fragments; Greig 1993b) in colonization of primary forest gaps needs further study. Finally, disturbance and habitat heterogeneity clearly play important roles in maintaining high local diversity in this *Piper* flora.

How do the *Piper* bats respond to this mosaic of *Piper* diversity and density? During our La Selva transect and gap study, we also captured phyllostomid bats in mist nets set at ground level in a variety of habitats (Fleming 1991). Bats were marked with ball–chain necklaces bearing a numbered aluminum bird band before being released (Fleming 1988). In a 2-month period (late May–late July), we recorded 819 captures and recaptures of three species of *Carollia* (252 *C. brevicauda*, 300 *C. castanea*, and 267 *C. perspicillata*). Proportion of fecal samples that contained *Piper* seeds ranged from 0.54 (in *C. perspicillata*) to 0.87 (in *C. castanea*). Reflecting the higher density of pipers in second growth forest, capture rates of *Carollia* in that habitat were 3.4 times higher than in the primary forest (second growth: 1.79 ± 0.43 SE captures per net-hour; primary forest: 0.53 ± 0.09 captures per net-hour). The smallest species (*C. castanea*) was most common in the second growth forest, and the largest species (*C. perspicillata*) was most common in the primary forest, a pattern that occurs elsewhere in these bats (Fleming 1991).

The distribution of recapture distances of these bats provides some insight into their movement patterns and the potential distances they can disperse *Piper* seeds. Fifty-five percent of the 80 *Carollia* recaptures occurred at the site where the bat was originally captured, but two individuals were recaptured 1.7 km from their original capture sites. In general, the distribution of recapture distances was leptokurtic and with a long tail (Fig. 4.7), a pattern also seen in my estimate of *Piper* seed dispersal distances in tropical dry forests (Fig. 4.5). These netting results suggest that the *Carollia* bats at La Selva are relatively sedentary but that they move among habitats and produce a mix of seed dispersal distances. For example, the recapture data show that the proportion of bats recaptured in a different habitat from the one in which they were first captured was 0.25 and 0.41 for bats first captured in the secondary and the primary forest, respectively. Thus, although most of the seeds they ingest are likely to be dispersed short distances (<100 m) from parent plants, some seeds can be dispersed substantial distances (1 km or more) and into different habitats from where they were produced.

4.4. COEVOLUTIONARY ASPECTS OF BAT–*Piper* INTERACTIONS

Compared with plant–pollinator mutualisms, most fruit–frugivore interactions are relatively unspecialized (Wheelwright and Orians 1982, Janzen 1983). The relationship between *Carollia* bats and Neotropical *Piper* plants, however, appears to be quite specialized. In this system, specialization seems to be higher on the plant side than on the frugivore side. Except for a few common large gap species (e.g., *P. aduncum* and *P. auritum*), Neotropical pipers appear to be nearly exclusively dispersed by *Carollia* bats (at least on the mainland). *Carollia* dependence on *Piper* varies predictably among species, with the smaller *C. castanea* being a stronger *Piper* specialist than the larger *C. perspicillata* (Fleming 1991).

Incidentally, although Old World frugivorous bats of the family Pteropodidae rarely eat the fruits of native pipers (Mickleburgh *et al*. 1992), certain species (e.g., *Cynopterus brachyotis*, *Syconycteris australis*) avidly eat and disperse the seeds of New World early successional species such as *P. aduncum* (Winkelmann *et al.* 2000). New World pipers can clearly attract bats to their infructescences, presumably using both visual and olfactory cues. Old World pipers appear to be dispersed primarily by birds (Snow 1981).

Conditions promoting a high degree of specialization in mutualisms are generally thought to involve reliability and effectiveness (e.g., McKey 1975, Howe 1984, Fleming and Sosa 1994, Waser *et al*. 1996). On the plant side, reliability usually refers to spatio-temporal predictability in resource availability, and effectiveness refers to providing a suitable nutritional reward. On the animal side, reliability refers to predictability of visitation or resource use, and effectiveness refers to treatment of pollen or seeds in a nonharmful fashion. In fruit–frugivore systems, effectiveness ultimately involves deposition of seeds in suitable recruitment sites.

On each of these counts, the *Piper*–*Carollia* dispersal system seems to meet conditions favoring specialization. As mentioned earlier, *Piper* flowering and fruiting phenology provides year-round resources for pollinators and frugivores in many Neotropical forests. Although the nutritional characteristics of *Piper* fruit are relatively unstudied, Herbst (1985) found that fruits of *P. amalago* are rich enough in protein, often a limiting resource in fruit pulp, to support wet season pregnancies and lactation in *C. perspicillata*. Dinerstein (1986) reported that protein levels in fruits of eight species of *Piper* plus *Potomorphe peltata* (now *Piper peltata*) averaged 7.2% (dry weight) and were higher than most bat fruits in a Costa Rican cloud forest. *Carollia* bats are avid consumers of *Piper* fruits, and nearly all fruits are removed and eaten as soon as they ripen (cf. certain species of *Ficus* trees that suffer enormous fruit wastage; Kalko *et al*. 1996). Although the probability of any given *Piper* seed being deposited in a site that is suitable for immediate germination is vanishingly small, the sheer numbers of seeds dispersed nightly across a varied landscape by roosts of several hundred *Carollia* bats is enormous (i.e., in tens of thousands) (Fleming 1988). Coupled with dormancy mechanisms and photoblastic responses, this "broadcast" dispersal is sufficient for *Piper* to quickly colonize both large and small habitat disturbances whenever they occur (e.g., see Chapter 10 in Fleming 1988). This mutual specialization has made *Piper* plants and *Carollia* bats very common in lowland Neotropical forests.

Despite a high degree of specialization between *Piper* plants and *Carollia* bats, their spatio-temporal distributions are not congruent. In terms of geographic distributions, diversity and abundance of both taxa are highest in the lowlands of the mainland Neotropics (especially in northwestern South America and southern Central America) and decrease with increasing latitude and altitude (e.g., Fleming 1986). Except for Grenada, *Carollia* bats are restricted to the mainland of Mexico and Central and South America, whereas pipers occur widely in the Greater and Lesser Antilles. *Sturnira lilium* is more broadly distributed in the Caribbean than *Carollia* bats and occurs as far north as Dominica in the Lesser Antilles; the endemic *S. thomasi* is known from Montserrat and Guadeloupe (Rodriguez-Duran and Kunz 2001). *Sturnira* bats presumably eat *Piper* fruits in the Lesser Antilles. In the Greater Antilles, *Erophylla bombifrons* (Phyllostomidae: Phyllonycterinae) eats *Piper* fruit in Puerto Rico (T. Fleming, pers. obs.) and probably elsewhere in the Greater Antilles.

In terms of geological history, *Piper* and Piperaceae are much older than phyllostomid bats. Pollen of Piperaceae, for example, is known from the Eocene (Muller 1970). Current geographic distributions and morphological and molecular data indicate that Piperaceae evolved early in angiosperm history, before the breakup of Gondwana and therefore probably before the Cenozoic. In contrast, the earliest phyllostomid bat (*Notonycteris*; probably not a frugivore) dates from the Miocene (Koopman 1984). On the basis of allozyme data, Straney *et al.* (1979) suggested that the Phyllostomidae evolved in the early Oligocene [ca. 30 million years ago (mya)]. Using DNA restriction site data, Lim and Engstrom (1998) concluded that *C. castanea* is the basal member of the genus and evolved in South America, presumably ≤20 mya. If this is true, then pipers and *Carollia* bats have been interacting for no more than 20 million years—long after the initial radiation of pipers around the world. This temporal mismatch raises the question: Who were the dispersers of *Piper* seeds before the evolution of *Piper* bats? What animals (pollinators and dispersers) originally selected for *Piper* inflorescence and infructescence characteristics?

Given its exceptionally high species richness in the Neotropics, we can further ask: What role, if any, have *Carollia* bats played in *Piper* speciation? As discussed by Hamrick and Loveless (1986), the foraging behavior of both pollinators and seed dispersers can affect the genetically effective size (N_e) of plant populations and the extent of gene flow between populations. Because seeds are diploid (cf. haploid pollen) and, in the absence of strong selective barriers, "foreign" seeds can establish as readily in local populations as "local" seeds, gene flow by seed dispersal can theoretically have a greater effect on genetic structure than gene flow by pollination. Long-distance seed dispersal will reduce levels of population subdivision and short-distance dispersal will have the opposite effect.

As described above, foraging by *Carollia* bats produces a mixture of seed dispersal distances, and these bats therefore serve as agents of both local and long-distance dispersal. Most of the seeds they ingest are deposited in clumps (of full or half-sibs) under night roosts located short distances from parent plants and will recruit locally, if they recruit at all. In contrast, some seeds move considerable distances (1–2 km) when bats change feeding areas and can potentially recruit in new habitats. Local recruitment of closely related seedlings, coupled with short-distance movements by insect pollinators, will produce high levels of genetic subdivision and potentially small values for N_es (Hamrick and Loveless 1986). The occasional colonization of recently disturbed habitats by a small number of closely related seedlings will also produce new populations with small N_es. Thus, dispersal of seeds by *Carollia* bats has the potential to produce genetically subdivided *Piper* populations, an important first step toward speciation.

Do *Piper* populations show high or low levels of genetic subdivision? Unfortunately, only one study appears to address this question. Heywood and Fleming (1986) studied genetic diversity and population structure using allozymes in three species of *Piper* in Costa Rican tropical dry forest. They found very low levels of protein polymorphism within populations (proportion of polymorphic loci ranged from 0 in *P. pseudo-fuligineum* to 0.095 in *P. amalago* for 20–24 loci), significant levels of genetic subdivision between populations of *P. amalago* ($F_{st} = 0.103$ at one locus, where a value of 0 indicates panmixia), and unusually high genetic distances (0.45–0.99 out of 1.0) between the three species. High genetic distances between congeners suggest that this genus is relatively old. Although more studies are needed to assess the role of pollinators and seed dispersers in determining the genetic structure of *Piper* populations, it appears that gene flow may indeed be limited in these bat-dispersed plants.

In his extensive review of Neotropical floral diversity, Gentry (1982) identified Piperaceae as a member of the Gondwanan, Andean-centered group of epiphytes, shrubs, and palmetto-like monocots that collectively accounts for nearly one-half of this diversity. Unlike Gentry's "Amazonian-centered trees and lianas," whose diversity is likely to be the result of allopatric speciation, Andean plants have undergone "explosive speciation and adaptive radiation, almost certainly most of it sympatric" (Gentry 1982). Gentry (1982) proposed that this speciation has occurred in small local populations and has involved constant recolonization of habitats separated by mountains, local rainshadows, shifting vegetation zones (in response to global climate fluctuations), and frequent landslides. Similarly, in his review of the ecological distributions of pipers and other species-rich genera in Costa Rica, Burger (1974) pointed out that many species have narrow ecological boundaries and that closely related species often co-occur in the same area but live in different habitats. Both scenarios suggest that genetic isolation occurs frequently in pipers. But the extent to which this is the result of limited pollen or seed movement remains to be seen. Genetic studies using both chloroplast and nuclear markers will help to clarify this issue.

4.5. CONCLUSIONS

To judge from its high diversity and abundance, the genus *Piper* has been extremely successful in the Neotropics. Its current diversity encompasses a wide range of life forms, including subshrubs, vines, shrubs, and treelets, that occupy a variety of habitats ranging from early successional disturbances to mature forests. Many species appear to reproduce most often by sexual means, and seed dispersal is accomplished using a small clade of fruit-eating bats of the genus *Carollia* (Phyllostomidae, Caroliinae). Reliable, year-round availability of relatively nutritious fruit has enabled *Carollia* bats to specialize on *Piper* for most of their diet in many places in the lowland Neotropics. These common and relatively sedentary bats nightly disperse tens of thousands of *Piper* seeds into a variety of habitats. Although most seeds are deposited close to parent plants, some can move 1–2 km into new habitats. This mixture of short- and long-distance dispersal is likely to create genetically subdivided populations of pipers. Genetic subdivision, in turn, can set the stage for speciation, especially in areas of high habitat diversity such as the foothills of the Andes in northern South America and southern Central America. It is likely that *Piper*'s relatively specialized dispersal ecology has played an important role in its evolutionary success.

4.6. ACKNOWLEDGMENTS

I thank Lee Dyer for an invitation to write this chapter and Elisabeth Kalko for a copy of her unpublished *Piper* manuscript, coauthored with Wiebke Thies. Financial support for my studies of *Piper* and *Carollia* in Costa Rica was provided by the U.S. National Science Foundation and the Center for Field Research. I was assisted by J. Maguire and several teams of Earthwatch volunteers in the La Selva study.

REFERENCES

Baker, R. J., Solari, S., and Hoffmann, F. G. (2002). A new Central American species from the *Carollia brevicauda* group. Occasional Papers of the Museum of Texas Tech University, 217: 1–16.

Bizerril, M. X. A., and Raw, A. (1998). Feeding behaviour of bats and the dispersal of *Piper arboreum* seeds in Brazil. *Journal of Tropical Ecology* 14:109–114.

Bonaccorso, F. J., and Gush, T. J. (1987). An experimental study of the feeding behaviour and foraging strategies of phyllostomid bats. *Journal of Animal Ecology* 56:907–920.

Burger, W. C. (1974). Ecological differentiation in some congeneric species of Costa Rican flowering plants. *Annals of the Missouri Botanical Garden* 61:297–306.

Charles-Dominique, P. (1986). Inter-relations between frugivorous vertebrates and pioneer plants: *Cecropia*, birds, and bats in French Guyana. In: Estrada, A., and Fleming, T. H. (eds.), *Frugivores and Seed Dispersal*. Dr. W. Junk Publishers, Dordrecht, The Netherlands, pp. 119–135.

Charles-Dominique, P. (1991). Feeding strategies and activity budget of the frugivorous bat *Carollia perspicillata* (Phyllostomidae) in French Guiana. *Journal of Tropical Ecology* 7:243–256.

Charles-Dominique, P. (1993). Tent-Use by the bat *Rhinophylla pumilio* (Phyllostomidae, Carolliinae) in French-Guiana. *Biotropica* 25:111–116.

Charles-Dominique, P., and Cockle, A. (2001). Frugivory and seed dispersal by bats. In: Bongers, F., Charles-Dominique, P., Forget, P.-M., and Thery, M. (eds.), *Nouragues. Dynamics and Plant–Animal Interactions in a Neotropical Rainforest*. Kluwer Academic Publishers, Dordrecht, The Netherlands, pp. 207–215.

Denslow, J. S., and Moermond, T. C. (1985). The interaction of fruit display and the foraging strategies of small frugivorous birds. In: D'Arcy, W., and Correa, M. D. (eds.), *The Botany and Natural History of Panama*. Missouri Botanical Garden, St. Louis, Missouri, pp. 245–253.

Dinerstein, E. (1986). Reproductive ecology of fruit bats and the seasonality of fruit production in a Costa Rican cloud forest. *Biotropica* 18:307–318.

Fleming, T. H. (1981). Fecundity, fruiting pattern, and seed dispersal in *Piper amalago* (Piperaceae), a bat-dispersed tropical shrub. *Oecologia* 51:42–46.

Fleming, T. H. (1986). Opportunism versus specialization: The evolution of feeding strategies in frugivorous bats. In: Estrada, A., and Fleming, T. H. (eds.), *Frugivores and Seed Dispersal*. Dr. W. Junk Publishers, Dordrecht, The Netherlands, pp. 105–118.

Fleming, T. H. (1988). *The Short-Tailed Fruit Bat: A Study in Plant–Animal Interactions*. University of Chicago Press, Chicago.

Fleming, T. H. (1991). The relationship between body size, diet, and habitat use in frugivorous bats, genus *Carollia* (Phyllostomidae). *Journal of Mammalogy* 72:493–501.

Fleming, T. H., and Heithaus, E. R. (1981). Frugivorous bats, seed shadows, and the structure of tropical forests. *Biotropica* 13(Suppl.):45–53.

Fleming, T. H., and Heithaus, E. R. (1986). Seasonal foraging behavior of *Carollia perspicillata* (Chiroptera: Phyllostomidae). *Journal of Mammalogy* 67:660–671.

Fleming, T. H., and Sosa, V. J. (1994). The effects of mammalian frugivores and pollinators on plant reproductive success. *Journal of Mammalogy* 75:845–851.

Fleming, T. H., and Williams, C. F. (1990). Phenology, seed dispersal, and recruitment in *Cecropia peltata* (Moraceae) in Costa Rican tropical dry forest. *Journal of Tropical Ecology* 6:163–178.
Fleming, T. H., Heithaus, E. R., and Sawyer, W. B., (1977). An experimental analysis of the food location behavior of frugivorous bats. *Ecology* 58:619–627.
Fleming, T. H., Williams, C. F., Bonaccorso, F. J., and Herbst, L. H. (1985). Phenology, seed dispersal, and colonization in *Muntingia calabura*, a Neotropical pioneer tree. *American Journal of Botany* 72:383–391.
Galindo-Gonzalez, J., Guevara, S., and Sosa, V. J. (2000). Bat- and bird-generated seed rains at isolated trees in pastures in a tropical forest. *Conservation Biology* 14:1693–1703.
Gentry, A. H. (1982). Neotropical floristic diversity: Phytogeographical connections between Central and South America, Pleistocene climatic fluctuations, or an accident of the Andean orogeny? *Annals of the Missouri Botanical Garden* 69:557–593.
Gentry, A. H. (ed.). (1990). *Four Neotropical Rainforests*. Yale University Press, New Haven, Connecticut.
Gentry, A. H., and Emmons, L. H. (1987). Geographical variation in fertility, phenology, and composition of the understory of Neotropical forests. *Biotropica* 19:216–227.
Gorchov, D. L., Cornejo, F., Ascorra, C. F., and Jaramillo, M. (1995). Dietary overlap between frugivorous birds and bats in the Peruvian Amazon. *Oikos* 74:235–250.
Greig, N. (1993a). Predispersal seed predation on five *Piper* species in tropical rainforest. *Oecologia* 93:412–420.
Greig, N. (1993b). Regeneration mode in Neotropical *Piper*: Habitat and species comparisons. *Ecology* 74:2125–2135.
Hamrick, J. L., and Loveless, M. D. (1986). The influence of seed dispersal mechanisms on the genetic structure of plant populations. In: Estrada, A., and Fleming, T. H. (eds.), *Frugivores and Seed Dispersal*. Dr. W. Junk Publishers, Dordrecht, The Netherlands, pp. 211–223.
Hartshorn, G. S. (1983). Plants: Introduction. In: Janzen, D. H. (ed.), *Costa Rican Natural History*. University of Chicago Press, Chicago, pp. 118–183.
Heithaus, E. R., and Fleming, T. H. (1978). Foraging movements of a frugivorous bat, *Carollia perspicillata* (Phyllostomidae). *Ecological Monographs* 48:127–143.
Heithaus, E. R., Fleming, T. H., and Opler, P. A. (1975). Patterns of foraging and resource utilization in seven species of bats in a seasonal tropical forest. *Ecology* 56:841–854.
Herbst, L. H. (1985). The role of nitrogen from fruit pulp in the nutrition of a frugivorous bat, *Carollia perspicillata*. *Biotropica* 18:39–44.
Heywood, J. S., and Fleming, T. H. (1986). Patterns of allozyme variation in three Costa Rican species of *Piper*. *Biotropica* 18:208–213.
Howe, H. F. (1984). Constraints on the evolution of mutualisms. *American Naturalist* 123:764–777.
Howe, H. F. (1989). Scatter- and clump-dispersal by birds and mammals: Implications for seedling demography. *Oecologia* 79:417–426.
Janzen, D. H. (1970). Herbivores and the number of tree species in tropical forests. *American Naturalist* 104:501–528.
Janzen, D. H. (1983). Seed and pollen dispersal by animals: Convergence in the ecology of contamination and sloppy harvest. *Biological Journal of the Linnean Society* 20:103–113.
Jaramillo, M. A., and Manos, P. S. (2001). Phylogeny and patterns of floral diversity in the genus *Piper* (Piperaceae). *American Journal of Botany* 88:706–716.
Kalko, E. K. V., Herre, E. A., and Handley, C. O., Jr. (1996). The relation of fig fruit syndromes to fruit-eating bats in the New and Old World tropics. *Journal of Biogeography* 23: 565–576.
Koopman, K. F. (1984). Bats. In: Anderson, S., and Jones, J. K. (eds.), *Orders and Families of Recent Mammals of the World*. John Wiley & Sons, New York, pp. 145–186.
Koopman, K. F. (1993). Order Chiroptera. In: Wilson, D. E., and Reeder, D. M. (eds.), *Mammal Species of the World*, 2nd ed. Smithsonian Institution Press, Washington, D.C., pp. 137–241.
Laska, M. S. (1997). Structure of understory shrub assemblages in adjacent secondary and old growth tropical wet forests, Costa Rica. *Biotropica* 29:29–37.
Levey, D. J. (1987). Seed size and fruit-handling techniques of avian frugivores. *American Naturalist* 129:471–485.
Lim, B. K., and Engstrom, M. D. (1998). Phylogeny of Neotropical short-tailed fruit bats, *Carollia* spp., phylogenetic analysis of restriction site variation in mtDNA. In: Kunz, T. H., and Racey, P. A. (eds.), *Bat Biology and Conservation*. Smithsonian Institution Press, Washington, D.C., pp. 43–58.
Loiselle, B. A. (1990). Seeds in droppings of tropical fruit-eating birds: Importance of considering seed composition. *Oecologia* 82:494–500.
Marhino-Filho, J. S. (1991). The coexistence of two frugivorous bat species and the phenology of their food plants in Brazil. *Journal of Tropical Ecology* 7:59–67.
Marquis, R. J. (1988). Phenological variation in the Neotropical understory shrub *Piper arieianum*: Causes and consequences. *Ecology* 69:1552–1565.
McKey, D. (1975). The ecology of coevolved seed dispersal systems. In: Gilbert, L. E. and Raven, P. H. (eds.), *Coevolution of Animals and Plants*. University of Texas Press, Austin, Texas, pp. 159–191.

Medellin, R. A., and Gaona, O. (1999). Seed dispersal by bats and birds in forest and disturbed habitats of Chiapas, Mexico. *Biotropica* 31:478–485.
Mickleburgh, S. P., Hutson, A. M., and Racey, P. A. (eds.). (1992). *Old World Fruit Bats, an Action Plan for Their Conservation*. International Union for the Conservation of Nature and Natural Resources, Gland, Switzerland.
Muller, J. (1970). Palynological evidence on early differentiation of angiosperms. *Biological Reviews* 45:417–450.
O'Donnell, S. (1989). A comparison of fruit removal by bats and birds from *Piper hispidum* Sw. (Piperaceae), a tropical second growth shrub. *Brenesia* 31:25–32.
Opler, P. A., Frankie, G. W., and Baker, H. G. (1980). Comparative phenological studies of treelet and shrub species in tropical wet and dry forests in the lowlands of Costa Rica. *Journal of Ecology* 68:167–188.
Orozco-Segovia, A., and Vazquez-Yanes, C. (1989). Light effect on seed germination in *Piper* L. *Acta Oecologica* 10:123–146.
Palmeirim, J. M., Gorchov, D. L., and Stoleson, S. (1989). Trophic structure of a Neotropical frugivore community: Is there competition between birds and bats? *Oecologia* 79:403–411.
Perry, A. E., and Fleming, T. H. (1980). Ant and rodent predation on small, animal-dispersed seeds in a dry tropical forest. *Brenesia* 17:11–22.
Prance, G. T. (1990). The floristic composition of the forests of central Amazonian Brazil. In: Gentry, A. H. (ed.), *Four Neotropical Rainforests*. Yale University Press, New Haven, Connecticut, pp. 112–140.
Rinehart, J. B. (2002). Evidence of territoriality in the Neotropical tent-roosting bat, *Rhinophylla pumilio*, in eastern Ecuador. *Abstracts of the North American Symposium on Bat Research*.
Rodriguez-Duran, A., and Kunz, T. H. (2001). Biogeography of West Indian bats: An ecological perspective. In: Woods, C. A., and Sergile, F. E. (eds.), *Biogeography of the West Indies*. CRC Press, Boca Raton, Florida, pp. 355–368.
Simmons, N. B., and Voss, R. S. (1998). The mammals of Paracou, French Guiana: A Neotropical lowland rainforest fauna. Part I: Bats. *Bulletin of the American Museum of Natural History* 237:1–210.
Snow, D. W. (1981). Tropical frugivorous birds and their food plants: A world survey. *Biotropica* 13:1–14.
Sollins, P., Sancho M., F., Mata Ch., R., and Sanford, J. R. L. (1994). Soils and soil process research. In: McDade, L. A., Bawa, K. S., Hespenheide, H. A., and Hartshorn, G. S. (eds.), *La Selva, Ecology and Natural History of a Neotropical Rain Forest*. University of Chicago Press, Chicago, pp. 34–53.
Sorensen, T. (1948). A method of establishing groups of equal amplitude in plant sociology based on similarity of species content. *Det. Kong. Danske Vidensk. Selsk. Biol. Skr.* 8:1–34.
Straney, D. O., Smith, M. H., Greenbaum, I. F., and Baker, R. J. (1979). Biochemical genetics. In: Baker, R. J., Jones, J. K., and Carter, D. C. (eds.), *Biology of Bats of the New World Family Phyllostomatidae, Part III*. Special Publications of the Museum, Texas Tech University, Lubbock, Texas, No. 16, pp. 157–176.
Thies, W., Kalko, E. K. V., and Schnitzler, H.-U. (1998). The roles of echolocation and olfaction in two Neotropical fruit-eating bats, *Carollia perspicillata* and *C. castanea*, feeding on *Piper*. *Behavioral Ecology and Sociobiology* 42:397–409.
Thomas, D. W., Cloutier, D., Provencher, M., and Houle, C. (1988). The shape of bird- and bat-generated seed shadows around a tropical fruiting tree. *Biotropica* 20:347–348.
Vitousek, P. M., and Denslow, J. S. (1987). Differences in extractable phosphorus among soils of the La Selva Biological Station, Costa Rica. *Biotropica* 19:167–170.
Waser, N. M., Chittka, L., Price, M. V., Williams, N. M., and Ollerton, J. (1996). Generalization in pollination systems, and why it matters. *Ecology* 77:1043–1060.
Wetterer, A. L., Rockman, M. V., and Simmons, N. B. (2000). Phylogeny of phyllostomid bats (Mammalia: Chiroptera): Data from diverse morphological systems, sex chromosomes, and restriction sites. *Bulletin of the American Museum of Natural History* 248:1–200.
Wheelwright, N. T., and Orians, G. H. (1982). Seed dispersal by animals: Contrasts with pollen dispersal, problems of terminology, and constraints on coevolution. *American Naturalist* 119:402–413.
Winkelmann, J. R., Bonaccorso, F. J., and Strickler, T. L. (2000). Home range of the southern blossom bat, *Syconycteris australis*, in Papua New Guinea. *Journal of Mammalogy* 81:408–414.

5
Biogeography of Neotropical *Piper*

Robert J. Marquis
Department of Biology, University of Missouri—St. Louis, St. Louis, Misssouri

5.1. INTRODUCTION

The history and nature of diversification of *Piper* in the New World has been little investigated. The genus *Piper* is often among the top 10 most speciose genera in Neotropical forests (e.g., Gentry and Emmons, 1987). Superficially, however, *Piper* species show relatively little morphological variation. In view of the high species richness in the genus but relatively low morphological variability among species, Gentry (1989) suggested that much diversification in Neotropical *Piper* may have occurred relatively recently (within the last 2 million years or even more recently). The premise is that because diversification was recent, little time has been available for morphological diversification.

Studies of the timing and pattern of diversification of speciose tropical genera may help reveal the underlying processes that have led to high tropical diversity in general (Richardson *et al.* 2001). In the Neotropics to date, only tree taxa (Prance 1982) have been investigated with this goal in mind, and rarely has a single genus been the focus of study (*cus*: Machado *et al.* 2001, *Inga*: Richardson *et al.* 2001). Studies of taxa that are both speciose regionally and locally, such as *Piper*, may provide insight into how regional and local processes have contributed to diversification and to local community species richness and composition (Ricklefs and Schluter 1993).

At the regional level, study of the biogeography of Neotropical *Piper* is limited to mapping the number of species found in various regions within the Neotropics (Jaramillo and Manos 2001). Apparently there have been no comparisons of regions, and sites within regions, based on similarity in species composition. Such a comparison would provide insight into the factors that determine the size of the regional pool of species, and in so doing determine the upper limit to the number of species that might occur locally. Comparison of floras across regions would also provide hypotheses about the pattern of diversification of the genus.

At the local level, species richness may be determined by the number of available niches. For species-rich genera such as *Piper* (as many as 60+ *Piper* species may be found in a relatively restricted geographic area), an obvious question is how so many closely related species can coexist at a single location. A commonly proposed hypothesis is that the number of species at a site is limited by the number of available niches (Pianka 1994). In turn, the number of niches will increase with productivity (Pianka 1994, Rosenzweig and Abramsky 1993). Productivity has been estimated at relatively few tropical forest sites (Clark *et al.* 2001); however, correlates of productivity such as rainfall and/or temperature are available as surrogate estimators of productivity. Although much additional data are needed (Clark *et al.* 2001), preliminary estimates suggest that the distribution of productivity with rainfall in tropical forests is unimodal, peaking at approximately 3,500 mm annual rainfall (Clark *et al.* 2001). *Piper* species richness, if limited by the number of niches, would be predicted to follow a similar pattern. The number of species is also expected to decline with elevation, given that productivity should decline with elevation. Finally, most species of *Piper* in the New World are shrubs, but small trees, herbs, vines, and lianas also occur. We might then predict that sites with the highest richness would be those in which these other growth forms (which could exploit additional niches beyond that of the "shrub niche") are most frequently represented.

An alternative hypothesis to the niche–productivity hypothesis is that the genus is most diverse in lowland humid forests, because it first arose there and therefore is best adapted to the environmental conditions of those forests. The genus is classically considered to be most diverse in wet, low-elevation tropical forests (Burger 1971). According to this hypothesis, richness would decline with decreasing rainfall and increasing seasonality of that rainfall. This pattern has been found for shrub species in general in the Neotropics (Gentry and Emmons 1987) and for Neotropical trees (Gentry 1988a,b). Richness would also decline with elevation, a pattern found for trees in Ecuador (Boyle 1996), and for lianas (Gentry 1991).

My objectives within this chapter are twofold. First, I describe patterns of biogeographic variation in community composition. This analysis provides information on how the regional species pool varies in size and composition across the Neotropics, and possible hypotheses regarding diversification of the genus. I compare local floras rather than regional floras (e.g., the flora of Mesoamerica or Venezuela) because most regional floras are not taxonomically current, have been poorly collected, or are not yet available. In contrast, the floras of many local sites are now relatively well known. Second, I relate geographic variation in local species richness with latitude, and with variation in rainfall and elevation. In so doing, I test whether richness patterns are consistent with the prediction that the number of niches, and therefore the number of species, is highest at highest productivity (i.e., at intermediate levels of rainfall), or alternatively, increases monotonically with increasing rainfall. In addition, I review what is known about variation in habitat affinities and growth form within local sites, with the goal of understanding how partitioning by soil type/local habitat has contributed to local species coexistence. The range of growth forms and habitat affinities has not been previously documented for this genus.

5.2. METHODS

With five exceptions, species lists came from local florulas published in 1980 or later (including Web-based florulas) or their unpublished updates (Table 5.1, Fig. 5.1). The exceptions were five additional data sets which have yet to be published: one site

TABLE 5.1
Site Characteristics of Those Used as the Database for This Study

Site	Country	Latitude	Longitude	Elevation (m)	Annual Rainfall (mm)	Forest Type	Sample Area (ha)	No. Piper Species	References
El Cielo[1]	Mexico	22°50′–23°25′N	99°05–26′W	330–1,600	2,522	Semideciduous, montane mesophyll	144,530	3	Johnston et al. 1998, http://maya.ucr.edu/pril/reservas/elcielo/elcielo1.html
Chamela	Mexico	19°22′–19°39′N	104°56′–105°10′W	10–150	731	Deciduous	3,319	8	http://www.ibiologia.unam.mx/ebchamela/GENINFOCHAMELA.html, Lott et al. 2002
Manantlan	Mexico	19°20′–40′N	103°50′–104°27′W	350–2,025	600–1,000	Bosque tropical caducifolia	55,700	15	Vazquez et al. 1995
Los Tuxtlas	Mexico	18°34′N	95°04′W	150–530	4,725	Evergreen	700	11	Manríquez and Colin 1997
Chiquibul	Belize	16°44′N	88°59′W	500–700	2,250	Deciduous semievergreen; deciduous	174,400	20	N. Garwood, C. Whitfoord, pers. comm.
Lacondona	Mexico	16°04′–16°57′ N	90°45′–91°30′ W	100–900	2,500	Evergreen	600,000	37	http://maya.ucr.edu/pril/reservas/montesazules/montesazules2.html, Dirzo and Miranda 1991, Martínez et al. 1994
Santa Rosa	Costa Rica	10°45′–11°00′N	85°30′–85°45′W	0–300	1,750	Deciduous	10,300	11	Janzen and Liesner 1980, Hartshorn 1983
La Selva	Costa Rica	10°26′N	83°59′W	50	4,000	Evergreen	1,536	43	http://www.ots.ac.cr/en/laselva/species/vascular.shtml, McDade and Hartshorn 1994, Sanford et al. 1994, Sollins et al. 1994
Palo Verde	Costa Rica	10°21′N	85°21′W	50–100	1,250	Deciduous	4,757	3	http://www.ots.ac.cr/en/paloverde/species/plants.shtml, Hartshorn 1983
Monteverde	Costa Rica	10°15′–25′N	84°35′–49′W	1,100–1,850	2,700[2]	Montane wet	2,519	35/9[3]	Nadkarni and Wheelwright 2000, Clark et al. 2000b Haber 2000
Barro Colorado Island	Panama	9°09′N	79°51′W	0–171	2,600	Semideciduous	1,560	21	Croat 1978, Dietrich et al. 1982

Site	Country	Latitude	Longitude	Elevation	Rainfall	Forest type	Area	Species	Reference
Sirena	Costa Rica	8°29′N	83°36′W	0	5,000	Evergreen	750	36	http://www.utexas.edu/courses/zoo384l/sirena/species/plants/; L. Gilbert and N. Greig, pers. comm.
Nouragues	French Guiana	4°05′N	52°40′W	40–100	2,990	Evergreen	114,347	22	Bongers et al. 2001
San Carlos	Venezuela	01°56′N	67°03′W	120	3,565	Evergreen, bana, gallery forest, caatinga	30,000	14	Jordan 1989, Ellsworth and Reich 1998, Clark et al. 2000a
Caqueta	Colombia	0°20′–1°30′N	70°40′–73°30′W	150	3,059	Evergreen	1,000	25	Sanchez Saenz 1997
Rio Palenque	Ecuador	0°50′S	79°03′W	150	2,650	Evergreen	167	23	Dodson and Gentry 1978
Juaneche	Ecuador	1°03′S	79°07′W	300	1,854	Evergreen	80	15	Dodson et al. 1985
Jatun Sacha	Ecuador	1°4′S	77°36′W	450	5,000	Evergreen	2,000	19	http://mobot.mobot.org/W3T/Search/Ecuador/projjs.html
BDFF	Brazil	2°19′–26′S	59°48′–60°05′W	100	2,607	Evergreen	1,788	12	Nee 1995, Bierregaard et al. 1992, Lawrence 2001
Reserva Ducke	Brazil	2°53′S	59°95′W	120	2,607	Evergreen	10,000	30	Ribeiro et al. 1999, Lawrence 2001
Explomapo Camp	Peru	3°15′S	72°53′W	120–140	2,498	Evergreen tierra firme, igapo	1,725	21	Vásquez 1997
Explorama Lodge	Peru	3°25′S	72°54′W	106–110	2,498	Evergreen tierra firme, varzea	195	5	Vásquez 1997
Allpahuayo—Mishana	Peru	3°53′S	73°25′W	110–180	2,498	Evergreen tierra firme, igapo	2,750	14	Vásquez 1997
Pernambuco	Brazil	7°30′–9°00′S	36°00′–39°00′W	300–1,000	750–1,000	Evergreen, campo rupestre	2,235	4	Salas et al. 1998
Cocha Cashu	Peru	11°54′S	71°22′W	380	2,000	Evergreen	1,000	62	R. Foster, pers. comm., Terborgh 1990
Serra da Chapindinha	Brazil	12°25′–28′S	41°25′–28′W	800–1,100	1,375	Cerrado and mata de gratão	50	3	Guedes and Orge 1998
Tambopata Reserve	Peru	12°49′S	69°43′W	280	2,200	Evergreen	5,500	39	Conservation International 1995, C. Reynel, pers. comm.
Pico das Almas	Brazil	13°32′–34′S	41°57′–58′W	450–1,958	800	Caatinga, montane evergreen, cerrado, campo rupestre	15,213	2	Stannard 1995
Una B.S. Bahia[1]	Brazil	15°10′S	39°03′W	0	1,918	Mesophytic	3,500	2	W. W. Thomas, A. M. de Carvalho, A. Amorim, and J. Garrison: http://www.nybg.org/bsci/res/una.html
Res. Ecol. IBGE	Brazil	15°60′S	47°60′W	1,048–1,160	1,436	Gallery	7,000	9	Pereira et al. 1993

(cont.)

TABLE 5.1
(cont.)

Site	Country	Latitude	Longitude	Elevation (m)	Annual Rainfall (mm)	Forest Type	Sample Area (ha)	No. Piper Species	References
Linhares	Brazil	19°06–18′S	39°45′–40°19′W	28–65	1,383	Semideciduous	16,500	9	R. M. de Jesus, pers. comm., Rolim et al. 2001
Caratinga	Brazil	19°50′S	41°50′W	400–680	1,146	Semideciduous	880	13	Lombardi and Goncalves 2000
Serra da Canastra	Brazil	20°00′–20°45′S	46°15′–47°00′W	800–900	1,300–1,700	Cerrado, evergreen	200,000	5	http://www.serracanastra.com.br
Southern Minas Gerais	Brazil	21°00′–22°30′S	44°00′–45°30′W	750–1,800	1,400–1,800	Cerrado, evergreen	2,624,400	11	A. T. Filho-Oliveira, pers. comm.
Est. Exper. de Marília—María[1]	Brazil	22°01′S	49°55′W	440	1,571	Gallery forest	155	5	http://www.bdt.fat.org.br/ciliar/sp/especies
Macae de Cima	Brazil	22°21–28′S	42°27–35′W	880–1,720	2,128	Montane wet	7,200	14	Lima et al. 1997
Est. Ecol. de Assis	Brazil	22°33–36′S	50°22–23′W	520–590	1,400	Cerrado/gallery forest/wetland	1,312	6	Durigan et al. 1999
Fazenda São Luís—Tarumã[1]	Brazil	22°44′S	50°4′W	400	1,500	Mesophyll	60	2	http://www.bdt.fat.org.br/ciliar/sp/especies, G. Durigan, pers. comm.
Ilha do Cardosa	Brazil	25°03′–25°18′S	47°53′–48°05′W	0–800	3,000	Evergreen	22,500	12	Barros et al. 1991

[1] Not included in ordination analysis because of too few species.
[2] Atlantic side of the Continental Divide.
[3] First number is of number of species on Atlantic side of the Continental Divide and the second for the number of species on the Pacific side of the Divide.

FIGURE 5.1. Geographic distribution of the 40 sites used in this study. See Table 1 for site descriptions and sources.

in Belize (Chiquibul; Table 5.1); two sites in Nicaragua (Volcan Mombacho, 11°82'N, 85°97'W, elevation 400–500 m; Cerro El Hormiguero, 13°73'N, 84°98'W, elevation 800–900 m) based on the TROPICOS database of the Missouri Botanical Garden (mobot.mobot.org/W3T/Search/vast.html), randomly chosen from six sites for which a minimum 100 specimens of *Piper* had been collected (T. Consiglio, pers. comm.); two sites in the Department of Antioquia, Colombia [valley of the Rio Magdalena River (elevation 0–50 m) and that surrounding the Golfo de Uraba (elevation 0–50 m)], both of which have been extensively collected by R. Callejas (pers. comm.). Both *Pothomorphe peltatum* and *Po. umbellatum* were considered as *Piper* species (Jaramillo and Manos 2001). Species richness per area was calculated as the total number of species recorded for a site, divided by the area of the site sampled in hectares. For a limited number of sites (Manantlan and Serra da Chapadinha) it was clear that some portion of the entire site was not suitable habitat. Accordingly, the area was reduced to the area of appropriate habitat based on a published vegetation map, the map was scanned (Hewlett Packard Scan Jet 6300C) and the area calculated (Sigma Scan Pro 5.0).

Analysis of biogeographic patterns was through ordination by non-parametric multidimensional scaling of presence–absence data for 40 sites (Table 5.1), based on the Sørensen similarity index, using the program PC-ORD, version 4 (MJM Graphics). Random starting configurations were used, with one run, and three dimensions in the final solution (McCune and Grace 2003). The best separation of sites occurred after first limiting the total number of species found at one or more sites (296) to those that occurred in at least five sites (33 species), and to only those sites with at least four reported species (40 total). Partial regression analyses using rainfall, latitude, and elevation as predictors of species richness were conducted in SAS (SAS 1995). Data on habitat affinities and growth form were gathered from published florulas.

5.3. RESULTS

5.3.1. Biogeographic Affinities and Regional Species Pools

Most species occurred in relatively few sites. Of the total 296 species found at the 40 sites, 170 (57%) occurred in only 1 site, whereas only 33 (11.1%) were recorded at 5 or more sites (Fig. 5.2). *Piper* species occur in practically all habitats available within the range of the genus. They are most common in wet humid and deciduous forest, but also occur in more extreme habitats: Brazilian cerradão (*P. tuberculatum* Jacq. and *P. mollicomum* Kunth; Durigan *et al.* 1999); bana (*P. aduncum* L.) and caatinga (*P. hermannii* Trel. & Yunck. and *P. holtii* Trel. & Yunck.; Clark *et al.* 2000a); in flooded varzea forests (*P. coruscans* H.B.K. and *P. heterophyllum* R. & P.; Vásquez 1997); swamp forests [*P. coruscans* H.B.K.; Dodson *et al.* 1985]; open wetlands (*Piper* sp.; Durigan *et al.* 1999); campo rupestre of Brazil (*P. lhotzkyanum* Kunth.; Guedes and Orge 1998); sandstone mesas [*P. arieianum* C.DC., *P. demeraranum* (Miq.) C.DC., and *P. macrotrichum* C.DC.; Sanchez Saenz 1997], and to elevations as high as 3,500 m (*Piper nubigenum* Kunth, Colombia; R. Callejas, pers. comm.). *Piper* species apparently are not found in cerrado *sensu stricto* or in less woody cerrado derivatives (Durigan *et al.* 1999, and references therein) or in restinga.

Ordination showed five major regions whose separation is based on presence–absence of the 33 most common *Piper* species (Fig. 5.3): (1) dry (<2,000 mm rainfall per year; mean ± SE = 1,266 ± 274 mm/year) Central America and Mexico sites; (2) wet (>2,200 mm rainfall per year; mean ± SE = 3,340 ± 434 mm/year) Central America and Mexico sites; (3) northwest Colombia; (4) Amazon, including Nouragues of French Guiana; and (5) Atlantic Forest sites. Andean sites broadly overlap with those of wet Central America and Mexico, northwest Colombia, and the Amazon. The wet and dry Central America and Mexico sites fall close together as might be expected, with the northwest Colombia sites mapping close to the latter. Axes 1, 2, and 3 accounted for 17.2, 34.9, and 20.8% of the total variation in data set, respectively, for a total of 72.9%.

The total number of species in the regional pool is highest in the three wettest regions: Central America/Mexico wet, Andes, and the Amazon (Fig. 5.4). The observed number of species (at five sampled sites) for the two drier regions (Central America/Mexico dry and Atlantic Forests) ($x = 30$ species) is approximately 66% less than that for the three

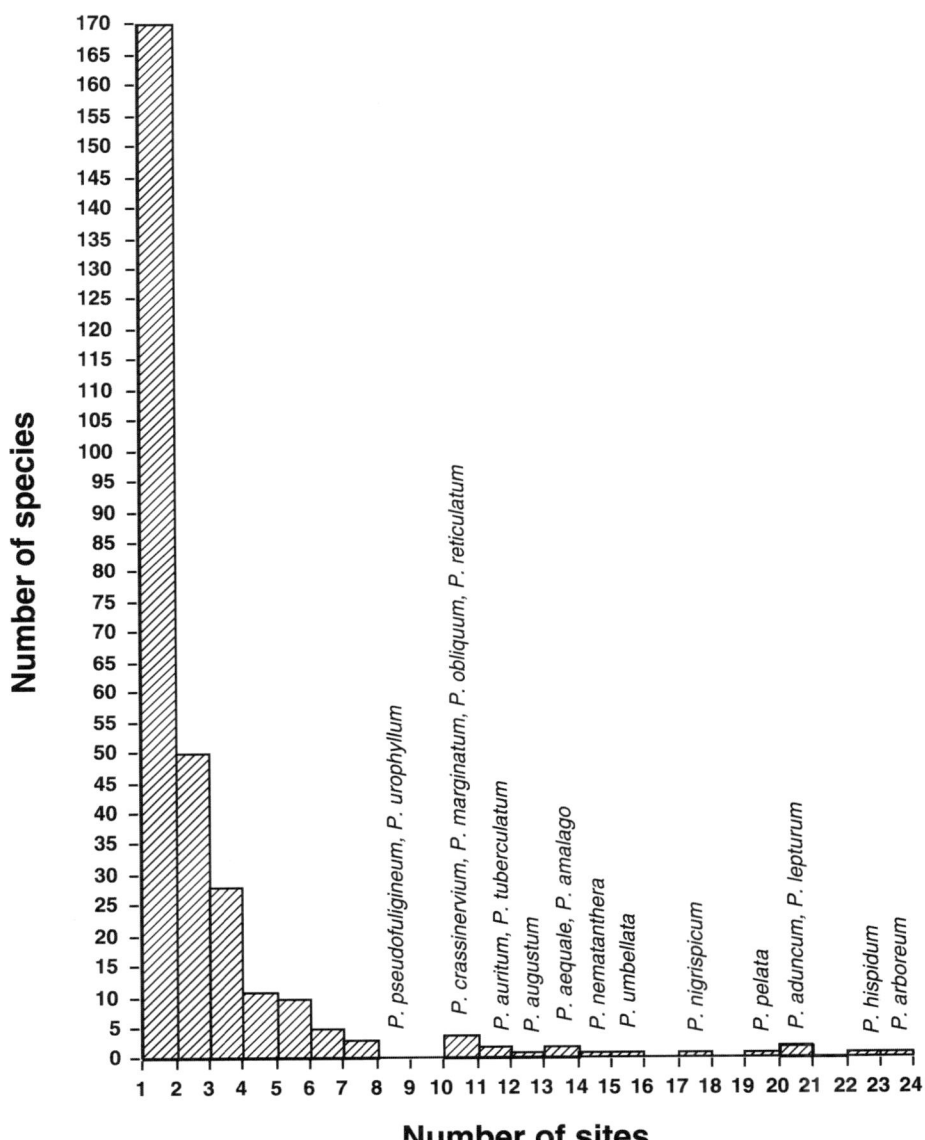

FIGURE 5.2. Distribution of the number of *Piper* species (total 296) based on the number of sites in which they have been recorded (maximum possible = 40).

wet regions ($x = 86.6$ species) (Fig. 5.4). Sites in the three wet regions receive on average 1,400 mm rainfall per year more than the sites of the two dry regions ($2,992 \pm 226$ vs. $1,541 \pm 145$; one-way ANOVA, $F = 29.72$, $P = 0.0001$; Pearson product correlation between rainfall and estimated number of species at five sites, $r = 0.87$, $P = 0.055$).

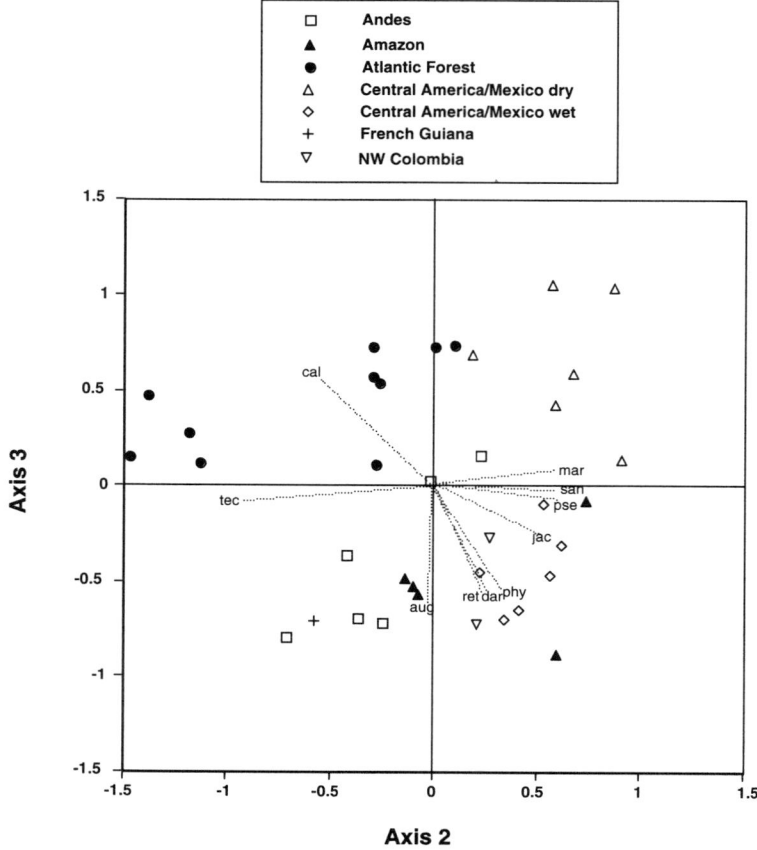

FIGURE 5.3. Variation in regional species composition: plot of distance in ordination space of 40 sites based on non-metric multidimensional scaling analysis. The top 10 *Piper* species in terms of weighting on the axes are indicated: aug = *P. augustum*; cal = *P. caldense*; dar = *P. dariense*; jac = *P. jacquemontanum*; mar = *P. marginatum*; phy = *P. phytolaccaefolium*; pse = *P. pseudofuligineum*; ret = *P. reticulatum*; san = *P. sanctum*; tec = *P. tectonii*.

5.3.2. Correlates of Local Species Richness

The highest species richness per area sampled occurred in Andean sites, more than twice as high as the next most diverse regional set of sites (Fig. 5.5; one-way ANOVA, $F = 5.76$, $P = 0.0012$). Species richness per area sampled (natural log) was marginally significantly greater at sites with higher rainfall for Central America and Mexico sites (Fig. 5.6(A); $P = 0.075$, partial $r^2 = 0.239$) but not for South America sites (Fig. 5.7(A); $P = 0.40$). Latitude was a significant negative predictor of species richness per area (natural log) sampled in Central America and Mexico (Fig. 5.6(B); $P = 0.0421$, partial $r^2 = 0.316$) and in South America (Fig. 5.7(B); $P = 0.0394$, $r^2 = 0.165$). For South America, species richness was highest at the equator, and lower to both the North and the South. Together, latitude and rainfall explained 59.8% of the total variation in species numbers per area

BIOGEOGRAPHY OF NEOTROPICAL *Piper* 87

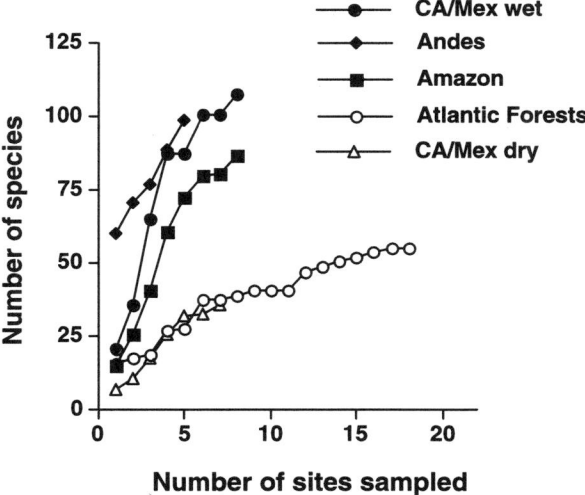

FIGURE 5.4. Species accumulation curves for the five geographic provinces as defined by the ordination. The average annual rainfall is significantly higher for the Central American/Mexican wet sites, Andes, and Amazon sites than for the Central American/Mexican dry sites and those of the Atlantic forests of Brazil.

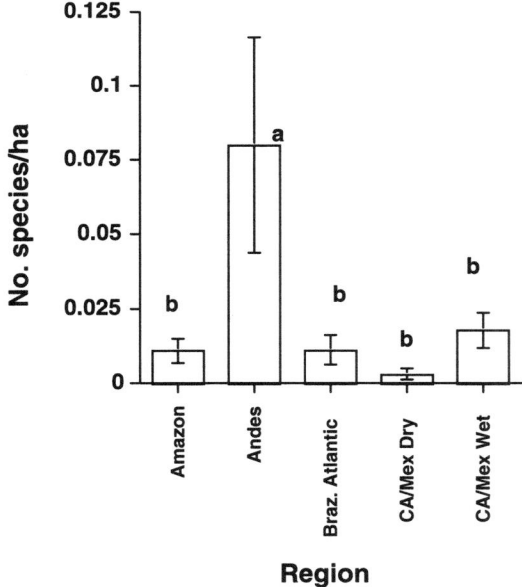

FIGURE 5.5. Variation in local species richness by region: *Piper* species density (number of species per ha) for five geographic provinces as defined by the ordination. Sites with different letters have significantly different numbers of species per hectare ($P < 0.05$).

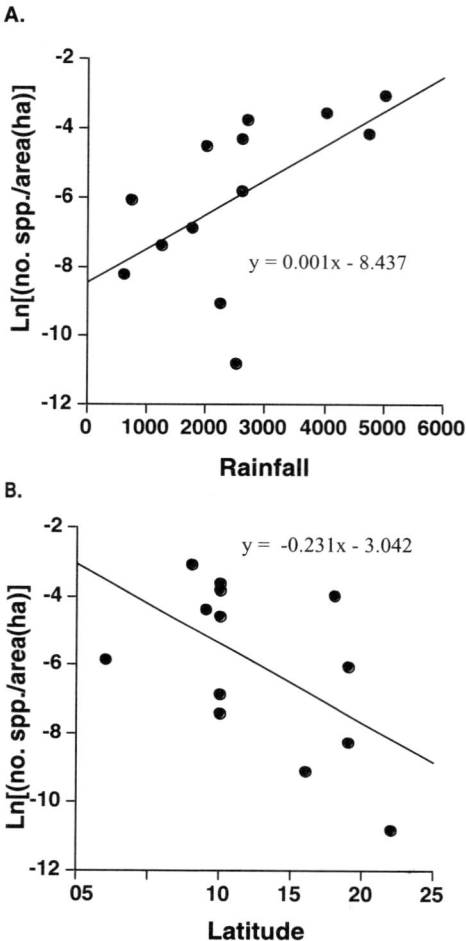

FIGURE 5.6. Correlates of local species richness: regression of *Piper* species density on rainfall (A) and latitude (B) for Central America and Mexico sites.

among Central America and Mexico sites. Mean local species richness for a region was not correlated with regional species richness (estimated for five sites) ($P = 0.184$). There was no significant relationship between elevation and species richness ($P = 0.87$), contrary to expectation. *Piper* diversity (species richness per area) in Colombia remains high even at relatively high elevations (Fig. 5.8), although no systematic sampling of diversity along elevational transects has been undertaken.

5.3.3. Variation in Growth Form and Habitat Affinity

Sites with other life forms in addition to shrubs (trees, vines, and/or herbs) had 5 times more species on average than those containing shrubs only: mean number of species

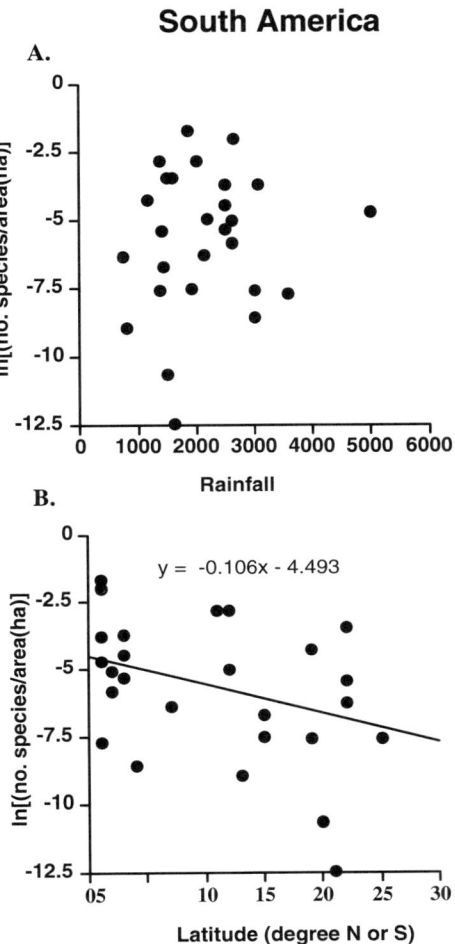

FIGURE 5.7. Correlates of local species richness: regression of *Piper* species density on rainfall (A) and latitude (B) for South America sites.

per hectare ± SE = 0.038 ± 0.013 versus 0.0075 ± 0.0028 (one-way ANOVA, $F = 5.07$, $P = 0.032$), respectively. Sites with multiple life forms on average receive more rain (2,718 mm/year ± 285) than those with a single life form (1,990 ± 261) (one-way ANOVA, $F = 3.55$, $P = 0.069$).

Although some species reach high canopy (*Piper arboreum* Aublet is 25–35 m tall at Nouragues Reserve in French Guiana; Bongers *et al.* 2001), most species are much smaller. At wet forest sites (Fig. 5.9), a full range of maximum heights occur up to 8 or 9 m. Unfortunately, there are no comparative data on maximum heights for low-rainfall sites. Given that *Piper* species are mostly shrubs in drier climates, maximum heights probably decrease with annual rainfall.

Colombia *Piper* floras include a number of small species (reproductive at less than 1 m) that appear to occur nowhere else. In addition, the flora of the Department of

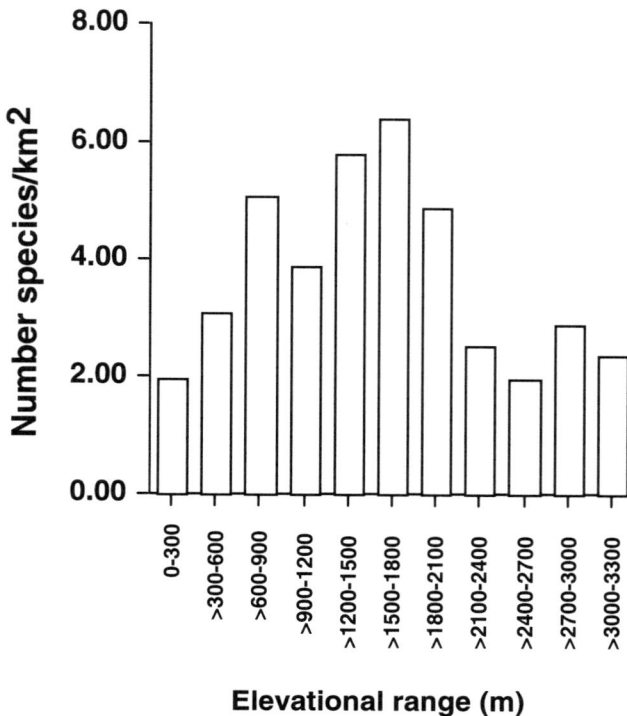

FIGURE 5.8. Effect of elevation on local species richness: number of *Piper* species collected with increasing elevation in the Department of Antioquia, Colombia (R. Callejas, pers comm.) standardized by the area represented by each elevation range. Total number of collections equals 773.

Antioquia, Colombia, includes a high concentration of species [*P. brachypodon* (Benth.) C.DC., *P. carpunya* Ruiz & Pav., *P. castroanum* Trel. & Yunck., *P. heterotrichum* C.DC., *P. lenticellosum* C.DC., *P. ottoniifolium* C.DC., *P. pinaresanum* Trel. & Yunck., and *P. oblongum* H.B.K.] that are shrubs early in their life but then switch to a vine/liana growth form. This growth form has not been previously reported in *Piper* but is known in the Dichapetalaceae, Combretaceae, Convolvulaceae, Malpighiaceae, Menispermaceae, and Sapindaceae (Lee and Richards 1991). Liana and vine *Piper* species were not found in any of the Atlantic Forest sites of Brazil. A single species of hemiepiphyte (*P. dactylostigmum* Yunck.) has been reported from the 40 sites surveyed here, and then only at the Reserve Ducke, Brazil (da S. Ribeiro *et al.* 1999).

Piper pseudobumbratum C.DC. (Marquis, pers. obs.) and *P. auritum* L. (Greig and Mauseth 1991) represent a unique growth form—reports of which appear to be limited, so far, to these two species. Both have underground rhizomes that give rise to aboveground shoots some distance from the parent plant. This results in large clonal patches of *P. auritum* in large, artificially cleared areas, and apparently only the largest of natural disturbances (Marquis, pers. obs.). In contrast, rhizome production and establishment of new plants from those rhizomes occurs in the deep shade of closed canopy in *P. pseudobumbratum* at the La Selva Biological Station, Costa Rica (R. Marquis and N. Greig, unpubl. data).

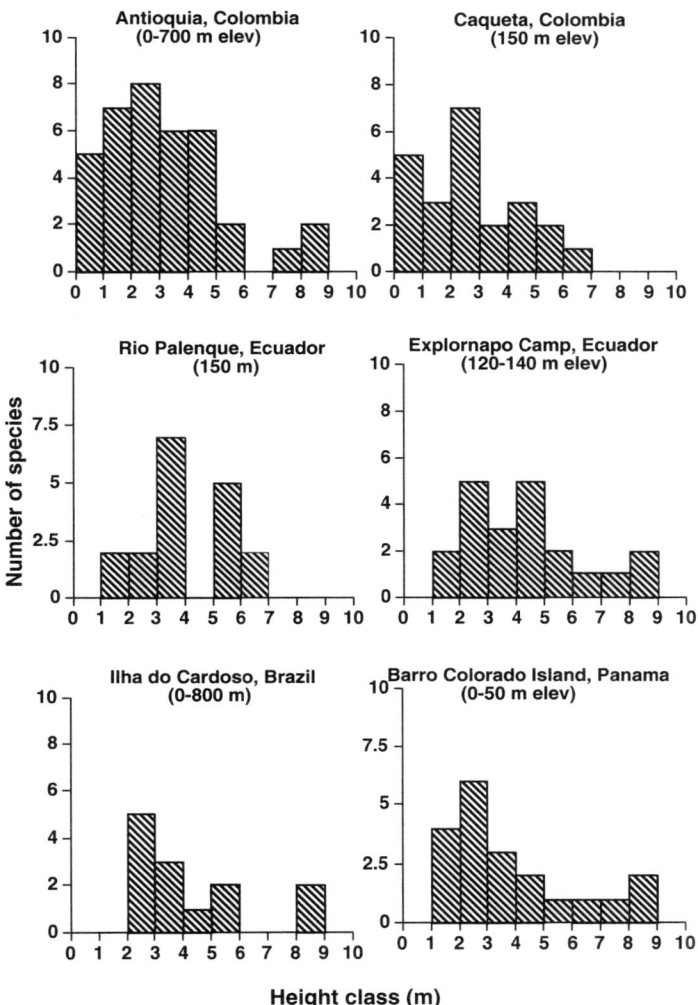

FIGURE 5.9. Height classes of *Piper* for six wet forest sites. Sources: Antioquia, Colombia (R. Callejas, pers. comm.); Caqueta, Colombia (Sanchez Saenz 1997); Rio Palenque, Ecuador (Dodson and Gentry 1978); Explornapa, Peru (Vásquez, 1997); Illha do Cardosa (Barros *et al.* 1991), and Barro Colorado Island (Croat 1978).

5.4. DISCUSSION

Ordination of the 40 sites revealed that there are three distinct biogeographic provinces: (1) Central America/Mexico, (2) the Amazon basin, and (3) Atlantic Forests of Brazil. The two sites of Northwest Colombia map closely to wet forests of Central America/Mexico, as has been found for other taxa (e.g., Prance 1982). Andean sites broadly overlap with wet forests of Central America/Mexico, the Amazon Basin, and the Colombian sites. On the basis of the currently available data, therefore, *Piper* communities of the Andes are not uniquely different from these other floras, but share elements of all three. Surprisingly, the flora of low-rainfall sites (<2,000 mm/year) in Central America and Mexico is distinct from that of high-rainfall sites (>2,200 mm/year) in that region. This pattern occurs despite the fact that these sites fall within the same geographic region. Thus the *Piper* flora of dry deciduous forests in this region is not entirely a subset of wet evergreen forests that comprises species that can live in both wet and dry climates but those that are unique to dry climates. Whether this pattern holds for South America remains to be seen. No data exist for relatively wet and dry sites occurring in relative geographic proximity (Table 5.1). Atlantic Forest sites receive significantly less rainfall than the Andean and Amazonian sites (1,536 mm/year ± 149 vs. 2,649 ± 349 and 2,959 ± 228, respectively; $F_{4,33} = 9.66$, $P = 0.0001$), but they are geographically isolated from both.

Although data are available for 40 sites, large areas of the range of Neotropical *Piper* are not as yet sampled (Fig. 5.1). Further sampling is required to understand the exact relationship of the floras of the Guianas, northern South America, Bolivia and Argentina, and the Amazon Basin to those presented here. The fact that the floras of Andean sites broadly overlap with those of other regions suggests that the Andes may have been the source pool for many of these other regions. Alternatively, given the relatively recent age of the Andes, these other regions may have served as the source pool for the Andean flora. Under this scenario, Central America contributed its species before the rise of the Andes. Clearly, a phylogeographic study is called for to determine the evolutionary relationships among species from the different regions.

The highest regional species richness occurs in the three high-rainfall regions, corroborating patterns of high diversification in this region for other taxa (Gentry 1988b). For the montane regions (in this study, the volcanic chain of Central America and the Andes), the proposed mechanism for the generation of such high diversity is the highly dissected nature of the terrain in combination with steep elevational gradients. The highly dissected terrain results in restricted gene flow over short distances, whereas dissection of the terrain and the elevational gradient together lead to variable climatic patterns, resulting in different selective regimes (Burger 1985, Gentry 1988b). This process would not appear to apply to the same degree in much of the Amazon, although relief even there apparently contributes to species coexistence (Tuomisto and Poulsen 1996). An alternative explanation, and one that is consistent for all sampled regions in this study, is that the genus is adapted to high-rainfall climates, resulting in more species in the source pool.

Species richness at the local level increases monotonically with increasing rainfall in Central America and Mexico (Fig. 5.6(A)). Although this pattern was not found for South America (Fig. 5.7(A)), only one site with rainfall over 4,000 mm of rain annually was sampled. This leaves open the possibility that the association of increasing rainfall with increasing species richness may hold throughout the tropics. Regardless of the universality of the pattern, there is no evidence for a peak in richness at intermediate levels of annual rainfall (Figs. 5.6(B) and 5.7(B)). If productivity shows an intermediate peak with increasing rainfall in tropical forested habitats, as suggested by initial data (Clark *et al.* 2001), then *Piper* diversity at the local level does not appear to be limited by the number of niches as determined by productivity. In contrast, total number of species and growth form diversity is highest at high-rainfall regions, each contributing to local diversity. High rainfall results in more niches, at least in allowing the coexistence of additional growth forms.

There is little evidence that local species richness may be constrained by the size of the regional pool. Across all regions, richness values at the local and regional level are not significantly correlated. Thus, the regional pool is low for the Atlantic Forests, but species richness per area for those forests is as high as that for Amazon forests and those of Central America and Mexico. In addition, although the Andes regional species pool appears to be no larger than that for the Amazon and Central America/Mexico wet sites (Fig. 5.4), Andean sites sample 4–5 times as many species from their regional pool as those from the other two high-rainfall regions (Fig. 5.5). The cause for such differences is not obvious.

These broad geographic patterns demonstrate not only that the number of species available in different regional *Piper* species pools varies by regions, but also representation of "unique kinds" of species varies among regions. Thus, small-statured species and shrub/liana forms appear to be highly concentrated in Colombia, hemiepiphytes have been reported only from the Amazon Basin, and species that clonally spread by underground rhizomes are reported only from the northern portion of the Neotropics. More patterns may emerge as the biology of the genus becomes better known. All three "kinds" of species would contribute to niche diversity in the genus.

On the basis of descriptions of habitat and growth form from flora lists reviewed in this study, most *Piper* species are shrubs and treelets, rarely growing above 10 m in height. In addition, most species commonly occur in forest habitats on well-drained soils. Some species, apart from a small percentage of the total Neotropical flora and that of any given site, are restricted to second growth forests, are found on poorly drained or low-nutrient soils, or are vines or lianas. Detailed studies of the distributions of individual species in local communities with respect to soil type and nutrient level (e.g., Tuomisto *et al.* 2002) and characterization of physiologies with respect to light environments (Chazdon *et al.* 1988) and water use efficiency are needed to determine how coexistence of many species of apparently similar habitat requirements can coexist.

A preliminary hypothesis, based on the evidence presented here, is that two evolutionary pathways have led to high local species richness: (1) evolution of additional growth forms from a shrub prototype, and (2) evolution of the ability to grow on unique combinations of soil type and soil moisture level. In both pathways, the abiotic environment defines the number of niches. The biotic portion of the environment could also potentially contribute to the number of niche axes. In fact, early suggestions regarding niche limitation in Neotropical forests focused on biotic factors; the number of flowering phenology niches [for Bignoniaceae (Gentry 1974) and Heliconiaceae (Stiles 1978)] or leaf shape niches [*Passiflora* (Passifloraceae); Gilbert 1975] placed an upper limit on the number of species that occur in any one Neotropical site. There is little evidence at this time, however, that the diversification in *Piper* has been driven by biotic interactions or that the number of *Piper* species at a site is limited by biotic interactions. *Piper* species show little or no variation in pollination mode (all or most are small insect-pollinated; see Chapter 3). In turn, leaf shape is more or less consistent across the genus, certainly much less so than that seen in *Passiflora* [compare leaf shape variation depicted by Burger (1971) with that by Gilbert (1975)]. In addition, the herbivore fauna of *Piper* does not include visually acute species (Marquis 1991) as does the *Passiflora* herbivore fauna, which are hypothesized to have driven the diversification of leaf shape in *Passiflora* (Gilbert 1975). Finally, the seeds of *Piper* are dispersed by a generalized bat fauna (Chapter 4). These hypotheses can be tested by mapping morphological traits, traits that determine habitat preference, and traits influencing biotic interactions onto molecular-based phylogenies.

5.5. ACKNOWLEDGMENTS

I thank Karina Boege, Ricardo Callejas, Jaime Cavelier, Howard Clark, Kleber Del Claro, Trisha Consiglio, Alberto Duran, Giselda Durigan, Robin Foster, Gordon Frankie, Nancy Garwood, Larry Gilbert, Nancy Greig, Arturo Mora, Renato de Molim de Jesus, Carl Jordan, Susana León, Lúcia Lohmann, Patricia Morellato, Ary Oliveira-Filho, Rosa Ortiz-Gentry, Cristian Samper, Doug Stevens, Alberto Vicentini, and Caroline Whitefoord for help in finding data sets, updates to published data, and access to unpublished data. Rebecca Forkner helped to prepare Fig. 5.1, and Trish Consiglio provided the area data for Fig. 5.8. The staff at the Missouri Botanical Garden library were especially helpful in finding materials. The manuscript was improved greatly by comments from Beatriz Baker, Karina Boege, Grace Chen, Rebecca Forkner, June Jeffries, Alejandro Masís, Rodrigo Rios, Kimberly Schultz, two anonymous reviewers, and the statistical help of Kurt Shultz.

REFERENCES

Barros, F., Melo, M. M. R. F., Chiea, S. A. C., Kirizawa, M., Wanderley, M. G. L., and Jung-Mendacolli, S. L. (1991). *Flora Fanerogamica da Ilha do Cardoso. Caràterização geral da vegetação e listagem das especias ocorrentes.* Instituto de Botânica, Sao Paulo, Brazil.

Bierregaard, R. O., Jr., Lovejoy, T. E., Kapos, V., Dos Santos, A. A., and Hutchings, R. W. (1992). The biological dynamics of tropical rainforest fragments. *BioScience* 42:859–866.

Bongers, F., Charles-Dominique, P., Forget, P.-M., and Thery, M. (eds.). (2001). *Nouragues, Dynamics and Plant–Animal Interactions in a Neotropical Rainforest.* Monographiae Biologicae. Kluwer Academic Publishers, Dordrecht, The Netherlands.

Boyle, B. L. (1996). *Changes on Altitudinal and Latitudinal Gradients in Neotropical Montane Forests.* Ph.D. dissertation, Washington University, St. Louis, Missouri.

Burger, W. C. (1971). Piperaceae. Flora Costaricensis. *Fieldiana, Botany* 35:5–227.

Burger, W. C. (1985). Why are there so many kinds of flowering plants in Costa Rica? In: D'Arcy, W. G. (ed.), *The Botany and Natural History of Panama.* Missouri Botanical Garden, St. Louis, Missouri, pp. 125–136.

Chazdon, R. L., Williams, K., and Field, C. B. (1988). Interactions between crown structure and light environment in five rain forest *Piper* species. *American Journal of Botany* 75:1459–1471.

Clark, D. A., Brown, S., Kicklighter, D. W., Chambers, J. Q., Thomlinson, J.R., Ni, J., and Holland, E. A. (2001). Net primary production in tropical forests: An evaluation and synthesis of existing field data. *Ecological Applications* 11:371–384.

Clark, H. L., Liesner, R., Berry, P. E., Fernandez, A., Aymard, G., and Maquirino, P. (2000a). Catalogo anotado de la flora del area de San Carlos de Rio Negro, Venezuela. *Scientia Guaianae* 11:101–333.

Clark, K. L., Lawton, R. O., and Butler, P. R. (2000b). The physical environment. In: Nadkarni, N. M., and Wheelwright, N. T. (eds.), *Monteverde, Ecology and Conservation of a Tropical Cloud Forest.* Oxford University Press, New York, pp. 15–34.

Conservation International. (1995). *Reporte Tambopata.* Universidad Nacional Agraria La Molina, Lima, Peru.

Croat, T. B. (1978). *Flora of Barro Colorado Island.* Stanford University Press, Stanford, California.

da S. Ribeiro, J. E. L., Hopkins, M. J. G, Vicentini, A., Sothers, C. A., da S. Costa, M. A., de Brito, J. M., de Souza, M. A. D., Martins, L. H. P., Lohmann, L. G., Assunção, P. A. C. L., da Silva, C. F., Mesquita, M. R., and Procópio, L. C. (1999). *Flora da Reserva Ducke.* INPA-DFID, Manaus, Brazil.

Dietrich, W. E., Windsor, D. M., and Dunne, T. (1982). Geology, climate, and hydrology of Barro Colorado Island. In: Leigh, E. G., Jr., Rand, A. S., and Windsor, D. M. (eds.), *The Ecology of a Tropical Forest, Seasonal Rhythms and Long-Term Changes.* Smithsonian Institution, Washington, D.C., pp. 21–46.

Dirzo, R., and Miranda, A. (1991). Altered patterns of herbivory and diversity in the forest understory: A case study of the possible consequences of contemporary defaunation. In: Price, P. W., Lewinsohn, T. M., Fernandes,

G. W., and Benson, W. W. (eds.), *Plant–Animal Interactions, Evolutionary Ecology in Tropical and Temperate Regions*. John Wiley & Sons, New York, pp. 273–287.
Dodson, C. H., and Gentry, A. H. (1978). *Flora of the Rio Palenque Science Center: Los Rios, Ecuador*. Selbyana. The Marie Selby Botanical Gardens, Sarasota, Florida.
Dodson, C. H., Gentry, A. H., and Valverde, F. M. (1985). *La flora de Jauneche, Los Ríos, Ecuador*. Florulas de las zonas de vida del Ecuador. Banco Central del Ecuador, Quito, Ecuador.
Durigan, G., Bacic, M. C., Franco, G. A. D. C., and Siqueira, M. F. (1999). Inventario floristico do cerrado na Estação Ecologica de Assis, SP. *Hoehnea (São Paulo)* 26:149–172.
Ellsworth, D. S., and Reich, P. B. (1996). Photosynthesis and leaf nitrogen in five Amazonian tree species during early secondary succession. *Ecology* 77:581–594.
Gentry, A. H. (1974). Flowering phenology and diversity in tropical Bignoniaceae. *Biotropica* 6: 64–68.
Gentry, A. H. (1988a). Changes in plant community diversity and floristic composition on environmental and geographical gradients. *Annals of the Missouri Botanical Garden* 75:1–34.
Gentry, A. H. (1988b). Tree species richness of upper Amazonian forests. *Proceedings of the National Academy of Sciences U.S.A.* 85:156–159.
Gentry, A. H. (1989). Speciation in tropical forests. In: Holm-Nielsen, L. B., Nielsen, I. C., and Balslev, H. (eds.), *Tropical Forests, Botanical Dynamics, Speciation and Diversity*. Academic Press, London pp. 113–134.
Gentry, A. H. (1991). The distribution and evolution of climbing plants. In: Putz, F. E., and Mooney, H. A. (eds.), *The Biology of Vines*. Cambridge University Press, Cambridge, England, pp. 3–49.
Gentry, A. H., and Emmons, L. H. (1987). Geographical variation in fertility, phenology, and composition of the understory of Neotropical forests. *Biotropica* 19:216–227.
Gilbert, L. E. (1975). Ecological consequences of a coevolved mutualism between butterflies and plants. In: Gilbert, L. E., and Raven, P. H. (eds.), *Coevolution of Animals and Plants*. University of Texas Press, Austin, Texas, pp. 210–240.
Greig, N., and Mauseth, J. D. (1991). Structure and function of dimorphic prop roots in *Piper auritum* L. *Bulletin of the Torrey Botanical Club* 118:176–183.
Guedes, M. L. S., and Orge, M. D. R. (1998). *Cheklist das Espécies Vasculares do Morro do Pai Ináico (Palmeiras) e Serre da Chapadinha (Lençóis) Chapada Diamantina, Bahia—Brasil*. Insituto de Biologia da UFBA, Salvador, Brazil.
Haber, W. A. (2000). Appendix 1. Vascular plants of Monteverde. In: Nadkarni, N. M., and Wheelwright, N. T. (eds.), *Monteverde, Ecology and Conservation of a Tropical Cloud Forest*. Cambridge University Press, New York, pp. 457–518.
Hartshorn, G. S. (1983). Plants, introduction. In: Janzen, D. H. (ed.), *Costa Rican Natural History*. University of Chicago Press, Chicago, pp. 118–183.
Janzen, D. H., and Liesner, R. (1980). Annotated checklist of plants of lowland Guanacaste Province, Costa Rica, exclusive of grasses and non-vascular plants (cryptograms). *Brenesia* 18:15–90.
Jaramillo, M. A., and Manos, P. S. (2001). Phylogeny and patterns of floral diversity in the genus *Piper* (Piperaceae). *American Journal of Botany* 88:706–716.
Johnston, M. C., Nixon, K., Neson, G. L., and Martínez, M. (1998). Listado de plantas vasculares conocidas en la Sierra de Guatemala, Gomez Farias, Tamaulipas, Mexico. *BIOTAM* 10:25–40.
Jordan, C. F. (1989). *An Amazonian Rain Forest*. Man and the Biosphere Series. UNESCO and Parthenon Publishing Group, Paris, France.
Lawrence, W. F. (2001). The hyper-diverse flora of the central Amazon: An overview. In: Bierragaard, J. R. O., Gascon, C., Lovejoy, T. E., and Mesquita, R. (eds.), *Lessons from Amazonia; Ecology and Conservation of a Fragmented Forest*. Yale University Press, New Haven, Connecticut, pp. 47–53.
Lee, D. W., and Richards, J. H. (1991). Heteroblastic development in vines. In: Putz, F. E., and Mooney, H. A. (eds.), *The Biology of Vines*. Cambridge University Press, Cambridge, England, pp. 205–243.
Lima, M. P. M. d., Guedes-Bruni, R. R., and Lima, H. D. d. (1997). Reserva ecológica de Macaé de Cima, Nova Friburgo, RJ : aspectos florísticos das espécies vasculares. Jardim Botânico de Rio de Janeiro, Rio de Janeiro, Brazil.
Lombardi, J. A., and Goncalves, M. (2000). Composição florística de dois remanescentes de Mata Atlântica do sudeste de Minas Gerais, Brasil. *Revista Brasiliera de Botanica*. 23:255–282.
Lott, E. J. (2002). Lista anotada de las plantas vasculares de Chamela-Cuixmala. In: Noguera, F. A., Vega Rivera, J. J., Garcia Aldreto, A. N., and Quesada A., M. (eds.), *Historia Natural de Chamela*. Unversidad Nacional Autonoma de Mexico, Mexico City, Mexico, pp. 99–136.
Machado, C. A., Jousselin, E., Kjellberg, F., Compton, S. G., and Herre, E. A. (2001). Phylogenetic relationships, historical biogeography and character evolution of fig-pollinating wasps. *Proceedings of the Royal Society of London B* 268: 685–694.
Manríquez, G. I., and Colín, S. S. (1997). Fanerógramas. In: Soriano, E. G., Dirzo, R., and Vogt, R. C. (eds.), *Historia Natural de Los Tuxtlas*. Unversidad Nacional Autonoma de Mexico, Mexico City, Mexico, pp. 162–181.

Marquis, R. J. (1991). Herbivore fauna of *Piper* (Piperaceae) in a Costa Rican wet forest: Diversity, specificity and impact. In: Price, P. W., Lewinsohn, T. M., Fernandes, G. W., and Benson, W. W. (eds), *Plant–Animal Interactions, Evolutionary Ecology in Tropical and Temperate Regions.* John Wiley & Sons, New York, pp. 179–208.

Martínez, E., Ramos A., C. H., and Chiang, F. (1994). Lista floristica de la Lacondona, Chiapas. *Boletín Sociedad Botanica Mexicana* 54:99–177.

McCune, B., and Grace, J. B. (2003). *Analysis of Ecological Communities.* MJM Press, Gleneden Beach, Oregon.

McDade, L. A., and Hartshorn, G. S. (1994). La Selva Biological Station. In: McDade, L. A., Bawa, K. S., Hespenheide, H. A., and Hartshorn, G. S. (eds.), *La Selva, Ecology and Natural History of a Neotropical Rain Forest.* University of Chicago Press, Chicago, pp. 7–14.

Nadkarni, N. M., and Wheelwright, N. T. (2000). Introduction. In: Nadkarni, N. M., and Wheelwright, N. T. (eds.), *Monteverde, Ecology and Conservation of a Tropical Cloud Forest.* Cambridge University Press, New York, pp. 3–10.

Nee, M. (1995). *Flora Preliminar do Projeto Dinâmica Biológica de Fragmentos Florestais (PDBFF).* New York Botanical Garden and INPA/Smithsonian Projeto Dinâmica Biológica de Fragmentos Florestais, Manaus, Brazil.

Oliveira-Filho, A. T., and Fontes, M. A. L. (2000). Patterns of floristic differentiation among Atlantic forests in Southeastern Brazil and the influence of climate. *Biotropica* 32:793–810.

Pereira, B. A. S., da Silva, M. A., and de Mendonca, R. C. (1993). *Reserva Ecologica do IBGE, Brasilia (DF): Lista das Plantas Vasculares.* IBGE, Divisao de Geociencias do Distrito Federal, Rio de Janeiro, Brazil.

Pianka, E. R. (1994). *Evolutionary Ecology,* 5th ed. Harper Collins, New York.

Prance, G. T. (ed.). (1982). *Biological Diversification in the Tropics.* Columbia University Press, New York.

Richardson, J. E., Pennington, R. T., Pennington, T. D., and Hollinsworth, P. M. (2001). Rapid diversification of a species-rich genus of Neotropical rain forest trees. *Science* 293: 2242–2245.

Ricklefs, R. E., and Schluter, D. (1993). *Species Diversity in Ecological Communities.* University of Chicago Press, Chicago.

Rolim, S. G., Zarate do Couto, H. T., and Moraes de Jesus, R. (2001). Fluctuatiónes temporales en la composición floristica del bosque tropical Atlantico. *Biotropica* 33:12–22.

Rosenzweig, M. L., and Abramsky, Z. (1993). How are diversity and productivity related? In: Ricklefs, R. E., and Schluter, D. (eds.), *Species Diversity in Ecological Communities.* University of Chicago Press, Chicago, pp. 52–65.

Salas, M. F., Mayo, S. J., and Rodal, M. J. N. (1998). *Plantas Vasculares das Florestas Serranas de Pernambuco. Um Checklist da Flora Ameacada dos Brejos de Altitude.* Universidade Federal de Pernambuco, Pernambuco, Brasil. Recife, Brazil.

Sanchez Saenz, M. (1997). *Catalogo Preliminar Comentado de la Flora del Medio Caqueta.* Tropenbos Colombia, Santafé de Bogotá, D.C., Colombia.

Sanford, R. L., Paaby, P., Luvall, J. C., and Phillips, E. (1994). Climate, geomorphology, and aquatic systems. In: McDade, L. A., Bawa, K. S., Hespenheide, H. A., and Hartshorn, G. S. (eds.), *La Selva, Ecology and Natural History of a Neotropical Rain Forest.* University of Chicago Press, Chicago, pp. 19–33.

SAS. (1995). *Statistical Analysis Systems.* The SAS Institute, Cary, North Carolina.

Sollins, P., Sancho M., F., Mata Ch., R., and Sanford, R. L., Jr. (1994). Soils and soil process research. In: McDade, L. A., Bawa, K. S., Hespenheide, H. A., and Hartshorn, G. S. (eds.), *La Selva, Ecology and Natural History of a Neotropical Rain Forest.* University of Chicago Press, Chicago, pp. 34–53.

Stannard, B. L. (1995). *Flora of the Pico das Almas: Chapada Diamantina—Bahia, Brazil.* Kew Botanic Gardens, London.

Stiles, F. G. (1978). Coadapted competitors: The flowering seasons of hummingbird-pollinated plants in a tropical forest. *Science* 196:1177–1178.

Terborgh, J. (1990). An overview of research at Cocha Cashu Biological Station. In: Gentry, A. H. (ed.), *Four Neotropical Forests.* Yale University Press, New Haven, Connecticut, pp. 48–59.

Tuomisto, H., and Poulsen, A. D. (1996) Influence of edaphic specialization on pteridophyte distribution in Neotropical rain forests. *Journal of Biogeography* 23:283–293.

Tuomisto, H., Ruokolainen, K., Poulsen, A. D., Moran, R. C., Quintana, C., Ganas, G., and Celi, J. (2002). Distribution and diversity of pteridophytes and Melastomataceae along edaphic gradients in Yansuni National Park, Ecuadorian Amazonia. *Biotropica* 34:516–533.

Vazquez G., J. A., Cuevas G., R., Cochrane, T. S., Iltis, H. H., Santana M., F. J., and Guzman H., L. (1995). Flora de Manantlan. *Sida, Botanical Misc.* No. 13.

Vásquez, R. (1997). *Flórula de las Reservas Biológicas de Iquitos, Perú: Allpahuayo-Mishana, Explornapo Camp, Explorama Lodge.* Monographs in systematic botany from the Missouri Botanical Garden. Missouri Botanical Garden, St. Louis, Missouri.

6

Faunal Studies in Model *Piper* spp. Systems, with a Focus on Spider-Induced Indirect Interactions and Novel Insect–*Piper* Mutualisms

Karin R. Gastreich[1,2] *and Grant L. Gentry*[3]

[1]*Organization for Tropical Studies, San Pedro, Costa Rica*

[2]*Duke University, Durham, North Carolina*

[3]*Tulane University, New Orleans, Louisiana*

6.1. INTRODUCTION

Indirect interactions arise when the effect of one species on a second species is mediated through a third species (Polis *et al.* 1989, Abrams 1995, Abrams *et al.* 1996, Menge 1995, Wootton 1994, Snyder and Wise 2001). In recent years, indirect interactions have been repeatedly shown to have an important influence on the structure and dynamics of ecological communities (e.g., Menge 1995, Schmitz and Suttle 2001, Chen and Wise 1999, reviewed in Polis and Winemiller 1996). Two model systems, in particular, have proven exceptionally rich for studying indirect interactions, albeit in distinct contexts: antiherbivore mutualisms in *Piper* and predator–prey interactions with spiders (e.g., Letourneau and Dyer 1998a, Gastreich 1999, Schmitz and Suttle 2001, Wise 1994). In this chapter, we discuss two examples of *Piper* systems that provide a unique opportunity to integrate the study of model antiherbivore mutualisms with the study of model arthropod predators. This approach can provide important insights into the dynamics of indirect interactions, as well as shed some light on the role of top predators in the origin and maintenance of antiherbivore mutualisms.

Indirect interactions can be conceptually divided into two general categories: density-mediated indirect interactions (DMIIs) and trait-mediated indirect interactions (TMIIs) (Abrams 1995, Abrams *et al.* 1996, Werner and Anholt 1996). In DMIIs, the indirect effect is mediated through a change in population of the intervening species. In TMIIs, the indirect effect is mediated through a change in behavior of the intervening species. These two categories are not necessarily mutually exclusive. Distinguishing between them, however, is essential because they have different implications for the structure of natural communities. To date, many examples of DMIIs have been confirmed in various studies of both marine and terrestrial ecosystems (e.g., Strauss 1991, Billick and Case 1994, Menge 1995, Wootton 1994, Chen and Wise 1999, reviewed in Polis and Winemiller 1996). Recent attention to TMIIs has resulted in the identification of various examples in natural communities, although TMIIs continue to be less well understood both theoretically and empirically (e.g., Peacor and Werner 2001, Schmitz 1998, Dukas 2001, Schmitz and Suttle 2001).

Application of the DMII/TMII concept to model multitrophic systems in the Piperaceae can contribute greatly to our understanding of the selective circumstances necessary for a plant to form a mutualism. In addition, it can elucidate the ecological and evolutionary forces such as parasitism of the mutualism (Letourneau 1990, Letourneau and Dyer 1998a,b) or intraguild predation (e.g., Polis *et al.* 1989) that may preclude mutualisms or cause them to break down. *Piper* species growing in similar habitats, in some cases in close proximity, exhibit all possible combinations of domatia and food body types, from hollow petiole domatia with food bodies inside, to no domatia and no food bodies (Chapter 9). They are also consistently associated with a wide array of arthropod herbivores and predators, which often interact in complex ways across multiple trophic levels. There are an unknown number of *Piper*–arthropod mutualisms, some of which could involve taxa other than ants.

Work with potential *Piper* mutualisms has focused on interactions between *Pheidole* ants, particularly *Pheidole bicornis*, and four *Piper* species that have hollow petioles, which the ants use as domatia. Food bodies are produced in these petioles and the ants use them as food (Risch *et al.* 1977, Letourneau 1983, 1998, Letourneau and Dyer 1998a,b, Dyer and Letourneau 1999a, Gastreich 1999, Fischer *et al.* 2002). Other *Piper* species have food bodies and domatia or domatia-like structures but only on the leaf surfaces. These structures could also allow the formation of mutualisms, yet mutualistic interactions have not yet been observed in these species.

As ubiquitous and dominant predators in many terrestrial ecosystems, spiders have been shown to alter the density and behavior of their prey in many contexts (e.g., McKay 1982, Gastreich 1999, Dukas 2001, Schmitz and Suttle 2001, reviewed by Wise 1994). More recently, these direct effects have been linked to both density-mediated and trait-mediated indirect interactions. Whether a given population of spiders generates TMIIs or DMIIs can depend on many factors, including the physical structure of the system, the capacity of potential prey to detect spider presence and respond accordingly, and the hunting strategy of the spider(s) in question. For example, Schmitz and Suttle (2001) found that the type of indirect interaction generated by spider presence depends on the hunting strategy of the spider in question. Dukas (2001) found that honeybees avoid flowers inhabited by crab spiders, thereby lowering pollination rates of occupied flowers. In *P. obliquum*, as discussed below, theridiid spiders alter the behavior of mutualist *Pheidole* ants such that levels of folivory are increased (Gastreich 1999). Spiders can therefore provide important model systems for understanding the relative dominance of TMIIs versus DMIIs, the interplay between the two types of interactions, and the conditions under which one or the other is most likely to arise. In the context of *Piper* communities, the study of indirect interactions involving spiders can be particularly productive given the large body of knowledge already generated about various species of *Piper*, as well as the diversity of contexts in which arthropod–*Piper* relationships exist.

Here we will use two *Piper* systems to discuss the implications of trait-mediated indirect interactions generated by spiders in the context of antiherbivore mutualisms. The two systems discussed in this section, *P. obliquum* and *P. urostachyum*, differ in terms of location, associated arthropod communities, and structure and morphology. Nonetheless, both have resident insects and associated spider predators that can serve as appropriate models for understanding the interplay between DMIIs, TMIIs, and the maintenance of mutualisms.

6.2. THE CASE OF *Piper obliquum*

Piper obliquum is one of a handful of *Piper* species that have a symbiotic mutualism with *Pheidole bicornis* ants (Risch et al. 1977, Risch and Rickson 1981, Risch 1982, Letourneau 1983). Found in the Pacific lowland wet forest of Costa Rica, it is morphologically similar to *P. cenocladum*, the most thoroughly studied *Piper* ant-plant to date. Work on *P. cenocladum*, especially in the Atlantic lowland forests of Costa Rica, has produced a tremendous body of information that has contributed substantially to our understanding of DMIIs, particularly in the context of the trophic cascade (e.g., Letourneau and Dyer 1998a,b, Dyer and Letourneau 1999b, Dyer et al. 2001). Like *P. cenocladum*, *P. obliquum* houses *Ph. bicornis* ants in hollow petioles, where it produces food bodies on which the ants feed. Ants protect *P. obliquum* from insect herbivores by removing eggs, larvae, and occasionally attacking adults found on leaves (Risch 1982, Letourneau 1983). *Piper obliquum* also produces food bodies on emerging leaves, which generally have the highest densities of ants compared with other areas of the plant surface (Gastreich, pers. obs.).

In the lowland Pacific rain forest of the Golfo Dulce Forest Reserve of the Osa Peninsula, Costa Rica, the most common arthropod predator found on *P. obliquum* is the Theridiid spider, *Dipoena schmidti* (= *D. banksii*, H. Levi, pers. comm.). Approximately 3–5 mm in length, *D. schmidti* builds small webs either at the base of the leaf or (less frequently) at the base of the petiole, and ambushes passing ants. *D. schmidti* is an efficient sit-and-wait predator that has never been observed to lose its prey once an attack is initiated. Although median daily consumption of ants is very low, the spider was seen successfully attacking up to a dozen *Pheidole* ants at once. Preference tests revealed that although *D. schmidti* can attack similar-sized ants of other species, it responds more quickly to the opportunity to attack *Ph. bicornis* and is more likely to reject ants of other species (Gastreich 1996). In addition, an intensive survey of *Piper* plants in the study area showed that these minute spiders are found only on *Piper* ant-plants (see below). These observations are consistent with the hypothesis that *D. schmidti* is a plant-ant specialist that may have evolved to exploit the particular conditions of the *Piper–Pheidole* relationship.

Although most spiders are generalist predators, monophagy has been reported in several families, with a number of social insect specialists and ant predators in the Theridiidae (e.g., Levi 1967, Hölldobler 1970, Carico 1978, Porter and Eastmond 1981, Nentwig 1986, Eberhard 1991). Interestingly, specialist spiders appear to have restricted their diet either to social insects (ants, termites, bees, and wasps) or to other spiders (Nentwig 1986). Although the chemical and behavioral defenses of social insects can be formidable for a small arthropod predator, they have nonetheless been proposed as an optimal diet for spiders because they occur locally in high density (*sensu* Pyke *et al*. 1977, Pyke 1984, Futuyma and Moreno 1988). Spiders able to overcome social insect defenses therefore gain access to a relatively unexploited and predictable food source. Although the genus *Dipoena* has not been well studied, different *Dipoena* species are typically collected with ant prey (William Eberhard, pers. comm.). It is therefore likely that ant preference is a primitive trait for the genus, predisposing its members to the formation of highly specialized associations such as that observed between *D. schmidti* and *Ph. bicornis*.

A 4-year study (from 1991 through 1994) of the *Piper–Pheidole–Dipoena* system on the Osa Peninsula produced data supporting both top–down and bottom–up scenarios for TMIIs in the system. In the top–down scenario, spiders were found to influence ant behavior in such a way that spider presence reduced the number of ants found on *P. obliquum* leaves (Gastreich 1999). This behavioral response depended on the ants' ability to detect spider presence via contact with support strands of silk extending from the web. Of 73 individual ants introduced to petioles occupied by *D. schmidti*, 60% contacted web silk before coming into range of attack, three-fourths of which chose not to continue up the petiole or onto the leaf. In conjunction with this, plants with spiders were shown to suffer higher levels of folivory during the first 30 days of leaf emergence when compared with plants without spiders (Gastreich 1999). Moreover, *P. obliquum* plants that had both resident ants and spiders suffered higher rates of folivory than *P. obliquum* with no ants at all. Spiders had no effect on overall ant densities on the plant, making it unlikely that this particular interaction included a DMII. In a separate study, Letourneau and Dyer (1998a) also found no evidence for spider-generated, top–down DMIIs in *Piper* ant-plants.

Evidence for a bottom–up TMII scenario was also compelling, but less conclusive. Key to this scenario is web site selection and preference in *D. schmidti*, which was shown through both experimental and observational data to strongly prefer not only *Piper* ant-plants, but also regions of high ant density within *Piper* ant-plants (Gastreich 1996). In a field survey of 234 plants representing 16 species of *Piper*, 11 species (59 individuals) were found with ants considered small enough to serve as prey for *D. schmidti*. Of these, 2 species (4 individuals) were *Piper* ant-plants with resident colonies of ants. (Eight *Piper* ant-plants were found, but only 4 with resident colonies.) Only *Piper ant-plants* with resident colonies of *Ph. bicornis* also housed *D. schmidti*. A separate survey of web site occupancy within 31 *P. obliquum* plants revealed that spiders occupy significantly higher portions of available web sites in young versus mature regions of the plant, with the proportion of occupied sites being on average 2 times higher in regions of new growth ($p = 0.02$, $N = 31$ plants; Wilcoxon signed-ranked test). In a behavioral assay, individual *D. schmidti* were introduced to *Piper* plants with and without ants. Behavioral responses to plants with and without ants differed markedly in many aspects, with spiders introduced to ant-plants being significantly more likely to encounter and attack prey, build webs at exposed sites, and remain at their web site for more than 1 day following introduction (Gastreich 1996).

Web site selection and preference in *D. schmidti* appears linked to the availability of its most common prey item, *Ph. bicornis*. Although many factors can interact to determine web site selection and preference in spiders, the fundamental role of prey availability in this process has been noted by several investigators (reviewed by Wise 1994).

Timed observations of *P. obliquum* petioles showed that on average 1.6 arthropods enter and exit a new petiole each minute, and that 95% of the potential prey items that cross the petioles are *Ph. bicornis* (Gastreich 1996). Mature petioles, in contrast, tend to have much lower densities of *Ph. bicornis* (Risch 1982, Letourneau 1983, Gastreich, pers. obs.). Consistent with these observations, *D. schmidti* occupying young petioles tend to have higher capture rates than spiders on mature petioles.

If one considers the possibility that food body production by *P. obliquum* indirectly determines the density and distribution of the system's top predator, web site selection by *D. schmidti* could be part of a bottom–up TMII scenario. The intermediary between *P. obliquum* and *D. schmidti* in this case would be *Ph. bicornis*, by altering its behavior in response to varying food body density on the plant. Various observations consistent with this hypothesis resulted from work with the system. First, a survey of petioles from *Piper* ant-plants revealed that food body density was significantly higher near young versus mature leaves ($p = 0.01$, $N = 38$ petioles from 19 plants; Wilcoxon signed-rank test). If food body production is an incentive to ant presence, higher food body production on young leaves would be an appropriate antiherbivore strategy for *P. obliquum*, as average rates of folivory on young leaves tend to be more than threefold higher when compared with folivory rates on mature leaves (Gastreich 1996).

The missing link in this bottom–up scenario is a clear demonstration of a causal relationship between food body density and ant density on *P. obliquum*. Is food body density actually a tool for *P. obliquum* to manipulate relative ant density on the plant (see Hespenheide 1985), or is it in fact locally high ant density that induces a higher production of food bodies? *Piper ant-plants* produce food bodies facultatively and only in the presence of ants or certain parasites (Risch and Rickson 1981, Letourneau 1990). Nonetheless, it remains unclear whether food body density is caused by local ant density or vice-versa. Some preliminary evidence indicates that *P. cenocladum* may react to intermediate increases in folivory by increasing food body production near young leaves, which in turn induces higher local ant densities (Chris Martin, pers. comm.) However, until more conclusive data is available, the existence of this bottom–up TMII will remain in question.

In concluding this brief review of *P. obliquum*, we would like to point out that the combined evidence for top–down and bottom–up interactions in the system brings to light the intriguing possibility of a positive feedback loop between some of the dominant interactions observed to date (Fig. 6.1). Specifically, if high folivory on young leaves of *P. obliquum* leads to high food body production, this could encourage higher local densities of *Ph. bicornis*. Locally high ant density, as we have seen, leads to locally high densities of *D. schmidti*, as spiders choose web sites according to to prey availability. Once in place, however, *D. schmidti* presence has been demonstrated to reduce visitation rates of ants to young leaves. This could close the circle by resulting in sustained high rates of folivory and maintaining the necessity for an antiherbivore response on the part of *P. obliquum*.

The existence of this positive feedback loop depends crucially on a causal relationship between folivory levels, food body production, and ant density. As mentioned earlier, the evidence for such a relationship is at best inconclusive. Nonetheless, the dynamics and intricacies of this positive feedback loop, if shown to exist, would provide very rich ground for further research and could greatly enhance our understanding of the interplay between top–down and bottom–up effects in terrestrial ecosystems.

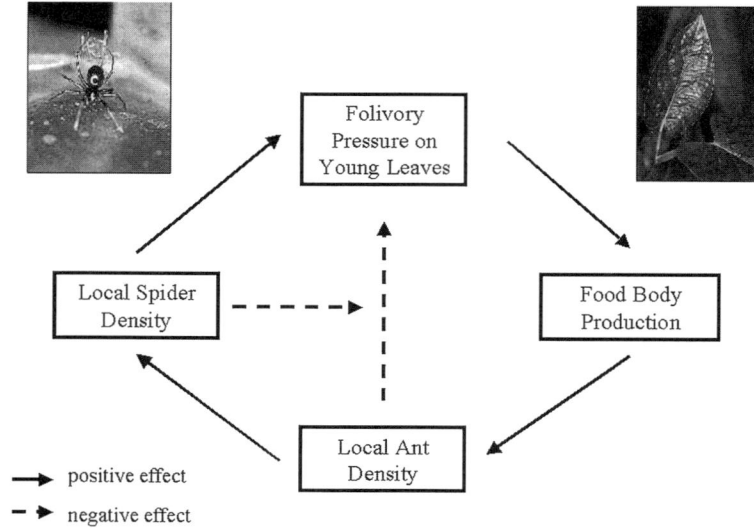

FIGURE 6.1. Proposed positive feedback loop for *P. obliquum*.

6.3. THE CASE OF *Piper urostachyum*

At La Selva Biological Station in the Atlantic lowland rain forest of Costa Rica, *P. urostachyum* is one of the most abundant *Piper* species in the understory (Chapter 4) and supports a remarkably diverse and interesting arthropod fauna. Unlike *P. obliquum*, *P. urostachyum* does not participate in an obligate mutualism with ants although it produces food bodies on its leaves and has large domatia structures that are occasionally used by a *Pheidole* sp. as a nesting site. With the presence of a seemingly stable food source it is curious that *P. urostachyum* does not have obvious facultative ant mutualists. Here, we present observational and experimental data that (1) describe the morphological characteristics of *P. urostachyum* that are conducive to mutualisms, (2) identify resident arthropods and their potential roles, and (3) indicate some possible dynamics of the spider–insect–*urostachyum* interaction.

6.3.1. Plant Characteristics that Encourage Mutualism

Piper urostachyum lacks the large, hollow, modified petioles of *P. obliquum* and *P. cenocladum*, but the leaves are covered in large trichomes, which are especially dense at the vein junctures. These vein junctures are also modified to form shallow pockets. Vein junctures with pockets and/or accompanying dense tufts of trichomes are typical characters for mite domatia (O'Dowd and Willson 1989, 1991, Walter 1996), although in *P. urostachyum* these characters are large for mite domatia, usually ranging from 0.4 to 0.7 cm^2 in length (Gentry, unpubl. data).

In addition to domatia, *P. urostachyum* has food bodies on the leaves (Fig. 6.2). In surveys where food bodies were measured and censused, 80% of *P. urostachyum* were found to have food bodies present on at least one leaf (Gentry and Rath, unpubl. data). Scanning electron micrographs reveal that foliar food bodies appear to be inducible and take the form of liquid-filled, translucent spheres that are probably modified trichomes as per da Silva and Machado (1999), as opposed to some specialized structure (Kelley and Rath, unpubl. data). They vary widely in size (0.1–0.87 mm, $n = 112$) and density (1–35 food bodies on a leaf; $N = 96$ leaves on 32 plants), and are located on the blade and veins of the undersides of the leaves, and on the floral and young fruiting spikes. They are present on both young and mature leaves and are particularly prevalent on leaves that have been damaged. Leaves cut from a plant often produce large numbers of food bodies (over 50) within a few hours, and leaves of young plants that have been grown from fragments are typically constantly covered with food bodies (Dyer, Kelley, pers. comm., Gentry, pers. obs.). Food bodies are produced continually although it is unknown whether they are reproduced in the same spot when removed. Because they are modified trichomes, this seems unlikely.

6.3.2. Resident Arthropods

Several types of faunal surveys were undertaken to learn about the resident fauna of *P. urostachyum* (Gentry and Gastreich, unpubl. data). Some of these were general surveys in which all insects on the first, third, and fifth leaf pairs of fixed numbers of *P. urostachyum* were identified at least to order. Other surveys were more specific, in which only the presence of targeted fauna was noted. On the basis of what is known about the life history of the taxa found, resident arthropods of *P. urostachyum* can be divided roughly into four general groups: potential herbivores, potential mutualists, potential parasites of the mutualism, and potential top predators. Each of these is discussed in turn below.

6.3.2a. Herbivores

The most common herbivores on *P. urostachyum* were scales (Coccidae: Heteroptera) and gall making ceccidomyiids (Diptera). New leaves were commonly attacked by weevils (Curculionidae: Coleoptera), and mature leaves were often subject to damage from leaf miners (these were long serpentine mines and so were probably made by microlepidopterans). Feeding damage from neonates and first instars of *Eois* sp. (Geometridae: Lepidoptera) was noted on five *P. urostachyum* leaves during insect surveys. Most of these observations (22 of 28) were of feeding damage typical of early instar *Eois* sp. (i.e., exactly the same, and adjacent to the damage caused by *Eois* sp. caterpillars that were present), although the caterpillars themselves were missing. The presumed predation events were not observed in these cases, but predation on early instar *Eois* sp. has been noted elsewhere and is discussed below. Interestingly, *Eois* sp., which specializes on pipers and is fairly abundant on other common species, such as *P. imperiale* and *P. cenocladum* (Gentry and Dyer, unpubl. data), was relatively scarce on *P. urostachyum*, even though *P. urostachyum* is the most common *Piper* at La Selva.

6.3.2b. Mutualist predators

Candidates for potential mutualists on *P. urostachyum* include various species of spiders and ants and a single, very common species of the family Miridae (Heteroptera).

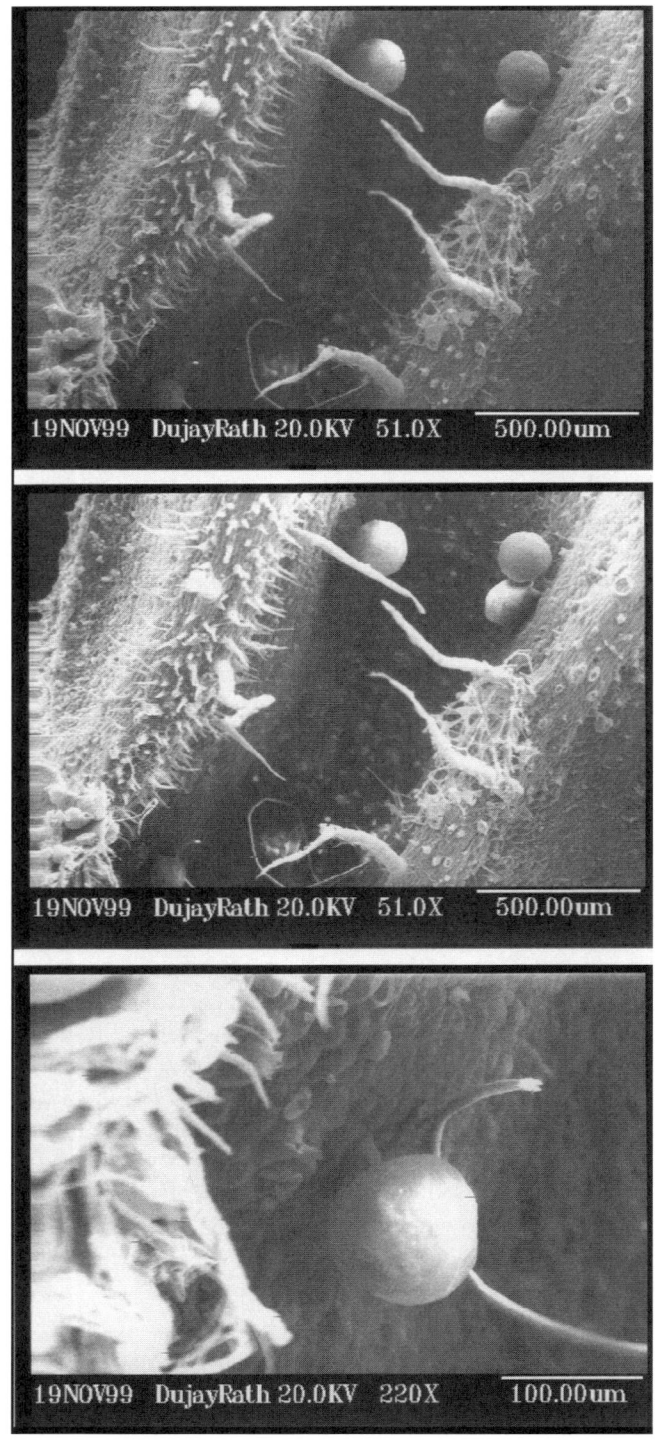

FIGURE 6.2. Photos of foliar food bodies (photo by Donna Rath and Rick Dujay).

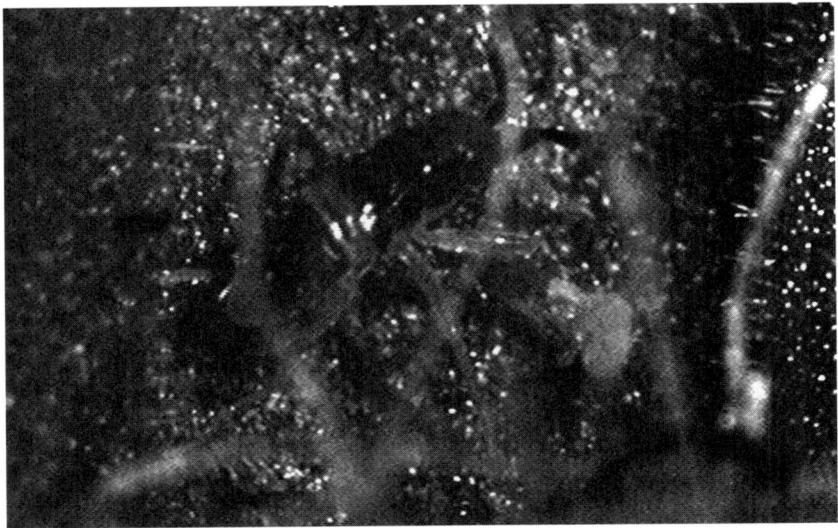

FIGURE 6.3. Mirid nymphs hiding behind trichomes on *P. urostachyum*.

Of several ant species surveyed on *P. urostachyum*, *Wasmannia* was the most common genus, with *Crematogaster* being the next most common. One species of *Wasmannia*, *Wasmannia scrobifera* (Kempf, 1961), makes small "jug"-type carton nests at the vein junctures of the leaves, usually near the petiole. These "jug" nests are fairly common and were found on the leaves of 7 of 32 surveyed *P. urostachyum* plants. They were not found on any other surveyed *Piper* plant, although a different type of carton nest was found on *P. imperiale* (as per Tepe *et al.*, Chapter 9). In two cases each, more than one "jug" nest was found per leaf and per plant. The nests contain brood and a small number of ants (up to 6 or 7). Resident ants were not observed harvesting food bodies, and examination of five nests found brood but no food bodies; so it is unknown if this *W. scrobifera* species utilizes foliar food bodies on *P. urostachyum*. In contrast, although they did not nest on *P. urostachyum*, ants in the genus *Crematogaster* were observed harvesting food bodies.

Although ants have been observed on *P. urostachyum* and may play an important role in this system, the most notable observation of this study was the constant presence of the adults and nymphs of a single species of predatory Miridae on the leaves. This mirid was found on 81% ($n = 60$) of surveyed *P. urostachyum*, yet was never seen on any other species of *Piper* (18 species surveyed, with >10 individuals per species) nor on surrounding vegetation (Gentry, unpubl. data). Both adults and nymphs spend much of their time sitting in the pockets formed by the vein junctures, where they are almost completely hidden by trichomes (Fig. 6.3). Apparently not confined to any particular leaf, they roam throughout the plant, including the flowers and fruits. Although no formal catch-and-release studies were performed to determine the extent of plant loyalty, when adults were disturbed or removed from the plant they consistently flew back to the same plant if released in the vicinity (Gentry, pers. obs.).

Resident mirids on *P. urostachyum* are omnivorous and feed on the plant and on insects. The nymphs had the most varied diet and in the field were observed feeding on food bodies produced by leaves, flowers, and fruits, and on eggs and scales. In the lab they fed on food bodies, tent coccinellid larvae if the tent was removed, and on first instar noctuid larvae. They would not feed on first instar larvae of an *Eois* species ($N = 5$). The adults have not been observed feeding on food bodies but have been seen feeding on scales in the field and early instar lepidopteran larvae in the lab.

Other occasionally encountered predators that could serve as mutualists were a beetle larva (possibly a staphylinid) that lived in a clear, flexible tube of silk; mites (various families); and thread-legged bugs (Tenthrenidae). Spiders were commonly observed on *P. urostachyum*, and have been seen preying on *Eois* sp. on *P. cenocladum* (Smilanich, pers. comm.). Because spiders could act either as mutualists or as top predators in the system, they are discussed separately below.

6.3.2c. Parasites of the mutualism?

A potential mutualism parasite in the *P. urostachyum* system is a coccinellid beetle whose larvae feed on food bodies, yet provide no discernable service to the plant. This species of coccinellid (and perhaps other species) is not restricted to *P. urostachyum* and occurs on other *Piper* species that produce food bodies on their leaves. The larvae construct "tents" in the vein junctures of the leaves, possibly from silk or saliva, during their first instar (Schaefer, unpubl. data). Food body size and density were much higher under the tents than on the leaf surface (Rath and Gentry, unpubl. data). If the plant is altering existing trichomes to make food bodies, the high density of tent food bodies is probably due to the high density of trichomes at the vein junctures. The tents have no openings and the larvae complete their entire life cycle under the tent, feeding on food bodies. It is unknown how the larvae cause the plant to produce so many food bodies under its tent. Mechanical damage may suffice to induce food body production, or the coccinellid larvae or adults (while ovipositing) may use some chemical signal, as the clerid *Tarsobaenus letourneau* does in *P. cenocladum* (Dyer and Letourneau, pers. comm.).

6.3.2d. Top predators

The spider fauna of *P. urostachyum* includes many families, the most common of which were the Anyphaenidae, Ctenidae, Pholcidae, Salticidae, Scytodidae, and Theridiidae. In general, these families are characterized by a variety of life histories ranging from generalist predators to ant- and even spider-specialists (Nentwig 1986, Levi 1990, Wise 1994). On *P. urostachyum*, they exploit a diversity of potential web sites and hunting strategies, and can be found in various stages of their life cycle. In contrast, the spider fauna of *P. obliquum*, discussed earlier in this chapter, consists almost exclusively of the ant predator *Dipoena schmidti*, a relatively specialized spider that exploits a very narrow range of web sites and prey items (Gastreich 1996).

The family most commonly represented on *P. urostachyum* was Theridiidae; the small, cobweb-building members of this family made up about 51% of the spiders surveyed ($N = 100$ of 194 spiders on 261 leaves of 32 plants; Gentry, unpubl. data). Similar in morphology to *D. schmidti*, these spiders have been observed preying on ants in the genera *Crematogaster*, *Paratrechina*, and *Pheidole* (but primarily *Crematogaster*), although the

FAUNAL STUDIES IN MODEL *Piper* spp. SYSTEMS

degree of preference and specialization is not known. Salticids, theridiids, and anyphaeniids have been observed preying on ants (*Crematogaster* and *Ectatoma* spp.), resident mirids, ceccidomyiids, ciccadellids, and fulgorids (Ahlstrom, Gastreich, Gentry, and Hodson, pers. obs.).

6.3.3. Possible Mutualisms and the Effects of Spiders

P. urostachyum poses an interesting conundrum because it possesses structures typical of other ant-mutualist pipers but does not appear to support stable ant mutualisms. The *Crematogaster* spp., foraging for food bodies or jug-nesting *Pheidole*, may act as facultative mutualists. However, they are not common residents or visitors on *P. urostachyum*; they are certainly not as common as ants acting as facultative mutualists on extrafloral nectary plants (as per Hespenheide 1985). Instead there are a wide range of predators on the plant that eat various types of prey, including herbivores and other predators. The most ubiquitous of these are mirid adults and nymphs, as well as the spider guild (Fig. 6.4).

In an effort to describe the relationships among the different players in this system, in June 2002 a general survey was conducted of 66 *P. urostachyum* plants in the forest understory at La Selva Biological Station. Total numbers of arthropods found were recorded for each of the following categories: spiders, mirid adults, mirid nymphs, coccinellid (tent) beetles, and other insects. Densities were determined for each of these categories by dividing the total number of arthropods by the total number of new and emerging leaves on each plant. In addition, percent leaf area lost was measured on new leaves (up to three leaves per plant) by tracing the leaves and scanning the tracings with a Li-Cor model 3100 area meter.

Resulting data was analyzed using two separate multiple regression models. The first model was a top–down scenario that examined the correlation between levels of folivory and densities of resident arthropods. In this model, percent leaf area missing was

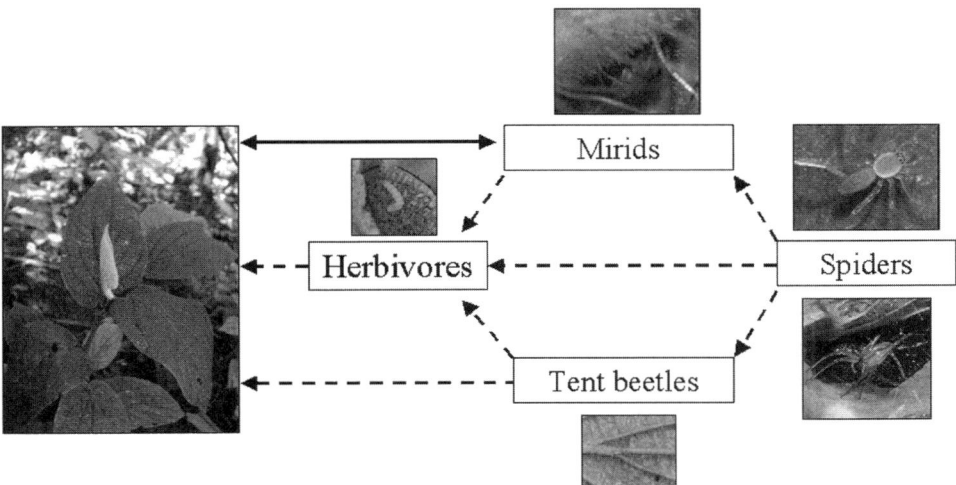

FIGURE 6.4. Summary of some main players in the *P. urostachyum* system.

TABLE 6.1
Results from Top–Down Multiple Regression Analysis

Variable (Densities)	Partial Correlation Coefficient	R^2	p
Spiders	0.004	0.26	0.97
Mirid adults	−0.27	0.36	0.03
Mirid nymphs	0.10	0.27	0.43
Tent beetles	0.29	0.05	0.02
Other insects	−0.13	0.08	0.33

Note: Dependent variable = average folivory; multiple $R = 0.43$, multiple $R^2 = 0.18$, $p < 0.03$, $N = 65$ plants. Results of subsequent partial correlations listed in table.

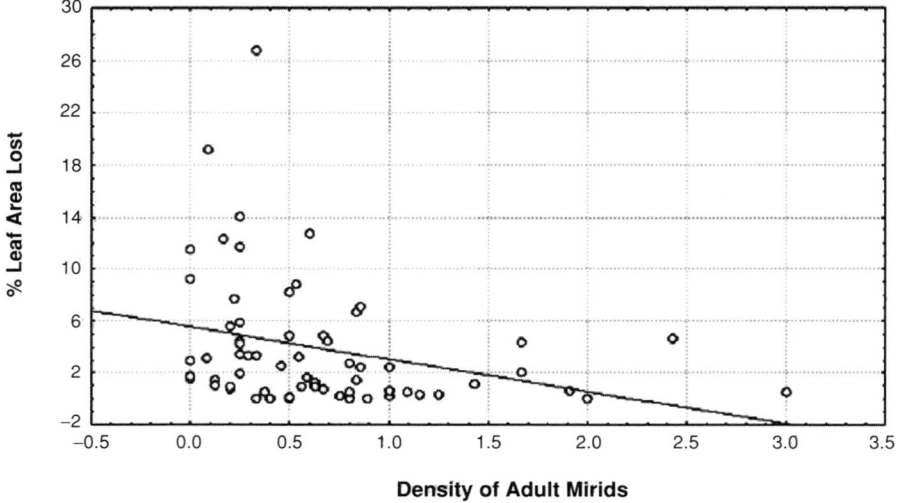

FIGURE 6.5. Correlation between mirid density and folivory.

hypothesized to result from top–down effects generated by covarying densities of arthropods occupying niches on higher trophic levels. Percent leaf area missing was therefore designated as the dependent variable, and correlated against the covariates of mirid adult density, mirid nymph density, tent beetle density, spider density, and the density of other insects. The second model was a bottom–up scenario. In this model, only the density relationships of resident arthropods were included. Resident spider density was hypothesized to result from bottom–up effects generated by covarying densities of potential prey occupying niches on lower trophic levels (Wise 1994). In this analysis, spider density was designated as the dependent variable, and correlated against the covariates of mirid adult density, mirid nymph density, coccinellid density, and the density of other insects.

The top–down scenario indicated a significant, negative relationship between folivory and mirid adult density, such that as mirid density increased, percent leaf area lost decreased (Table 6.1, Fig. 6.5). This analysis also indicated a statistically significant,

TABLE 6.2
Results from Bottom–Up Multiple Regression Analysis

Variable (Densities)	Partial Correlation Coefficient	R^2	p
Mirid adults	0.46	0.19	0.0002
Mirid nymphs	0.07	0.23	0.59
Tent beetles	0.15	0.03	0.24
Other insects	−0.15	0.04	0.18

Note: Dependent variable = spider density; multiple $R = 0.53$, multiple $R^2 = 0.28$, $p < 0.0004$, $N = 65$ plants. Results of subsequent partial correlations listed in table.

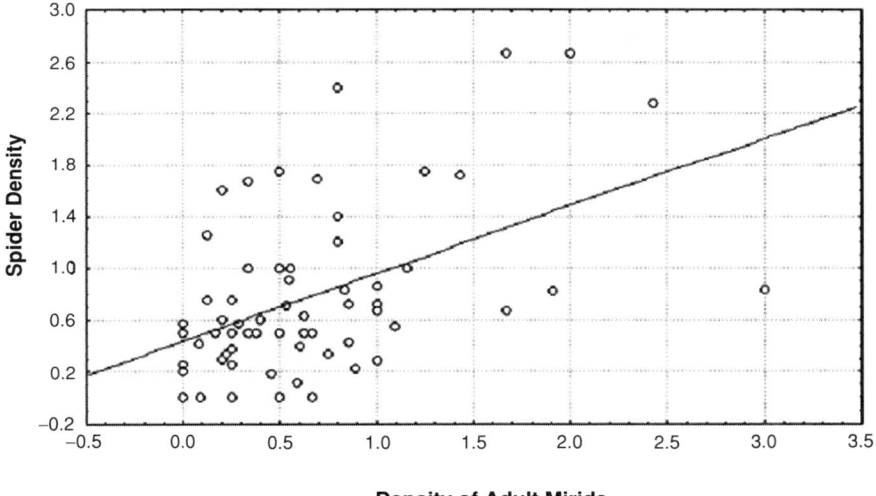

FIGURE 6.6. Correlation between spider density and mirid density.

although weak, positive correlation between tent beetle density and levels of folivory. The bottom–up model indicated a significant, positive relationship between mirid adult density and spider density, such that as mirid density increased, spider density increased (Table 6.2, Fig. 6.6).

Together, these results are consistent with, though not demonstrative of, the hypothesis of a mutualistic relationship between mirids and *P. urostachyum*, and a predator–prey relationship between spiders and mirid bugs. Another type of interaction that might result in the same positive correlation between spider and mirid density would be the use of spider presence by mirids as a form of enemy-free space. Indeed, both possibilities may be at work depending on the spider species in question. For example, anyphaeniid spiders have been observed preying on mirid adults and nymphs in the field and in laboratory feeding experiments (Gastreich and Hodson, unpubl. data). However, mirid adults have also been observed taking shelter under the webs of these same spiders, which weave sheets of silk on the underside of *P. urostachyum* leaves (Gentry, Hodson, pers. obs.).

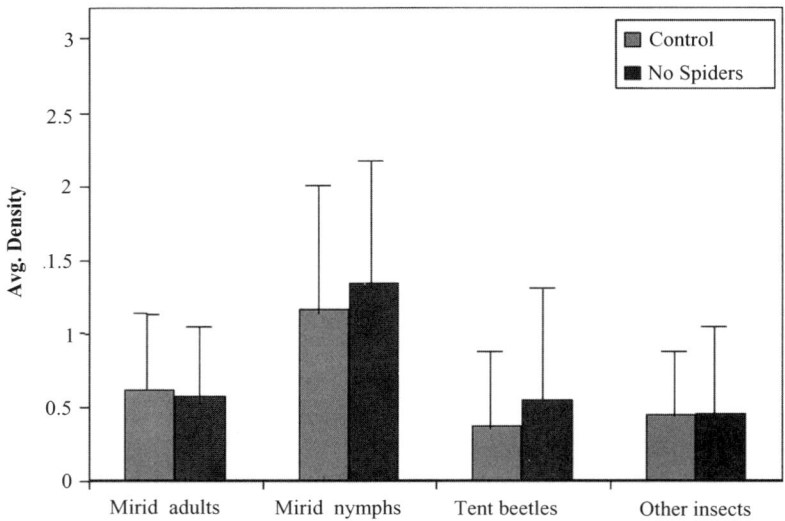

FIGURE 6.7. Mean arthropod densities, spider removal experiment.

To begin experimentally testing the relationships suggested by the above correlation analysis, a spider removal experiment was conducted with *P. urostachyum* at La Selva. Sixty-six plants between 0.5 and 2 m in height were located in the field. To ensure a similar range of plant sizes in experimental and control groups, plants were sorted by height, number of leaves, and number of stems. Similar plants were then paired, with pairs randomly divided between control and experimental groups. At the beginning of the manipulation, all new or emerging leaves were identified and marked individually. Baseline folivory levels on new leaves were recorded by tracing leaves, and then estimating baseline percent leaf area missing to the nearest 1% using a Li-Cor model 3100 area meter. (Baseline levels of folivory for emerging leaves were recorded at 0%.) All spiders and their webs were then removed from plants in the experimental group, which were monitored every 2 days for 5 weeks in order to keep them free from colonizing spiders.

At the end of the 5-week period, numbers of spiders, mirid adults, mirid nymphs, tent beetles, and other insects were recorded for each plant. All marked leaves were traced, and final percent leaf area missing was estimated from tracings to the nearest 1% using a Li-Cor model 3100 area meter. Percent leaf area lost to folivory during the manipulation was then calculated by subtracting [baseline percent leaf area missing] from [final percent leaf area missing]. Average percent area lost per leaf was then calculated for each plant. Arthropod densities were calculated separately for each plant by dividing the number of arthropods by the number of new leaves at the end of the study. Arthropod densities and average percent leaf area lost were then compared between control (with spiders) and experimental (without spiders) groups using a one-way ANCOVA model, with average percent leaf area lost as the dependent variable and arthropod densities as covariates.

Data from the 5-week manipulation included folivory levels of 321 leaves from 65 plants. The final census included 311 mirid adults, 589 mirid nymphs, 116 tent beetles, 127 other insects, and 240 spiders (with 88% of the spiders on control plants).

TABLE 6.3
Results from One-Way ANCOVA of Spider Removal Experiment

Comparison	df	MS Effect	df Error	MS Error	F	p
Summary of all effects	1	64.4	59	10.9	5.89	0.018
Univariate comparisions						
Average folivory	1	72.1	63	10.6	6.77	0.012
Mirid adults	1	0.03	63	0.30	0.11	0.742
Mirid nymphs	1	0.52	63	0.78	0.66	0.419
Tent beetles	1	0.33	63	0.56	0.95	0.333
Other insects	1	0.00	63	0.28	0.00	0.968

Note: Dependent variable = average folivory; Covariates = mirid adults, mirid nymphs, tent beetles, other insects.

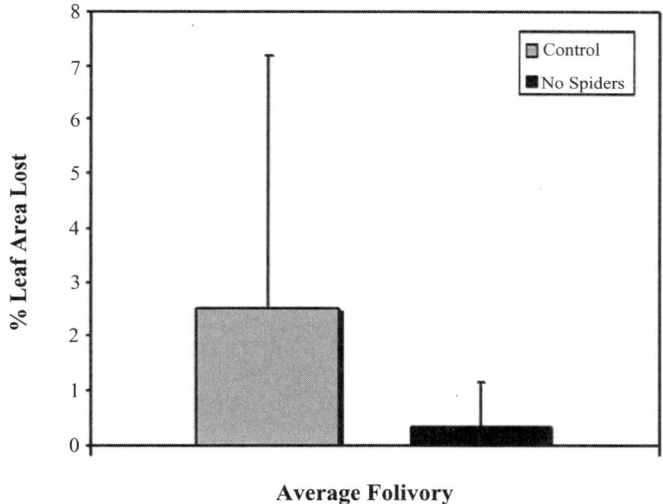

FIGURE 6.8. Mean folivory, spider removal experiment.

Mean insect density, including mirid adult and nymph densities, was the same for both control and experimental groups (Fig. 6.7). However, on plants without spiders, folivory was reduced sixfold within 5 weeks (Fig. 6.8). The ANCOVA model indicated that the effect of spider absence on percent leaf area lost was statistically significant (Table 6.3). However, the model indicated no significant relationship between spider presence and any of the other arthropod densities measured (Table 6.3).

The results of this brief study are very interesting. While detecting a strong indirect negative effect of spider presence on levels of folivory, the results give no indication that spiders have any direct impact on the densities of resident arthropods (such as mirids or herbivores) that should be instrumental in transmitting this effect. This result closely parallels what was observed with the spider removal experiments in *P. obliquum*, and provides intriguing support for the possibility of a spider-generated, top–down TMII in the

P. urostachyum system. Like *Ph. bicornis* on *P. obliquum*, in the absence of spiders mirids may alter their behavior such that they become more effective in antiherbivore defense.

Despite these results, at least two important questions about the *P. urostachyum* system remain unanswered. First, do mirids exhibit a predator avoidance response that is compatible with the TMII model? Preliminary behavioral experiments indicate that overall insect density does increase on *P. urostachyum* leaves within 24 h after spider removal (Ahlstrom, unpubl. data), but the specific response of mirid adults and nymphs is still unknown. Moreover, if spider removal results in increased insect density *coupled* with increased mirid activity, the net effect of this response could be complex and context-dependent. Much work remains to be done here.

Second, and more fundamentally, do mirids protect *P. urostachyum* plants from potential herbivores? Current evidence supports the hypothesis of a mirid–*urostachyum* mutualism; in essence, all the pieces are in place for such a relationship to exist. In addition, preliminary evidence from more recent experiments indicates that removal of mirids results in increased folivory (Hodson and Gastreich, unpubl. data). However, the experimental data needed to confirm this relationship are still incomplete.

6.4. SUMMARY AND CONCLUSIONS

In this chapter, we have presented two systems within the *Piper* group that at first glance appear qualitatively different, but in fact show very similar patterns in terms of both direct and indirect interactions. *P. obliquum*, found in the south Pacific lowlands of Costa Rica, provides food bodies and shelter for resident ant mutualists who, in turn, protect the plant against herbivores. Antiherbivore protection is impeded, however, by the presence of a single specialized species of spider that does not affect ant populations, but does alter ant behavior so as to leave the plant more vulnerable to attack. *P. urostachyum*, found in the Atlantic lowlands of Costa Rica, produces food bodies and shelter utilized by resident mirid bugs, whose presence is correlated with lowered levels of folivory. Antiherbivore protection of *P. urostachyum* is impeded by the presence of a generalist guild of spiders, composed of representatives from many families, who do not appear to affect resident arthropod populations, but apparently do alter some other trait of the mirid–*urostachyum* interaction so as to leave the plant more vulnerable to attack. Although many questions remain unanswered about both systems, the strong similarities detected in the potential impact of spiders is very compelling and is, in fact, what motivated this chapter.

*Piper ant-plant*s and *P. urostachyum* are not unique in the sense that many *Piper* species have been observed to have food bodies and domatia or domatia-like structures that could potentially allow the formation of mutualisms. Nonetheless, these species have not been observed to have mutualistic interactions. This could be due to several factors, including a lack of appropriate character combinations that would allow mutualisms to form with arthropods, the presence of predators or parasites that disrupt the formation of mutualisms, or lack of study of alternative types of mutualisms that either exhibit qualitatively different dynamics and/or do not involve ants.

Pipers such as *P. obliquum* and *P. cenocladum*, which produce food bodies in large, sheltered, and relatively defensible domatia structures such as hollow petioles, may be better able to form strong mutualist associations with ants. Providing this degree of protection to the resident mutualist may allow greater buffering from interference by top predators like spiders. In the case of *P. obliquum* on the Osa Peninsula, *D. schmidti* was by far the most dominant spider species found on the plants, accompanied only occasionally by certain Salticids or other *Dipoena* (Gastreich, unpubl. data). This spider was therefore the only major predator with which *Ph. bicornis* had to contend and even then it was present at low densities and avoided with fairly simple behavioral responses. In the case of *P. cenocladum* at La Selva, spiders are relatively nonexistent on individual plants, and when found are generally also ant specialists. These two examples contrast strongly with *P. urostachyum*, which only provides limited shelter for defense against a large variety of resident, generalist spiders and other predators.

That predation pressure from spiders may mediate the establishment and maintenance of plant–insect mutualisms is an intriguing hypothesis, particularly if the mechanism of mediation is through TMIIs. For a plant experiencing the disruption of a mutualistic interaction, the resulting increase in folivory may be more costly when it happens through a TMII, as opposed to a DMII, depending on the relative strength of each indirect effect. Assuming both phenomena generate equal increases in folivory, under a TMII the plant may continue to sustain populations of mutualist partners, despite increasing folivory. In a DMII, declining populations of mutualist partners could release the plant from some of the costs of maintaining those mutualists, thereby mitigating, at least in part, the effects of increased folivory.

In *P. obliquum*, it was observed that plants without ants actually had less folivory than plants with ants and spiders (Gastreich 1999). This is consistent with the hypothesis that in populations subject to TMIIs, *P. obliquum* plants may do better by not investing in the maintenance of ant colonies for antiherbivore defense. In the case of *P. urostachyum*, high spider density coupled with limited options for effective shelter may prevent ants from commonly establishing nests or foraging lines. This may greatly reduce the importance of ants as even facultative mutualists for *P. urostachyum*. Because facultative mutualisms by necessity include a large behavioral component (i.e., varied potential mutualists have to be constantly enticed to visit the plant because none actually live there), TMIIs may be sufficient to destabilize or preclude the mutualism. At the same time, this would increase the importance and utility of mirids or other predatory insects that can make better use of the limited shelter provided, and can also serve as facultative mutualists.

The emphasis on TMIIs in this chapter is by no means meant to undercut the importance of DMIIs in *Piper* systems in general, and in *Piper* mutualisms in particular. On the contrary, it is likely that both are ubiquitous, and that the predominance of one or the other in any given situation depends on the predator(s) in question. Spiders, to date, appear to be associated primarily with TMIIs. Of the three published studies that have looked for spider-generated DMIIs in *Piper* systems, none have found evidence supporting their existence (Gastreich 1999, Letourneau and Dyer 1998a). On the other hand, a large body of evidence supports the existence of DMIIs generated by other top predators in *Piper* systems, such as the parasitic beetle *Tarsobaenus letourneauae* in *P. cenocladum* (e.g., Letourneau and Dyer 1998a,b, Dyer *et al*. 2001). An opportunity offered by *Piper* systems, then, is the study of coexisting DMIIs and TMIIs, which may interact in complex ways to produce observed density patterns in terrestrial systems.

One important conclusion from the study of these two systems is that not all antiherbivore mutualisms exhibit the same characteristics and follow the same rules, even within the same genera of plants. Other *Piper* species such as *P. urostachyum* with exposed foliar food bodies may also form associations with potential mutualists that are not ants. That mirid bugs may serve as mutualists on *P. urostachyum* is a novel idea but not particularly surprising given observations of predatory Hemiptera in other systems. Harpactorine reduviids can develop on Mullerian and pearl bodies of *Cecropia*, and other hemipterans utilize extrafloral nectar, membracid honeydew, and sap flows as food (Berenger and Pluot-Sigwalt 1997). Bugs in the Geocoridae have been posited as mutualists of cotton (Agrawal *et al.* 2000). That predators of some type are constantly present on *P. urostachyum* is suggested by the fact that the resident coccinellid larvae build shelters for protection during development. Mirids readily eat these larvae when the tents are removed, although they do so only if they manage to find them.

Piper plants that provide food bodies but limited shelter for potential mutualists may be more susceptible to impacts from the presence of top predators such as spiders. These systems may develop first as de facto insectaries, with production of foliar food bodies creating the same watering hole effect as that observed at extrafloral nectaries (Atsatt and O'Dowd 1976, Hespenheide 1985, Gentry 2003). In these systems, many different enemies congregate at the resource and in the course of utilizing it also provide the plant with protection. Given the wide range of morphological characteristics that apparently predispose different *Piper* species to this effect, further comparative studies of *Piper*–arthropod associations could contribute greatly to our understanding of what plant characters and arthropod behaviors are necessary to facilitate the transition from a facultative watering hole effect to a stronger mutualism, and the role of arthropod predators in this process.

6.5. ACKNOWLEDGMENTS

Thanks to Donna Rath, Walt Kelley, Amy Schaefer, Ashby Merton-Davies, Orlando Vargas, Malia Fincher, Cheryl Pacheco, Lee Dyer, John Longino, Mariamalia Araya, Heather Rosenberg, Betsey Ahlstrom, and Amanda Hodson for assistance with data collection, development of a conceptual model of the *Piper urostachyum* system, and comments and assistance with this manuscript. Special thanks go to Lee Dyer for staying on our backs until we finished this chapter. Parts of this study were made possible by the NSF-funded Research Experiences for Undergraduates Program run by the Organization for Tropical Studies at La Selva Biological Station. Funding was also provided by the National Geographic Society and the Earthwatch Institute.

REFERENCES

Abrams, P. A. (1995). Indirect effects arising from optimal foraging. In: Kawanabe, H., Cohen, J., and Iwasaki, K., (eds.), *Mutualism and Community Organization: Behavioral, Theoretical and Food Web Approaches.* Oxford University Press, New York, pp. 255–279.

Abrams, P., Menge, B. A., Mittlebach G. G., Spiller, D., and Yodzis, P. (1996). The role of indirect effects in food webs. In: Polis, G. A., and Winemiller, K. O. (eds.), *Food Webs: Integration of Patterns and Dynamics.* Chapman & Hall, New York, pp. 371–395.

Agrawal, A. A., Karban, R., and Colfer, R. G. (2000). How leaf domatia and induced plant resistance affect herbivores, natural enemies and plant performance. *Oikos* 89:70–80.

Atsatt, P. R., and O'Dowd, D. J. (1976). Plant defense guilds. *Science* 193:24–29.

Berenger, J. M., and Pluot-Sigwalt, D. (1997). Special relationships of certain predatory Heteroptera (Reduviidae) with plants. First known case of a phytophagous Harpactorinae. *Comptes Rendus De L' Academie Des Sciences Serie III—Sciences De La Vie—Life Sciences* 320(12):1007–1012.

Billick, I., and Case, T. J. (1994). Higher order interactions in ecological communities: What are they and how can they be detected? *Ecology* 75:1529–1543.

Carico, J. E. (1978). Predatory behavior in *Euryopsis funebris* (Hentz) (Araneae: Theridiidae) and the evolutionary significance of web reduction. *Symposium of the Zoological Society of London* 42:51–58.

Chen, B., and Wise, D. H. (1999). Bottom–up limitation of predaceous arthropods in a detritus-based terrestrial food web. *Ecology* 80:761–772.

da Silva, E. M. J., and Machado, S. R. (1999). Ultrastructure and cytochemistry of the pearl gland in *Piper regnellii* (Piperaceae). *Nordic Journal of Botany* 19:623–634.

Dukas, R. (2001). Effects of perceived danger on flower choice by bees. *Ecology Letters* 4:327–333.

Dyer, L. A., and Letourneau, D. K. (1999a). Trophic cascades in a complex terrestrial community. *Proceedings of the National Academy of Sciences U.S.A.* 96:5072–5076.

Dyer, L. A., and Letourneau, D. K. (1999b). Relative strengths of top–down and bottom–up forces in a tropical forest community. *Oecologia* 119(2):265–274.

Dyer, L. A., Dodson, C. D., Beihoffer, J., and Letourneau, D. K. (2001). Trade-offs in antiherbivore defenses in *Piper cenocladum*: Ant mutualists versus plant secondary metabolites. *Journal of Chemical Ecology* 27:581–592.

Eberhard, W. G. (1991). *Chrosiothes tonala* (Araneae, Theridiidae): A web-building spider specializing on termites. *Psyche* 98:7–19.

Fischer, R. C., Richter, A., Wanek, W., and Mayer, V. (2002). Plants feed ants: Food bodies of myrmecophytic *Piper* and their significance for the interaction with *Pheidole bicornis* ants. *Oecologia* 133:186–192.

Futuyma, D. J., and Moreno, G. (1988). The evolution of ecological specialization. *Annual Review of Ecological Systems* 19:207–233.

Gastreich, K. R. (1996). *Interactions between the Ant Predator* Dipoena sp. *and the* Piper–Pheidole *Mutualism*. Ph.D. dissertation, University of Texas at Austin. UMI, Ann Arbor, MI.

Gastreich, K. R. (1999). Trait-mediated indirect effects of a Theridiid spider on an ant–plant mutualism. *Ecology* 80:1066–1070.

Gentry, G. L. (2003). Multiple parasitoid visitors to the extrafloral nectaries of *Solanum adherens*. Is *S. adherens* an insectary plant? *Basic and Applied Ecology* 4:405–411.

Hespenheide, H. A. (1985). Insect Visitors to Extrafloral Nectaries of *Byttneria aculeata* (Sterculiaceae)—Relative Importance and Roles. *Ecological Entomology* 10:191–204.

Hölldobler, B. (1970). *Steatoda fulva* (Theridiidae), a spider that feeds on harvester ants. *Psyche* 77:202–208.

Letourneau, D. K. (1983). Passive aggression: An alternative hypothesis for the *Piper–Pheidole* association. *Oecologia* 60:122–126.

Letourneau, D. K. (1990). Code of an ant–plant mutualism broken by a parasite. *Science* 248:215–217.

Letourneau, D. K. (1998). Ants, stem-borers, and fungal pathogens: Experimental tests of a fitness advantage in *Piper* ant-plants. *Ecology* 79:593–603.

Letourneau, D. K., and Dyer, L. A. (1998a). Density patterns of *Piper* ant-plants and associated arthropods: Top-predator trophic cascades in a terrestrial system? *Biotropica* 30:162–169.

Letourneau, D. K., and Dyer, L. A. (1998b). Experimental test in lowland tropical forest shows top–down effects through four trophic levels. *Ecology* 79:1678–1687.

Levi, H. W. (1967). The Theridiid spider fauna of Chile. *Bulletin of the Museum of Comparative Zoology* 136:1–20.

Levi, H. W. (1990). *Spiders and Their Kin*. Golden Press, New York.

McKay, W. P. (1982). The effect of predation of western widow spiders (Araneae: Theridiidae) on harvester ants (Hymenoptera: Formicidae). *Oecologia* 53:406–411.

Menge, B. A. (1995). Indirect effects in marine rocky intertidal interaction webs: Patterns and importance. *Ecological Monographs* 65:21–74.

Nentwig, W. (1986). Non–web building spiders: Prey specialists or generalists? *Oecologia* 69:571–576.

O'Dowd, D. J., and Willson, M. F. (1989). Leaf domatia and mites on Australasian plants—ecological and evolutionary implications. *Biological Journal of the Linnean Society* 37:191–236.

O'Dowd, D. J., and Willson, M. F. (1991). Associations between mites and leaf domatia. *Trends in Ecology and Evolution* 6:179–182.

Peacor, S. D., and Werner, E. E. (2001). The contribution of a trait-mediated indirect effects to the net effects of a predator. *Proceedings of the National Academy of Sciences U.S.A.* 98:3904–3908.

Polis, G. A., and Winemiller, K. O. (eds.). (1996). *Food Webs: Integration of Patterns and Dynamics*. Chapman & Hall, New York.

Polis, G. A., Myers, C. A., and Holt R. D. (1989). The Ecology and Evolution of intraguild predation—Potential comptetitors that eat each other. *Annual Review of Ecological Systems* 20:297–330.

Porter, S. D., and Eastmond, D. A. (1981). *Euryopsis coki* (Theridiidae), a spider that preys on *Pogonomyrmex* ants. *Journal of Arachnology* 10:275–277.

Pyke, G. H. (1984). Optimal foraging theory: A critical review. *Annual Review of Ecological Systems* 15:523–575.

Pyke, G. H., Pulliam, H. R., and Charnov, E. L. (1977). Optimal foraging: A selective review of theory and tests. *Quarterly Review of Biology* 52:137–154.

Risch, S. (1982). How *Pheidole* ants help *Piper* plants. *Brenesia*, 19–20:545–548.

Risch, S. J., and Rickson, F. R. (1981). Mutualism in which ants must be present before plants produce food bodies. *Nature* 291:149–150.

Risch, S., McClure, M., Vandermeer, J., and Waltz, S. (1977). Mutualism between 3 species of tropical *Piper* (Piperaceae) and their ant inhabitants. *American Midlands Naturalist* 98:433–444.

Schmitz, O. J. (1998). Direct and indirect effects of predation and predation risk in old-field interaction webs. *American Naturalist* 151:327–342.

Schmitz, O. J., and Suttle, K. B. (2001). Effects of top predator species on direct and indirect interactions in a food web. *Ecology* 82:2072–2081.

Snyder, W. E., and Wise, D. H. (2001). Contrasting trophic cascades generated by a community of generalist predators. *Ecology* 82:1571–1583.

Strauss, S. Y. (1991). Indirect effects in community ecology: Their definition, study and importance. *Trends in Ecology and Evolution* 6:206–210.

Walter, D. E. (1996). Living on leaves: Mites, tomenta, and leaf domatia. *Annual Review of Entomology* 41:101–114.

Werner, E. E., and Anholt, B. R. (1996). Predator-induced behavioral indirect effects: Consequences to competitive interactions in anuran larvae. *Ecology* 77:157–169.

Wise, D. H. (1994). *Spiders in Ecological Webs*. Cambridge University Press, New York.

Wootton, J. T. (1994). The nature and consequences of indirect effects in ecological communities. *Annual Review of Ecological Systems* 25:443–466.

7
Isolation, Synthesis, and Evolutionary Ecology of *Piper* Amides

Lee A. Dyer[1], *Joe Richards*[2], *and Craig D. Dodson*[2]
[1]*Tulane University, New Orleans, Louisiana*

[2]*Mesa State College, Grand Junction, Colorado*

7.1. INTRODUCTION TO *Piper* CHEMISTRY

The phytochemistry of the genus *Piper* is rich in terms of numbers of compounds discovered, but given the diversity of this genus and the intraspecific diversity of secondary metabolites in those species that have been examined, *Piper* chemistry has not been adequately investigated. The natural products chemistry that has been elucidated is well documented and has been the subject of extensive review (Sengupta and Ray 1987, Parmar *et al.* 1997). Since those reviews were published, 28 new species have been investigated (Benevides *et al.* 1999, Chen *et al.* 2002, Ciccio 1997, de Abreu *et al.* 2002, Dodson *et al.* 2000, dos Santos *et al.* 2001, Facundo and Morais 2003, Jacobs *et al.* 1999, Joshi *et al.* 2001, Martins *et al.* 1998, Masuoka *et al.* 2002, McFerren and Rodriquez 1998, Moreira *et al.* 1998, Mundina *et al.* 1998, Srivastava *et al.* 2000a, Stohr *et al.* 2001, Terreaux *et al.* 1998, Torquilho *et al.* 2000, Vila *et al.* 2001, 2003, Wu *et al.* 1997), and 69 compounds new to *Piper* have been discovered (Adesina *et al.* 2002, Alecio *et al.* 1998, Baldoqui *et al.* 1999, Banerji *et al.* 1993, 2002b, Boll *et al.* 1994, Chen *et al.* 2002, Ciccio 1997, Da Cunha and Chaves 2001, Das and Kashinatham 1998, daSilva *et al.* 2002, de Araújo-Júnior *et al.* 1997, Dodson *et al.* 2000, Gupta *et al.* 1999, Jacobs *et al.* 1999, Joshi *et al.* 2001, Martins *et al.* 1998, Masuoka *et al.* 2002, Menon *et al.* 2000, 2002, Moreira *et al.* 1998, 2000, Mundina *et al.* 1998, Navickiene *et al.* 2000, Pande *et al.* 1997, Parmar *et al.* 1998, Santos and Chaves 1999a,b, Santos *et al.* 1998, Seeram *et al.* 1996, Siddiqui *et al.* 2002, Srivastava *et al.* 2000a,b, Stohr *et al.* 1999, Terreaux *et al.* 1998, Wu *et al.* 2002a,b, Zeng *et al.* 1997).

Only about 10% of all *Piper* species (112 of 1,000+ known species worldwide) have been phytochemically investigated. These 112 species have been found to produce 667 different compounds distributed as follows: 190 alkaloids/amides, 49 lignans, 70 neolignans, 97 terpenes, 39 propenylphenols, 15 steroids, 18 kavapyrones, 17 chalcones/dihydrochalcones, 16 flavones, 6 flavanones, 4 piperolides (cinnamylidone butenolides) and 146 miscellaneous compounds that do not fit into the common major groups of secondary metabolites. The categories used here are not important from a biosynthetic perspective, they are simply consistent with the classifications of the most recent review (Parmar *et al.* 1997). Secondary metabolites have been found in all parts of the plant, including leaves, stems, roots, and inflorescences. Although all species investigated produce mixtures of secondary metabolites and some particular species contain very diverse suites of secondary metabolites (e.g., *P. nigrum*, *P. betle*, *P. auritum*), only a few studies have tested for the additive or synergistic interactions that may exist between these compounds (Haller *et al.* 1942, Scott *et al.* 2002, Dyer *et al.* 2003). Synergistic plant defenses are broadly defined as effects of multiple compounds that are greater than expected on the basis of projected additive values of each individual compound (Jones 1998), and we use that definition in this chapter.

The propenylphenols have relatively simple structures, with phenol (or their methyl ethers) plus a 3-carbon side chain (Fig. 7.1), and include active compounds commonly found in other plants, such as eugenol, safrole, and myristicine. Safrole is found in particularly high concentrations in *P. hispidinervium* (83–93% of the essential oils; Rocha and Ming 1999) and, like other *Piper* compounds, could act synergistically with other compounds found in the plant. Piperonyl butoxide, the well-known insecticidal synergist (Jones 1998), is derived from safrole, as are other important compounds, such as heliotropin, which is used as a fragrance and as a flavoring agent (Rocha and Ming 1999).

Other *Piper* compounds produced by the phenylpropanoid pathways include the lignans, neolignans, chalcones, dihydrochalcones, flavones, flavanones, piperolides, and kavapyrones (Fig. 7.1). All of these classes of compounds have been investigated for various bioactivities. For example, the lignan sesamin (Fig. 7.1) was shown to be a synergist. Hallar *et al.* (1942) found that mortality of flies exposed to pyrethrins plus sesamin (85% fly mortality) was greater than the additive effects of the pyrethrins (20%) and sesamin (5%). Neolignans, such as kadsurin A (Fig. 7.1), have been tested as inhibitors of platelet binding on receptor sites of mammal plasma membranes with some success (Parmar *et al.* 1997). The dihydrochalcone asebogenin (Fig. 7.1) demonstrates antiplasmodial activity (Jenett-Siems *et al.* 1999). The kavapyrones are unique to the genus *Piper* and have mostly been isolated from *P. methysticum*; they are probably an important component of the medicinal qualities of this species and are the focus of Chapter 8. The mevalonic acid pathway in *Piper* produces a wide variety of sesquiterpenes, monoterpenes, and steroids. Many of these terpenoids are also important compounds in other plant families, including camphor, linalool, and α-pinene (monoterpenes); guaiol (sesquiterpene); and cholesterol and sitosterol (steroids). Sitosterol is found in at least 37 species of *Piper* (Parmar *et al.* 1997, Facundo and Morais 2003).

This chapter focuses on *Piper* amides, which are abundant in this genus and have great ecological and economic importance. All of these compounds are neutral (to weakly acidic) as opposed to basic (as in the case of alkaloids) since in all cases the nitrogen is in an amide functionality and thus they cannot be isolated by standard acid/base procedures. The majority of these compounds are composed of an acid such as cinnamic acid forming an amide where the nitrogen is in a five- or a six-membered ring (e.g., piperine, Fig. 7.1) or on an isobutyl chain. A small number of oxoaporphine and *N*-formyl aporphines have also been

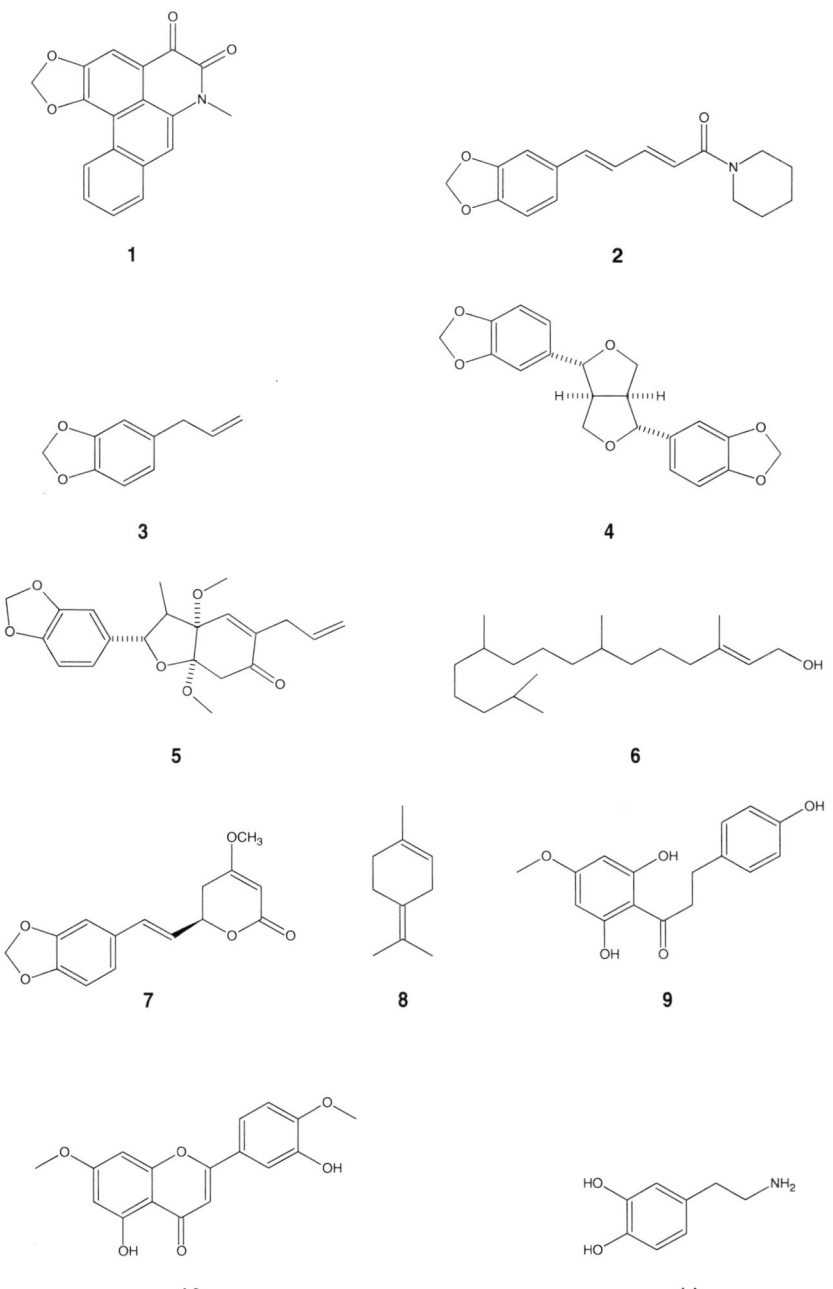

FIGURE 7.1. Examples of major classes of compounds found in *Piper* spp. Stereochemistry is depicted in all cases for which it is known. *Amides*: cepharadione A (**1**) and piperine (**2**); *Propenylphenol*: safrole (**3**); *Lignan*: sesamin (**4**); *Neolignan*: kadsurin A (**5**); *Terpenes*: transphytol (**6**), terpinolene (**8**); *Kawapyrone*: methysticin (**7**); *Dihydrochalcone*: asebogenin (**9**); *Flavone*: 7,4′-dimethoxy-5,3′-dihydroxyflavone (**10**); *Other*: dopamine (**11**).

reported. One of the most important of these amides is piperine (Fig. 7.1), which has been isolated from 20 Old World species of *Piper* (Parmar *et al.* 1997) but does not exist in the new world. Other amides that have been studied in an ecological context include pipericide, piplartine, 4'-desmethyl piplartine, and cenocladamide; the latter two compounds appear to be unique to *P. cenocladum*. There are two studies that have demonstrated strong insecticidal synergy between different amides (Scott *et al.* 2002, Dyer *et al.* 2003).

7.2. ISOLATION AND QUANTIFICATION OF *Piper* AMIDES

Diverse methods for isolation and quantification of amides are described in papers such as Pring (1982), Gbewonyo and Candy (1992), Bernard *et al.* (1995), Alecio *et al.* (1998), and da Silva *et al.* (2002). Most of the methods therein focus on isolating compounds with cytotoxic, insecticidal, fungicidal, or other economically important activities. Dodson and colleagues (Dodson *et al.* 2000, Dyer *et al.* 2001) developed methods that are useful for the analysis of amides in plants harvested from large field experiments. The general method for analyzing amides from *P. cenocladum* (Fig. 7.2) and *P. imperiale* (Dodson and Dyer, unpubl. data) should be useful for examining amides that may have roles in trophic dynamics associated with other rain forest *Piper* species.

For quantification of amides in *P. cenocladum* (**12, 13, 14**) and three amides from *P. imperiale* that were harvested from experiments, leaves were air-dried at room temperature and ground to a fine powder, and 1-g aliquots from each plant were extracted overnight, twice, at room temperature with 95% ethanol. The crude residue of the extract was resuspended in 3:1 water/ethanol and exhaustively extracted with chloroform in a separatory funnel. Combined chloroform extracts were dried in vacuo and the residue redissolved in 10 ml of methylene chloride. One-milliliter aliquots of the samples were transferred to autosampler vials. Each sample was quantitatively analyzed by GC/FID using the method of internal standardization. Piperine was used as the internal standard and was added to the samples at the 80-μg/ml level. Five point calibrations (50, 100, 200, 300, and 500 μg/ml; r^2 values for calibration were 0.99 or better) were prepared with synthetic piplartine, 4'-desmethylpiplartine, and cenocladamide (Dodson *et al.* 2000, 2001). These three amides are present in *P. cenocladum* at roughly equivalent (i.e., not statistically different) concentrations, and total percent dry weight of the three amides varies from 0.48 to 2.0% (Table 7.1). This method can be modified with good detection response and linearity in the range of 20–1,000 μg/ml.

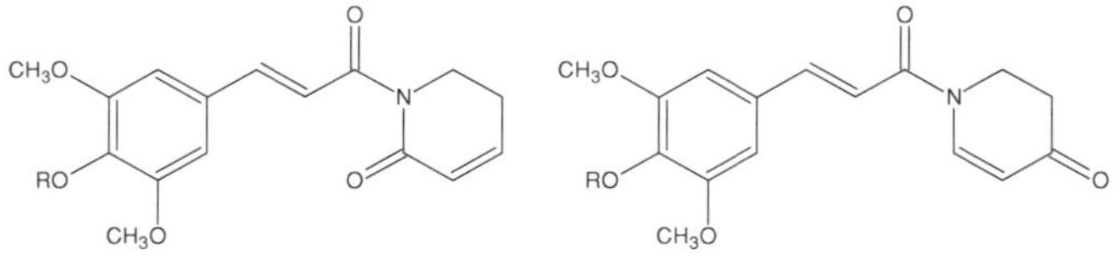

12 R = CH$_3$ 13 R = H   14

FIGURE 7.2. Amides from *Piper cenocladum*: piplartine (**12**), 4'-desmethylpiplartine (**13**), and cenocladamide (**14**).

TABLE 7.1
Mean Total amide Content (Piplartine + 4′-Desmethylpiplartine + Cenocladamide) in *Piper cenocladum* Shrubs and Fragments, Expressed as % Dry Weight of the Leaf Tissue (with 1 SE in Parentheses)

Treatment	Field Shrubs	Shadehouse	Field Fragments
Ant symbionts	0.48 (0.02)	1.4 (0.17)	1.4 (0.22)
Beetle symbionts	0.39 (0.03)	1.5 (0.20)	Not measured
No symbiont	1.36	2.0 (0.22)	1.9 (0.22)

Note: Values are from various experiments at La Selva Biological Station, Costa Rica, that were most recently described in Dyer *et al.* (2004). The "shadehouse" at La Selva is an outdoor structure, with walls and roof made of shade cloth and with tables for growing plants. Shadehouse plants were small fragments (3–15 leaves) grown in pots. Field fragments were planted as part of experiments in a manner that simulates natural fragmentation.

HPLC methods are also available and have been described by Gbewonyo and Candy (1992). More recently, Navickiene *et al.* (2003) developed a rapid reverse-phase HPLC method for the quantitative determination of amides in *P. tuberculatum*. The method resulted in well-resolved peaks with good detection response and linearity in the range of 15–3,000 μg/ml.

7.3. SYNTHESIS OF *Piper* AMIDES AND THEIR ANALOGS

Synthetic studies involving amides found in *Piper* species began in the late 19th century. The earliest syntheses were designed to confirm the structures of isolated compounds. This was especially important for isolations performed prior to the advent of modern spectroscopic techniques. Often, materials used in the syntheses were prepared from isolated compounds whose structures were still being elucidated. The first synthesis of piperine (**2**) was described in 1882 (Rügheimer 1882) and involved the condensation of piperidine with the acyl chloride of piperic acid that had been obtained via hydrolytic digestion of isolated piperine. The *total* synthesis of piperine was not described until 12 years later (Ladenburg and Scholtz 1894). As additional components of *Piper* extracts were isolated, their structures were also commonly elucidated by digestion and reconstitutive or total synthesis. Other examples of *Piper* amide syntheses (Fig. 7.3) described within the context of structural identification include the synthesis of piperettine, a second component of piperine extracts from *P. nigrum* (Spring and Stark 1950), piperstachine, an alkaloid isolated from *P. trichostachyon* C.DC. (Viswanathan *et al.* 1975), wisanidine and wisanine (**15**), piperine-type amides isolated from *P. guineense* (Addae-Mensah *et al.* 1977, Crombie *et al.* 1977), pipericide (**16**) and dihydropipericide (**17**), insecticidal amides first isolated from *P. nigrum* (Miyakado and Yoshioka 1979, Banerji *et al.* 1985, Rotherham and Semple 1998), *N-trans*-feruloyl tyramine (**18**), coumaperine (**19**), *N-trans*-feruloyl piperidine (**20**), *N*-5-(4-hydroxy-3-methoxyphenyl)-2*E*, 4*E*-pentadienoyl piperidine (**21**), and *N*-5-(4-hydroxy-3-methoxyphenyl)-2*E*-pentenoyl piperidine (**22**), phenolic amides isolated from *P. nigrum* L. (Nakatani *et al.* 1980, Inatani *et al.* 1981), *N*-isobutyl-4,5-dihydrox-2*E*-decenamide

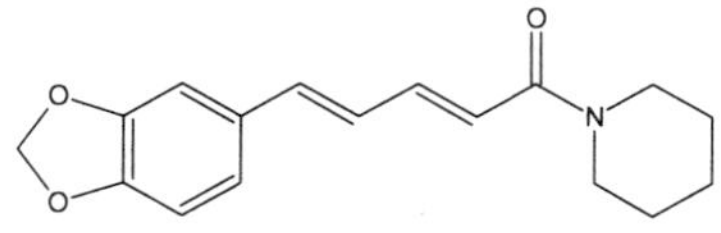

FIGURE 7.3. Structures of some synthesized *Piper* amides.

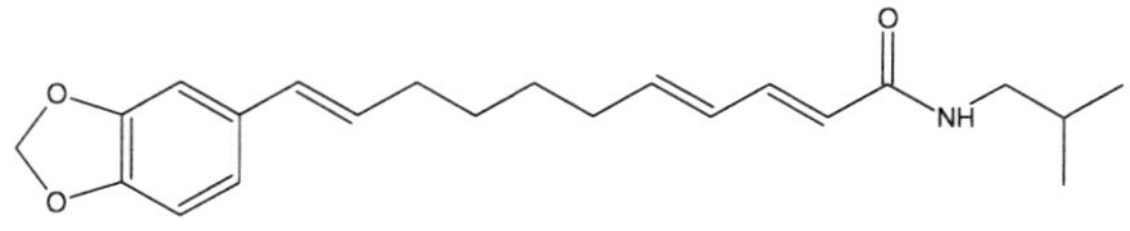

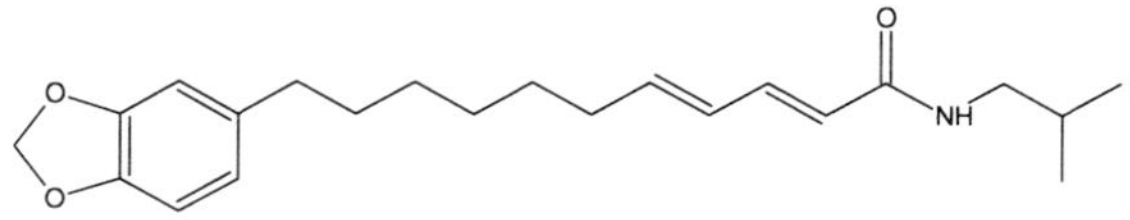

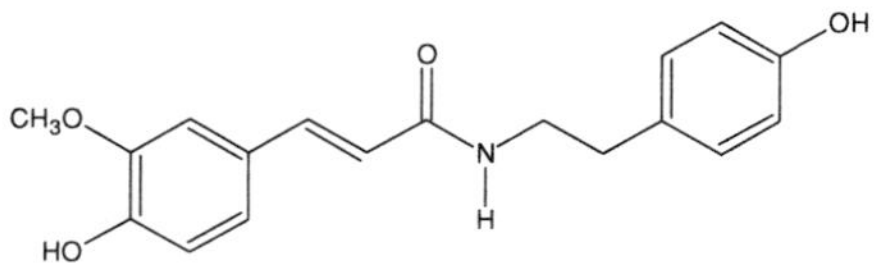

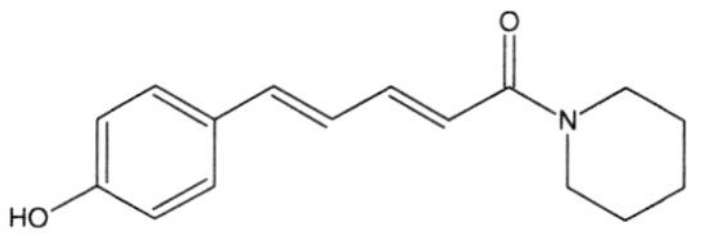

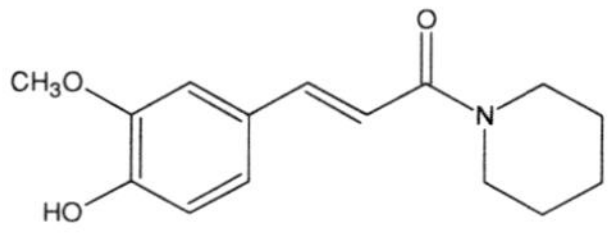

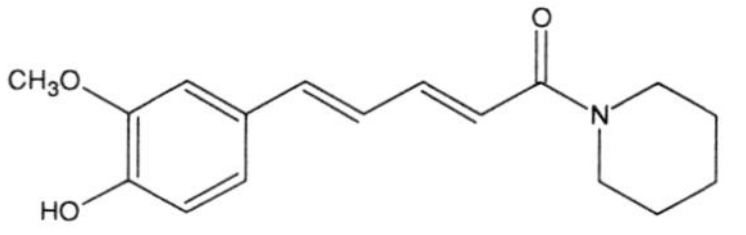

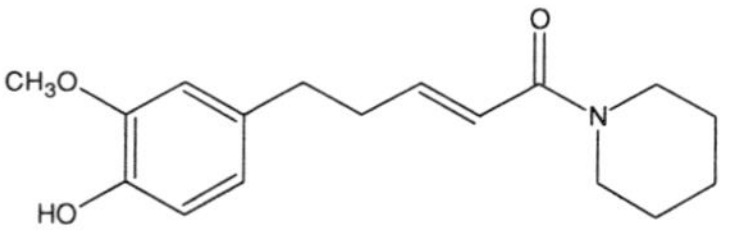

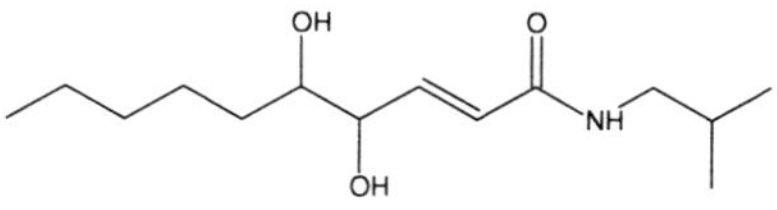

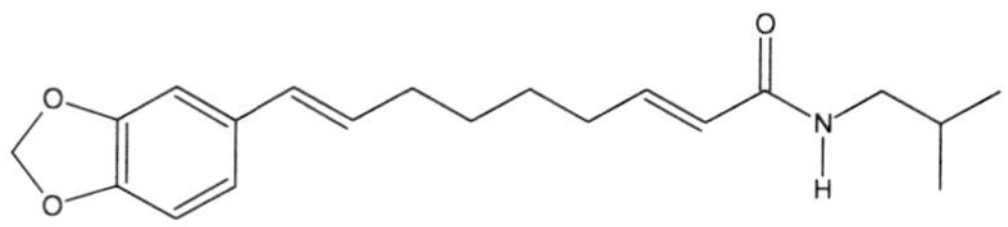

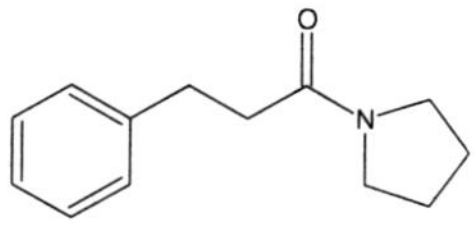

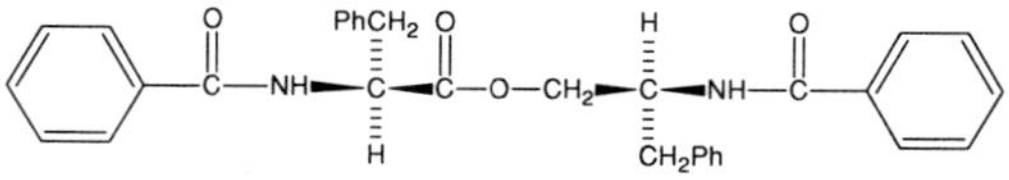

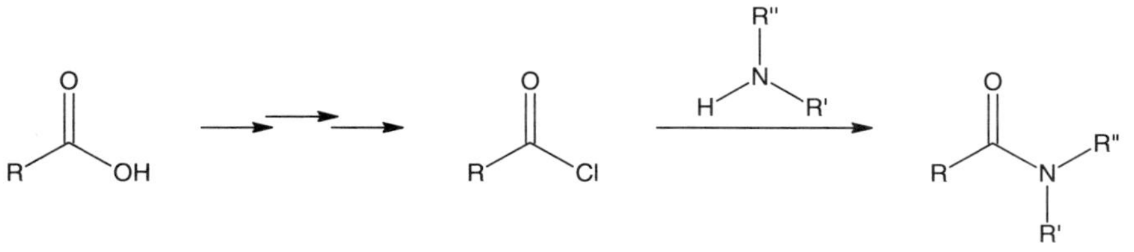

EVOLUTIONARY ECOLOGY OF *Piper* AMIDES

FIGURE 7.4. Typical generation of the amide linkage in the synthesis of *Piper* amides.

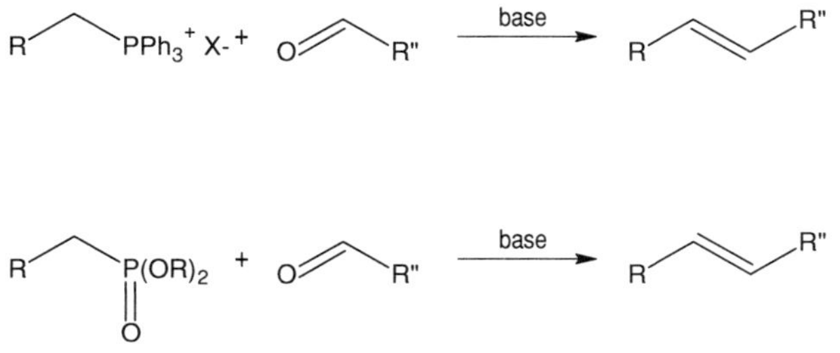

FIGURE 7.5. Typical introduction of unsaturation in the synthesis of *Piper* amides.

(sylvamide, **23**), an alkenamide isolated from *P. sylvaticum* Roxb. (Banerji and Pal 1983), retrofractamide C (**24**), an unsaturated amide from *P. retrofractum* (Banerji et al. 1985), N-(3-phenylpropanoyl)pyrrole (**25**), a pyrrole amide isolated from *P. sarmentosum* (Likhitwitayawuid and Ruangrungsi 1987), and aurantiamide benzoate (**26**), a phenylalanine-derived amide isolated from *P. aurantiacum* Wall. (Banerji et al. 1993).

The syntheses described in the references listed above generally involve very well-understood chemistry and share several synthetic features. Typically, the amide linkages are generated very late via the condensation of an appropriate amine with an acyl halide generated from the corresponding carboxylic acid (Fig. 7.4). In the cases where it is necessary to introduce unsaturation into the carbon framework of the acid moiety, the use of phosphonium salts or phosphonate esters in Wittig-type chemistry appears to have been the method of choice (Fig. 7.5).

Increased interest in the isolation, identification, and activity of amides found in *Piper* species have also made them attractive targets for new, efficient total syntheses and the exploration of new synthetic methodology. A method for the stereoselective introduction of double bonds by coupling of α-bromoacrylamides with alkenylboronates and 1-alkylboronates was recently described and applied to the total synthesis of the *Piper* amides piperine (**2**), piperlongumine (**27**), dehydropipernonaline (**28**), and pipernonaline (**29**) (Kaga et al. 1994; Fig. 7.6).

Strunz and Findlay exploited a Sakai aldol condensation–Grob fragmentation sequence to stereoselectively incorporate unsaturation during the synthesis of more than a dozen *Piper* amides on the basis of the piperonyl framework, including pipericide (**16**), piperolein A (**30**), piperstachine (**13**), retrofractamide A (**31**), dehydropipernonaline (**28**), and piperamide-C9:3($2E,4E,8E$) (**32**) (Strunz and Findlay 1994, 1996). The same papers describe an unrelated systematic approach used for the synthesis of six nonaromatic *Piper*

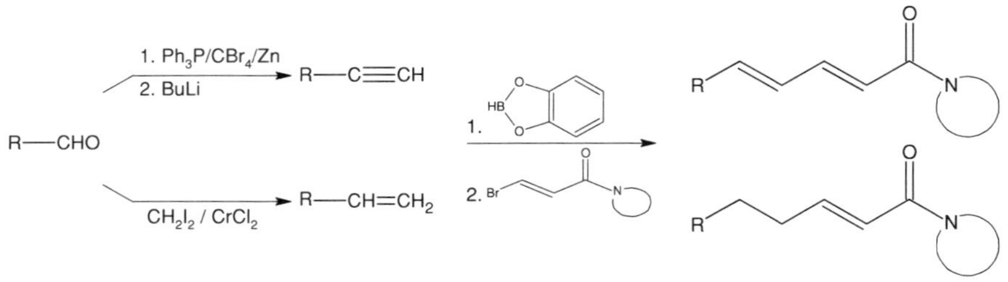

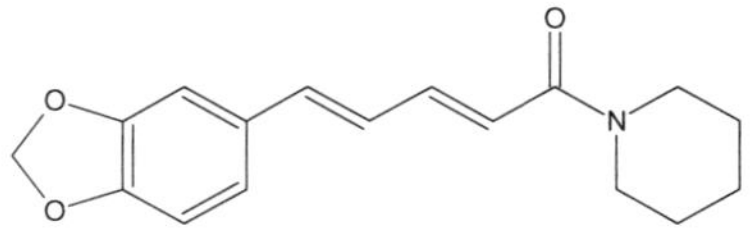

FIGURE 7.6. -Bromoacrylamide–organoboronate coupling used in the synthesis of piperine (**2**), piperlongumine (**27**), dehydropipernonaline (**28**), and pipernonaline (**29**).

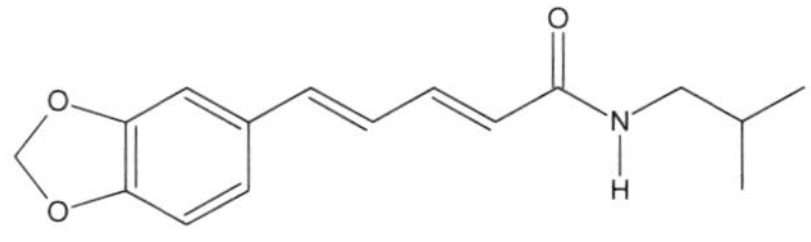

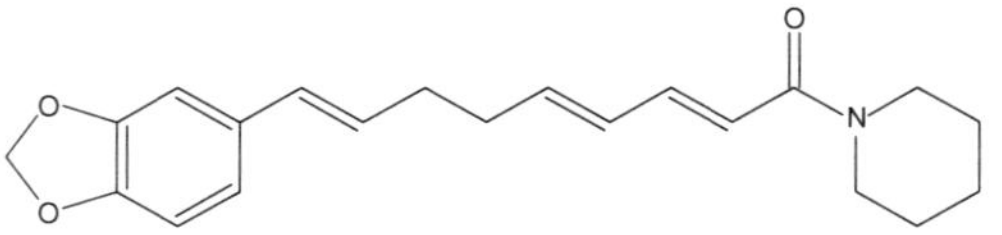

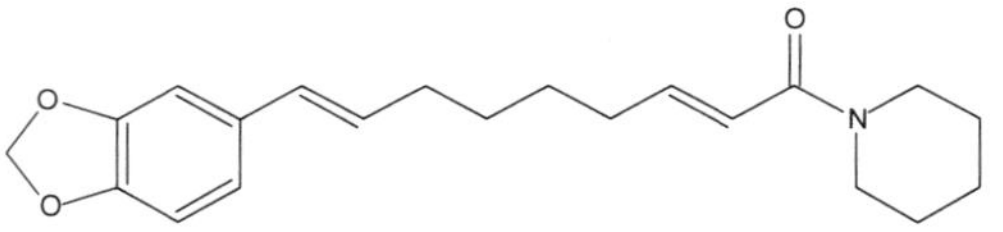

amides, including pellitorine, that is suggested to be an improvement over an earlier reported synthesis (Ma and Lu 1990; Fig. 7.7).

A palladium-catalyzed alkenylation originally applied to α,β-unsaturated aldehydes has been used to synthesize naturally occurring dienamides, including piperdardine (**33**), an amide isolated from *P. tuberculatum* (Schwarz and Braun 1999). This approach involved the alkenylation of a carbonate ester using a palladium(0) catalyst after the carbon framework and amide linkage had been generated (Fig. 7.8).

Similar, naturally occurring dienamides have also been prepared using a hypervalent iodine(III) reagent [phenyliodine(III) bis(trifluoroacetate) (PIFA)] to generate a Pummerer-type reaction intermediate (Kang *et al.* 2001). This intermediate can undergo an ene reaction with 1-alkenes to stereoselectively generate olefins with a nearby sulfide functionalization that can be used to generate a second olefin moiety (Fig. 7.9).

The total syntheses of piplartine (**12**), 4′-desmethylpiplartine (**13**), and cenocladamide (**14**)—amides containing unsaturation in the nitrogen heterocycle isolated from *P. cenocladum*—have also been accomplished (Richards *et al.* 2002). This represents an improvement over the original synthesis of these targets (Richards *et al.* 2001) and involved

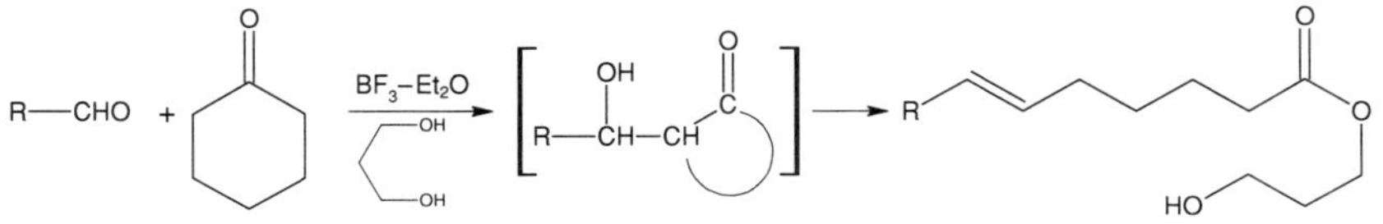

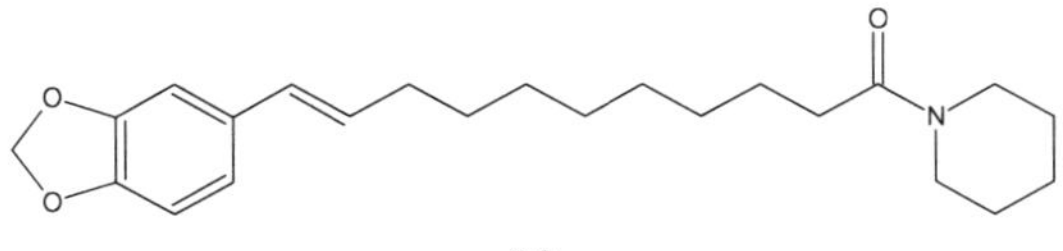

30

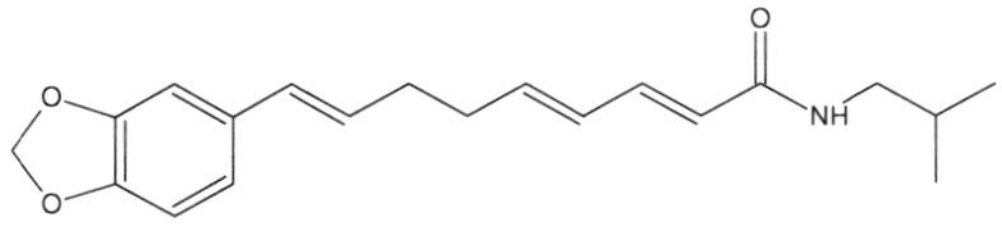

31

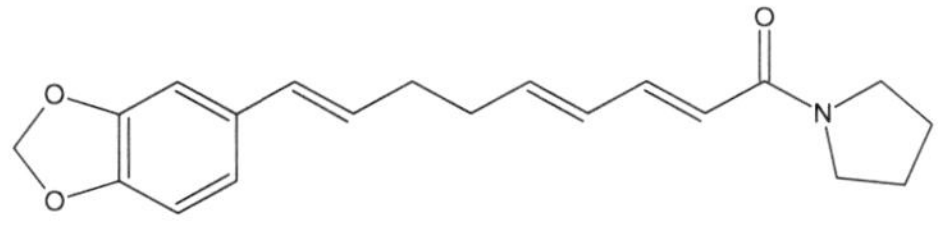

32

FIGURE 7.7. Sakai aldol condensation–Grob fragmentation used in the synthesis of several *Piper* amides.

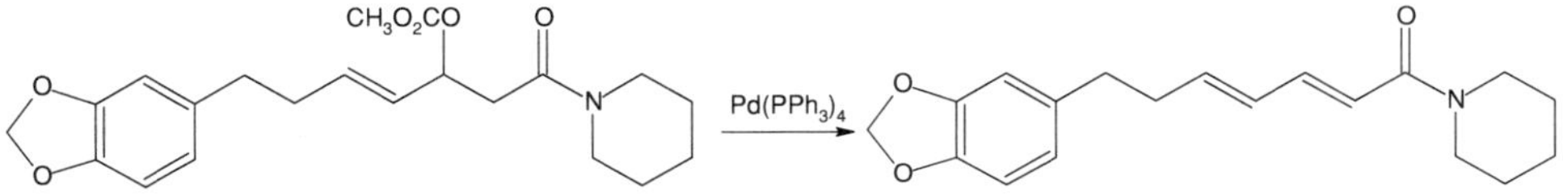

33

FIGURE 7.8. Palladium-catalyzed alkenylation used to prepare piperdardine (**33**).

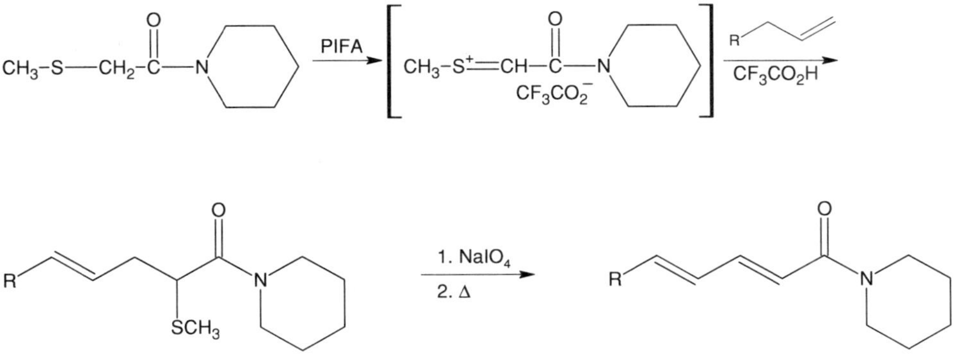

FIGURE 7.9. Generation of reactive intermediate using hypervalent iodine(III) and its use in an ene reaction during the synthesis of dieneamides.

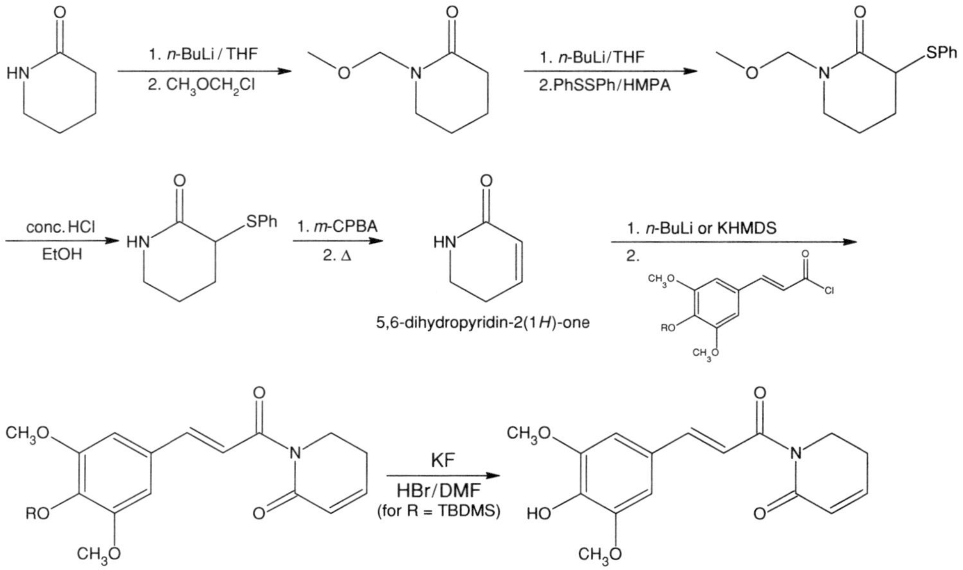

FIGURE 7.10. Synthesis of 5,6-dihydropyridin-2(1H)-one as a precursor to piplartine (**12**) and 4′-desmethylpiplartine (**13**).

early introduction of unsaturation in the heterocyclic moiety and an efficient, convergent approach. The syntheses of **12** and **13** involved the preparation of 5,6-dihydropyridin-2(1H)-one as an important common intermediate (Fig. 7.10).

Again exploiting structural similarity, the syntheses of **13** and **14** involved the preparation of (2E)-3-(4-t-butyldimethylsiloxy-3,5-dimethylphenyl)acryloyl chloride as a common intermediate (Fig. 7.11). This involved the direct conversion of a silyl ester into an acyl chloride in the presence of a silyl ether. The acyl halide was coupled with the anion of the appropriate unsaturated, nitrogen-containing heterocycle, yielding the natural products after deprotection.

The interesting biological activities exhibited by *Piper* amides have also spurred a significant number of synthetic studies designed to prepare analogs of naturally occurring *Piper* amides and to evaluate their potential for commercial and medical use against and with naturally occurring compounds. Synerholm et al. (1945) prepared a series of amides and esters from piperic acid and evaluated their toxicities toward houseflies. Similar series of synthetic and semisynthetic *Piper* amides and analogs based on the piperonyl or homologous frameworks have been prepared and described by Gokhale et al. (1945), Chatterjee and Dutta (1967), Joshi et al. (1968), Atal et al. (1977), Chandhoke et al. (1978), Gupta et al. (1980), Miyakado et al. (1989), Strunz and Finlay (1994), Kiuchi et al. (1997), Das and Madhusudhan (1998), de Mattos Duarte et al. (1999), de Araújo-Júnior et al. (1999, 2001),

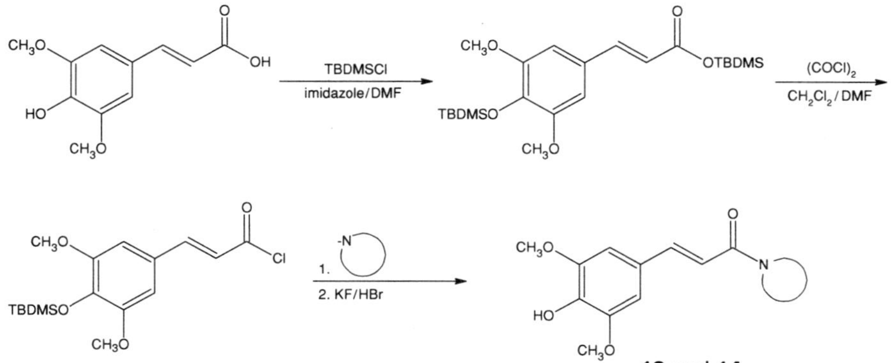

FIGURE 7.11. Synthesis of (2E)-3-(4-t-butyldimethylsiloxy-3,5-dimethylphenyl) acryloyl chloride as a precursor to 4′-desmethylpiplartine (**13**) and cenocladamide (**14**).

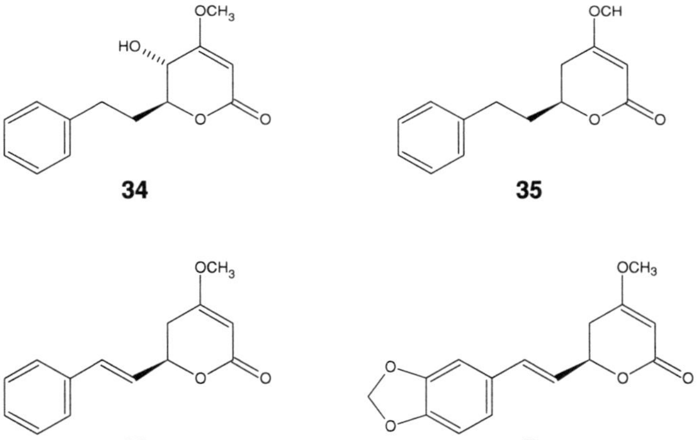

FIGURE 7.12. Structures of (+)-dihydrokawain (**34**), (+)-dihydrokawain-5-ol (**35**), kawain (**36**) and methysticin (**7**).

de Paula *et al.* (2000), Scott *et al.* (2002), and Banerji *et al.* (2002a). Since the focus of these studies was to examine the activity of biologically relevant and/or novel compounds, little effort was made to optimize the efficiency of the syntheses or to examine novel methodology.

Although this review has focused on the *Piper* amides, other classes of compounds isolated from *Piper* species have been targets of synthetic studies. Two notable examples include the syntheses of (+)-dihydrokawain (**34**) and (+)-dihydrokawain-5-ol (**35**), relatives of kawain (**36**) and methysticin (**7**) isolated from *P. methysticum* (Spino *et al.* 1996, Arai *et al.* 2000; Fig. 7.12). Although there are certainly many other examples, these syntheses are notable for their use of interesting enantioselective methodologies.

7.4. ECOLOGY OF *Piper* CHEMISTRY

Although adequate methods exist for isolation, quantification, and synthesis of *Piper* secondary compounds, the ecology of *Piper* chemistry is not well explored (Dyer *et al.* 2000, 2001, 2003). There are countless demonstrations of the strong effects on herbivores and pathogens for all the classes of *Piper* compounds, especially the amides (e.g., Bernard *et al.* 1995, Navickiene *et al.* 2000, Siddiqui *et al.* 2002, also reviewed by Parmar *et al.* 1997, Yang *et al.* 2002). These effects are likely to influence the population dynamics of *Piper* and associated herbivores and could even influence entire communities where *Piper* is found (Dyer and Letourneau 1999a), but effects beyond the cellular or organismal level have rarely been explored. Furthermore, virtually nothing is known about the role of secondary metabolites important for pollination dynamics (e.g., as pollinator rewards or attractants) or other mutualistic interactions. The focus on detrimental cellular and organismal effects of *Piper* compounds has been driven by the applied sciences, where ecologically relevant population or community effects are not important. Thus, most *Piper* compounds are tested in unrealistic biological settings and the tests are not very relevant to *Piper* evolution or ecology. For example, we know that the amides in the infructescence of *P. longum* are effective at killing mosquito larvae (Yang *et al.* 2002) and at vasodilating mammalian coronary tissue (Shoji *et al.* 1986), but nothing is known about the effects these amides might have on the population ecology of fruit-dispersing bats (see Chapter 4).

The bioassay-guided fractionation methods utilized by Wiemer and colleagues (Howard *et al.* 1988, Capron and Wiemer 1996, Green *et al.* 1999) have resulted in the isolation of ecologically important compounds from *Piper* (and other genera). This research focuses on compounds that are deterrent to leafcutting ants (*Acromyrmex* spp. and *Atta* spp.). As these authors point out, leafcutting ants are significant herbivores in most Neotropical forests and pose a serious threat to agriculture in Central and South America, due to the very large amount of plant material they harvest and the high diversity of plant species they attack. The deterrence to leafcutting ants alone highlights the potential for *Piper* chemistry to dramatically affect population dynamics of this genus and to indirectly affect other plants and herbivores. In any forest where leafcutting ants are a dominant herbivore, the species of *Piper* that are not attacked could be more abundant in the understory, and could support a diverse fauna of specialist herbivores that are released from competition with the plant-harvesting ants. This herbivore fauna could support a high diversity of associated predators and parasites. No study has formally examined these potential community-level consequences of *Piper* compounds that deter leafcutting ants.

Our own work at La Selva Biological Station, Costa Rica, with Letourneau and other colleagues has demonstrated the important roles that amides can play in community ecology. We have examined predictions of herbivore regulation models using arthropods associated with the understory shrub *P. cenocladum*, which is defended from herbivores by symbiotic ants (*Pheidole bicornis*) and three amides (Chapter 2 contains many details about the *P. cenocladum* system; Fig. 7.2 illustrates the amides; Table 7.1 summarizes concentrations of amides in shrubs and fragments). Another symbiont, the predatory beetle *Tarsobaenus letourneauae*, preys on *Ph. bicornis*, and both the beetle and ant induce increased production of protein- and lipid-rich food bodies in the petiole of the plant. Individuals housing ants have 2–5 times lower amide content versus those without ants; this is true whether ants are experimentally excluded or the plants are naturally found without ants (Dodson et al. 2000, Dyer et al. 2001; also see Table 7.1). *Pheidole bicornis* exerts strong negative effects on specialist caterpillars (*Eois apyraria* and *Eois* sp.), and enhanced resources cause biologically significant increases in plant biomass, but there are no indirect positive effects of enhanced plant resources on any arthropods associated with *P. cenocladum* (Dyer and Letourneau 1999b).

In a series of experiments we tested these hypotheses: (1) enhanced plant resources cause increases in amides, (2) the amides prevent primary and secondary consumers from proliferating, and (3) the amides have stronger effects on specialist versus generalist herbivores (Dyer et al. 2003, 2004). Experiments were conducted on full-size shrubs in the field and small fragments in the field and in outdoor shadehouses (greenhouses with shade cloth rather than glass). The experiments included manipulations of known and suspected sources of variation in secondary metabolites: presence and absence of symbionts, poor and rich soil (based on N, P, and K availability), and high and low light. We quantified amide concentration of plants from these experiments using the methods described above and tested for correlations between amide content and damage by the most common groups of generalist (orthopterans, leafcutters) and specialist (lepidopterans, coleopterans) folivores. For all experiments, resources had significant effects on amide content, as did the previously demonstrated effects of removing symbionts. Levels of amides were higher in balanced versus unbalanced resource (light and nutrients) conditions, and amides increased when symbionts were excluded and food bodies were not being produced (amide concentrations are summarized in Table 7.1). These results were consistent for shrubs or fragments, and in the field as well as the controlled environment of shade houses. The optimal conditions for production of amides were fertilized fragments in an outdoor shadehouse at low light conditions without symbionts ($2.8 \pm 0.31\%$ dry weight) and the conditions associated with the lowest levels of amides were poor soils, with beetles, and high light ($0.34 \pm 0.05\%$ dry weight in field shrubs). In general, *P. cenocladum* appears to be N-limited; if the protein-rich food bodies are not produced because symbionts are excluded, or if the C:N ratios are low because of "balanced" resource availability, N will be shifted to production of amides. Also, shrubs, which are much more likely to put energy into reproductive parts and roots, have much lower levels of amides than fragments (0.39–0.48% dry weight for shrubs and 1.3–2.0% for fragments).

One interesting aspect of the herbivore–plant chemistry interactions in this system is that production of amides is not induced by herbivores. There are three strong lines of evidence for lack of induction. First, amide production is significantly lower for shrubs in the field with beetles compared to those with ants, and beetle shrubs experience extremely high levels of herbivory by specialist geometrids, whereas ant-plants are infrequently attacked by geometrids (Letourneau and Dyer 1998, Dyer and Letourneau 1999a). Second, the removal of symbionts resulted in the same magnitude of increase in amides both in the presence (field experiments) and absence (shadehouse experiments) of herbivores. In fact, plants that were experimentally protected from herbivores (shadehouse experiments) had the highest levels of amides (Table 7.1). This constitutes strong experimental evidence for the absence of induction in this system. Measuring amides in plants randomly assigned to herbivore exposure versus herbivore exclusion (i.e., our study) is a much stronger test of induction than measuring amides on plants before and after exposure to herbivores. Third, cutting the plants in various experiments did not cause higher levels of amides (although we acknowledge there are problems with simulating herbivory). Thus, allocation of resources

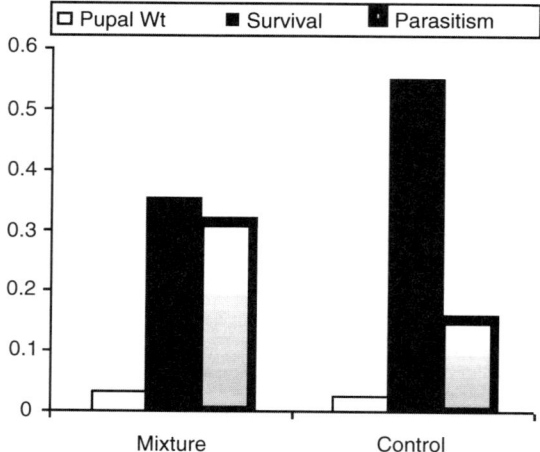

FIGURE 7.13. Pupal weights, survivorship, and percent parasitism for 250 caterpillars (*Eois* sp., Geometridae) that were field collected in early instars and reared on leaf diets spiked with a solution of *Piper cenocladum* amides ("mixture") or only solvent ("control") added to the leaf surface. GC analysis indicated that these caterpillars sequester the mixture of amides. Although there was no significant relationship between pupal weight and diet, survivorship was significantly lower ($\chi^2 = 17.1$, df = 1, $P < 0.0001$) and parasitism was significantly higher ($\chi^2 = 5.5$, df = 1, $P = 0.02$) for caterpillars on mixture versus control diets. Random assignment to treatments occurred after the caterpillars were collected, and they were not exposed to parasitoids during the rearing process, thus the differences observed in parasitism were due to changes after oviposition by parasitoids. It is likely that these patterns of parasitism were due to physiological differences in caterpillars or parasitoids that were caused by the differences in diets.

to amides in *P. cenocladum* is determined by symbionts and resource availability, regardless of the presence and absence of herbivores.

Our experiments demonstrated that control of herbivores by plant toxins is stronger against generalists than specialists. Although the top–down effects of mutualistic ants were much greater than effects of plant chemistry on specialist herbivores, the amides did have subtle effects on these animals. Specialist geometrids (*Eois* spp.) sequester all three amides found in *P. cenocladum*, possibly protecting them against predators. *Piper* amides negatively affected these caterpillars, by making them more susceptible to parasitism (Fig. 7.13), possibly via disruption of normal parasitoid encapsulation processes. This interesting result could enhance the existing efforts to uncover insecticidal properties of *Piper* secondary compounds, since it presents a new type of integrated pest management: utilizing natural products to enhance natural control by parasitoids.

Results from work with the *P. cenocladum* system could be unique to that "quirky" species and perhaps not relevant to general questions about the roles of secondary metabolites in terrestrial communities (Hunter 2001, Heil *et al.* 2002). Nevertheless, preliminary data suggest that similar effects of chemistry are found in the closely related species *P. imperiale*, which is not an ant-plant. Also, the results are consistent with chemical defense theory (e.g., Ehrlich and Raven 1964, Coley *et al.* 1985, Moen *et al.* 1993). It is likely that the secondary metabolites found in *Piper* species play an important role in tropical communities, but more ecological studies are obviously needed to determine the specifics and magnitude of these ecological roles.

7.5. EVOLUTION OF *Piper* CHEMISTRY

Not enough data are available to make significant conclusions about the evolution of secondary chemistry in *Piper*, since fewer than one-third of the species included in the most recent published phylogeny (Jaramillo and Manos 2001) have been investigated for phytochemicals (Fig. 7.14). In addition, most of the phytochemical studies that have been done within the genus have not been exhaustive investigations of all potential secondary metabolites. Thus, it is difficult to rule out the presence of specific compounds in any given species. However, if a few simple assumptions are made, some interesting patterns emerge. For example, amides are present in almost 75% (82 species) of the *Piper* species that have

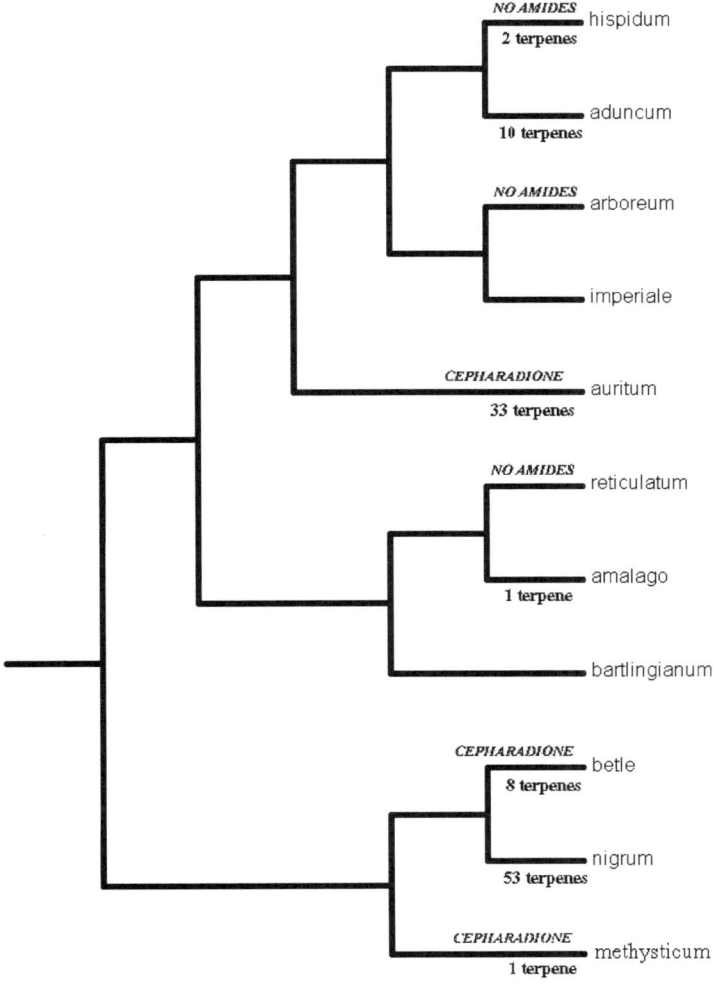

FIGURE 7.14. A phylogenetic hypothesis including all *Piper* species for which there are chemistry data (see References); the tree is based on one published by Jaramillo and Manos (2001). For clarity, the originally published tree was pruned to include only those species for which chemistry data were available.

been examined and may represent an ancestral character, with some instances of loss of amides (Fig. 7.14). There are some common amides that have evolved multiple times in both Old World and New World *Piper* species (e.g., cepharadione, Fig. 7.14), but most of the common amides that are present in multiple unrelated species have only evolved in Old World species (piperine, pipericide, and guineensine) or only in New World species (8,9-dihydropiplartine and piplartine), which are potentially monophyletic groups (Jaramillo and Manos 2001).

Some species have evolved very high diversities of amides, such as *P. amalago*, from which 31 different amides have been isolated. In contrast, *P. reticulatum* does not have any amides (Fig. 7.14). It is interesting to note that there are no apparent trade-offs between N- and C-based defenses. Species with high amide diversity or concentration are not necessarily devoid of terpenes (e.g., *P. betle* with five amides and eight terpenes), nor are there any striking patterns of evolution of shade-adapted N-based defensive systems versus gap-adapted terpene defensive systems. Despite our findings that the phenotypic plasticity of amide concentrations allows amides to respond to nutrient availability in accordance to predictions of nutrient availability hypotheses (Dyer *et al.* 2004), it certainly does not appear as if evolutionary predictions based on carbon–nutrient balance (Bryant *et al.* 1983) will be upheld once enough data are available for phylogenetically independent contrasts. Nor does it appear that herbivores that are susceptible to chemical defense (i.e., mostly generalists) will be influenced greatly by the phylogenetic history of the host plant (*sensu* Baldwin and Schultz 1988).

The majority of amides have evolved only once (e.g., cenocladamide in *P. cenocladum*; arboreumine in *P. arboreum*; aduncamide in *P. aduncum*), but in most instances these contain common moieties (e.g., isobutyl, pyrrolidine, dihydropyridone, piperidine) and only involve slight modifications of amides found in unrelated species. Although it is likely that herbivores and pathogens have provided strong selective pressures that are partly responsible for the diverse secondary chemistry in *Piper*, no formal macro- or microevolutionary hypotheses have been tested. One prediction of coevolutionary theory (Ehrlich and Raven 1964, Cornell and Hawkins 2003) is that generalists should be more susceptible than specialists to toxins such as amides. The one study that directly compares effects of amides on generalists versus that on specialists (Dyer *et al.* 2003) supports this prediction. The specialist herbivores in that study were two species in the genus *Eois*. Patterns of *Piper–Eois* interactions are ideal for studying coevolution, since most *Eois* species are probably specialized on one or two species of *Piper* (Dyer and Gentry 2002, unpubl. data) and amides are toxic to generalist herbivores. Experiments comparing performance of *Eois* on diets with amides from their specific *Piper* hosts versus amides from related *Piper* species could reveal whether or not predicted trade-offs occur in specialist herbivores' susceptibility to toxins. In addition, since several good *Piper* phylogenies exist (Jaramillo and Manos 2002; Chapters 9 and 10), an *Eois* phylogeny would allow for tests of parallel diversification of the plant and herbivore species (Farrell *et al.* 1992).

7.6. APPLIED *Piper* CHEMISTRY

Most *Piper* chemistry has been conducted to find potential pharmaceuticals or pesticides, and over 90% of the literature focuses on compounds that are cytotoxic, antifungal, antitumor, fragrant, or otherwise useful to humans. For example, safrol occurs in high concentrations in several species of *Piper*, particularly *P. hispidinervum*. This propenylphenol and its derivatives have been used successfully in powerful insecticides, as well as in fragrances, waxes, polishes, soaps, and detergents. Thus, *P. hispidinervum* has been cultivated for high levels of safrole and could contribute significantly to tropical economies and conservation efforts (Rocha and Ming 1999). A recent example that typifies the search for insecticidal properties of *Piper* amides is the work of Yang *et al.* (2002), in which the authors demonstrate the effectiveness of a piperidine amide (pipernonaline, extracted from *P. longum* infructescence) against *Aedes aegypti* mosquito larvae. This insecticidal activity has potential human importance because these mosquitoes are vectors for yellow fever. The most extensive review of *Piper* phytochemistry (Parmar *et al.* 1997) summarizes the bioactivity of *Piper* chemistry, and most examples are medicinal or pesticidal. The most common uses reported in the Parmar review and subsequent papers are anticarcinogens, insecticides, treatments for respiratory diseases, pain killers, mood enhancers, and treatments for gastrointestinal diseases. Current applied work particularly stresses the insecticidal effects.

The common economic uses of *Piper* spp. are as a spice (*P. nigrum* and *P. guineense*) and herbal medicine (*P. methysticum*, or "Kava"). These species have rich histories in economic botany and ethnobotany. For example, Kava has been used for centuries in the Pacific Islands to prepare intoxicating beverages and offerings to gods. Traditional preparations of the beverage involved spitting chewed Kava leaves into a communal bowl for eventual consumption; the salivary enzymes from the chewing putatively helped create a more potent beverage (Cotton 1996). Modern Kava beverages are prepared commercially, without expectoration, and many of the psychoactive constituents of Kava have long been known. Methysticin, one of the more potent physiologically active components of Kava (Parmar *et al.* 1997), was isolated in 1860 (see Chapter 8). What remains unknown are the specifics of the causes of inebriation or other physiological effects: What are the actual compounds involved, potential synergies, interactions with salivary enzymes, and other aspects of chemistry that are responsible for strong effects on animals?

Aside from the significant economic impact of *P. nigrum* chemistry throughout its history (e.g., Dove 1997), the actual contributions of *Piper* chemistry to any country's economy are relatively small. For example *P. methysticum* sales were only US$ 50 million worldwide in 1998 (Laird 1999). It is unlikely that any aspect of the bioactivity of *Piper* chemistry will significantly alter the economies of the tropical countries where *Piper* species are found, and there is always the potential for disaster when *Piper* species are exploited for economic purposes. One particular concern is that *Piper* species that are imported for cultivation for chemistry could easily become invasive, such as *P. auritum* (J. Denslow, pers. comm.) and *P. aduncum* (Novotny *et al.* 2003). Another potential problem is overharvesting of sensitive species (Balick *et al.* 1995), but this is unlikely and can be made even less likely with techniques such as synthesis, tissue culture production (see Chapter 8), and sustainable harvesting (Balick *et al.* 1996). Finally, there are some notable and important efforts to link economic benefits of sustainable use of *Piper* chemistry resources to conservation efforts (Laird 1999; other references).

7.7. FUTURE RESEARCH ON *Piper* CHEMISTRY

Future research on *Piper* chemistry should focus on the ecological effects of variation in secondary chemistry, chemical coevolution between *Piper* species and specialist herbivores, and synergistic interactions between different secondary metabolites. For all of these studies, enhanced isolation and synthesis techniques will be necessary, so these techniques should also be a focus of future research.

Probably the most interesting investigations will be the search for synergy. The relative scarcity of examples of synergistically acting secondary compounds is likely due to the lack of research on this topic or weak statistical methods (Nelson and Kursar 1999), and we have suggested these synergistic effects may be the rule rather than the exception (Dyer *et al.* 2003). Indeed, synergy may explain the apparent lack of defensive properties that have been indicated for a variety of plant secondary compounds, which have no known function (Harborne 1988, Ayres *et al.* 1997). This is particularly important for *Piper* chemistry, because many species have a plethora of defensive compounds that could potentially interact. The large number of research programs that test for pharmaceutical, pesticidal, or other activities of *Piper* plant secondary metabolites should at least be supplemented with appropriate tests (e.g., Jones 1998, Nelson and Kursar 1999) of pertinent mixtures and whole plant extracts.

7.8. ACKNOWLEDGMENTS

We thank G. Vega, H. Garcia, M. Tobler, R. Fincher, A. Smilanich, J. Searcy, J. Stireman, and A. Hsu for expert assistance in the field and laboratory. A. Smilanich, A. Palmer, and two anonymous reviewers provided helpful comments on the manuscript. D. Letourneau introduced us all to the genus *Piper*. This work was supported by NSF (DEB-0074806), Earthwatch Institute, Tulane University, and Mesa State College.

REFERENCES

Addae-Mensah, I., Torto, F. G., Dimonyeka, C. I., Baxter, I., and Sanders, J. K. M. (1977). Novel amide alkaloids from the roots of *Piper guineense*. *Phytochemistry* 16:757–759.

Adesina, S. K., Abebayo, A. S., Adesina, S. K. O., and Groning, R. (2002). GC/MS investigations of the minor constituents of *Piper guineense* stem. *Pharmazie* 57:622–627.

Alecio, A. C., Bolzani, V. D., Young, M. C. M., Kato, M. J., and Furlan, M. (1998). Antifungal amide from leaves of *Piper hispidum*. *Journal of Natural Products* 61:637–639.

Arai, Y., Masuda, T., Yoneda, S., Masaki, Y., and Shiro, M. (2000). Asymmetric synthesis of (+)-dihyrokawain-5-ol. *Journal of Organic Chemistry* 65:258–262.

Atal, C. K., Dhar, K. L., Gupta, O. P., and Gupta, S. C. (1977). Synergists for Pyrethrum: Synthetic analogues of some *Piper* compounds. *Indian Journal of Experimental Biology* 15(12):1230–1232.

Ayres, M. P., Clausen, T. P., MacLean, J., Redman, A. M., and Reichardt, P. B. (1997). Diversity of structure and antiherbivore activity in condensed tannins. *Ecology* 78:1696–1712.

Baldoqui, D. C., Kato, M. J., Cavalheiro, A. J., Bolzani, V. D., Young, M. C. M., and Furlan, M. (1999). A chromene and prenylated benzoic acid from *Piper aduncum*. *Phytochemistry* 51:899–902.

Baldwin, I. T., and Schultz, J. C. (1988). Phylogeny and the patterns of leaf phenolics in gap-adapted and forest-adapted *Piper* and *Miconia* understory shrubs. *Oecologia* 75:105–109.

Balick, M. J., Elisabetsky, E., and Laird, S. A. (1995). *Medicinal Resources of the Tropical Forest: Biodiversity and Its Importance to Human Health*. Columbia University Press, New York.

Banerji, A., and Pal, S. C. (1983). Total synthesis of sylvamide, a *Piper* amide. *Phytochemistry* 22(4):1028–1030.

Banerji, A., Bandyopadhyay, D., Sarkar, M., Siddhanta, A. K., Pal, S. C., Ghosh, S., Abraham, K., and Shoolery, J. N. (1985). Structural and synthetic studies on the refractamides—Amide constituents of *Piper retrofractum*. *Phytochemistry* 24(2):279–284.

Banerji, A., Ray, R., Bandyopadhyay, D., Basu, S., Maiti, S., Bose, A., and Majumder, P. L. (1993). Structure and synthesis of aurantiamide benzoate—A modified dipeptide. *Indian Journal of Chemistry Section B—Organic Chemistry Including Medicinal Chemistry* 32:776–778.

Banerji, A., Banerjee, T., Sengupta, R., Sengupta, P., Das, C., and Sahu, A. (2002a). Synthetic and spectroscopic studies of structural analogs of *Piper* amides—The 5-aryl-2E,4E-pentadienamides. *Journal of the Indian Chemical Society* 79(11):876–883.

Banerji, A., Sarkar, M., Datta, R., Sengupta, P., and Abraham, K. (2002b). Amides from *Piper brachystachyum* and *Piper retrofractum*. *Phytochemistry* 59:897–901.

Benevides, P. J. C., Sartorelli, P., and Kato, M. J. (1999). Phenylpropanoids and neolignans from *Piper regnellii*. *Phytochemistry* 52:339–343.

Bernard, C. B., Krishnamurty, H. G., Chauret, D., Durst, T., Philogene, B. J. R., Sanchezvindas, P., Hasbun, C., Poveda, L., Sanroman, L., and Arnason, J. T. (1995). Insecticidal defenses of piperaceae from the Neotropics. *Journal of Chemical Ecology* 21:801–814.

Boll, P. M., Parmar, V. S., Tyagi, O. D., Prasad, A., Wengel, J., and Olsen, C. E. (1994). Some recent isolation studies from potential insecticidal *Piper* species. *Pure and Applied Chemistry* 66:2339–2342.

Bryant, J. P., Chapin, F. S., III, and Klein, D. R. (1983). Carbon/nutrient balance of boreal plants in relation to vertebrate herbivory. *Oikos* 40:357–368.

Capron, M. A., and Wiemer, D. F. (1996). Piplaroxide an ant-repellent piperidine epoxide from *Piper tuburculatum*. *Journal of Natural Products* 59:794–795.

Chandhoke, N., Gupta, S., and Dhar, S. (1978). Interceptive activity of various species of *Piper*, their natural amides and semi-synthetic analogs. *Indian Journal of Pharmaceutical Sciences* 40(4):113–116.

Chatterjee, A., and Dutta, C. P. (1967). Alkaloids of *Piper longum* Linn-I structure and synthesis of piperlongumine and piperlonguminine. *Tetrahedron* 23:1769–1781.

Chen, J. J., Huang, Y. C., Chen, Y. C., Huang, Y. T., Want, S. W., Peng, C. Y., Teng, C. M., and Chen, I. S. (2002). Cytotoxic amides from *Piper sintenense*. *Planta Medica* 68:980–985.

Ciccio, J. F. (1997). Essential oil components in leaves and stems of *Piper bisasperatum* (Piperaceae). *Revista de Biologia Tropical* 45:35–38.

Coley, P. D., Bryant, J. P., and Chapin, F.S. (1985). Resource availability and plant antiherbivore defense. *Science* 230:895–899.

Cornell, H. V., and Hawkins, B. A. (2003). Herbivore responses to plant secondary compounds: A test of phytochemical coevolution theory. *American Naturalist* 161:507–522.

Cotton, C. M. (1996). *Ethnobotany: Principles and Applications*. John Wiley & Sons Ltd., West Sussex, England.

Crombie, L., Pattenden, G., and Stemp, G. (1977). Synthesis of wisanine, a new piperine amide from *Piper guineense*. *Phytochemistry* 16:1437–1438.

Da Cunha, E. V. L., and Chaves, M. C. D. (2001). Two amides from *Piper tuberculatum* fruits. *Fitoterapia* 72:197–199.

da Silva, R. V., Navickiene, H. M. D., Kato, M. J., Bolzania, V. D. S., Meda, C. I., Young, M. C. M., and Furlan, M. (2002). Antifungal amides from *Piper arboreum* and *Piper tuberculatum*. *Phytochemistry* 59:521–527.

Das, B., and Madhusudhan, P. (1998). Transformation of the conjugated dienamide system of some natural alkamides to the fi,fl-unsaturated amide function using Zn/HOAc. *Tetrahedron Letters* 39(49):9099–9100.

Das, B., and Kashinatham, A. (1998). Futoamide from *Piper longum*. *Fitoterapia* 69:548.

de Abreu, A. M., Sevegnani, L., Machicado, A. R., Zimermann, D., and Rebelo, R. A. (2002). *Piper mikanianum* (Kunth) Steudel from Santa Catarina, Brazil—A new source of safrole. *Journal of Essential Oil Research* 14:361–363.

de Araujo-Júnior, J. X., DaCunha, E. V. L., Chaves, M. C. D., and Gray, A. I. (1997). Piperdardine, a piperidine alkaloid from *Piper tuberculatum*. *Phytochemistry* 44:559–561.

de Araujo-Júnior, J. X., Barreiro, E. J., Parente, J. P., and Fraga, C. A. M. (1999). Synthesis of piperamides and new analogues from natural safrole. *Synthetic Communications* 29:263–273.

de Araújo-Júnior, J. X., de Mattos Duarte, C., de O. Chaves, M. C., Parente, J. P., Fraga, C. A. M., and Barreiro, E. J. (2001). Synthesis of natural amide alkaloid piperdardine and a new bioactive analog. *Synthetic Communications* 31(1):117–123.

de Mattos Duarte, C., de Araújo Júnior, J. X., Parente, J. P., and Barreiro, E. J. (1999). Synthesis of new hypotensive piperamides analogues. *Revista Brasileira de Farmácia* 80(1/2):35–37.

de Paula, V. F., Barbosa, L. C. D., Demuner, A. J., Pilo-Veloso, D., and Picanco, M. C. (2000). Synthesis and insecticidal activity of new amide derivatives of piperine. *Pest Management Science* 56:168–174.

Dodson, C. D., Dyer, L. A., Searcy, J., Wright, Z., and Letourneau, D. K. (2000). Cenocladamide, a dihydropyridone alkaloid from *Piper cenocladum*. *Phytochemistry* 53:51–54.

dos Santos, P. R. D., Moreira, D. D., Guimaraes, E. F., and Kaplan, M. A. C. (2001). Essential oil analysis of 10 Piperaceae species from the Brazilian Atlantic forest. *Phytochemistry* 58:547–551.

Dove, M. R. (1997). The banana tree at the gates: Perceptions of production of *Piper nigrum* (Piperaceae) in a seventeenth century Malay state. *Economic Botany* 51:347–361.

Dyer, L. A., and Gentry, G. L. (2002). Caterpillars and parasitoids of a tropical lowland wet forest. Available from: http://www.caterpillars.org.

Dyer, L. A., and Letourneau, D. K. (1999a). Trophic cascades in a complex, terrestrial community. *Proceedings of the National Academy of Sciences U.S.A.* 96:5072–5076.

Dyer, L. A., and Letourneau, D. K. (1999b). Relative strengths of top–down and bottom–up forces in a tropical forest community. *Oecologia* 119:265–274.

Dyer, L. A., Williams, W., Dodson, C., and Letourneau, D. K. (2000). A commensalism between *Piper marginatum* Jacq. (Piperaceae) and a coccinellid beetle. *Journal of Tropical Ecology* 15:841–846.

Dyer, L. A., Dodson, C. D., Beihoffer, J., and Letourneau, D. K. (2001). Trade offs in anti-herbivore defenses in *Piper cenocladum*: Ant mutualists versus plant secondary metabolites. *Journal of Chemical Ecology* 27:581–592.

Dyer, L. A., Dodson, C. D., Stireman, J. O., Tobler, M. A., Smilanich, A. M., Fincher, R. M., and Letourneau, D. K. (2003). Synergistic effects of three *Piper* amides on generalist and specialist herbivores. *Journal of Chemical Ecology* 29:2499–2514.

Dyer, L. A., Dodson, C. D., Letourneau, D. K., Tobler, M. A., Hsu, A., and Stireman, J. O., III. (2004). Ecological causes and consequences of variation in defensive chemistry of a Neotropical shrub. *Ecology* (in press).

Ehrlich, P. R., and Raven, P. H. (1964). Butterflies and plants: A study in coevolution. *Evolution* 18:568–608.

Facundo, V. A., and Morais, S. M. (2003). Constituents of *Piper aleyreanum* (Piperaceae). *Biochemical Systematics and Ecology* 31:111–113.

Farrell, B. D., Mitter, C., and Futuyma, D. J. (1992). Diversification at the insect–plant interface. Insights from phylogenetics. *Bioscience* 42:34–42.

Gbewonyo, W. S. K., and Candy, D. J. (1992). Chromatographic isolation of insecticidal amides from *Piper guineense* root. *Journal of Chromatography* 607:105–111.

Gokhale, V. G., Phalnikar, N. L., and Bhidde, B. V. (1945). Synthetic mosquito larvacides: Part I. *Journal of the University of Bombay, Science: Physical Sciences, Mathematics, Biological Sciences and Medicine* 16(5):32–36.

Green, T. P., Treadwell, E. M., and Wiemer, D. F. (1999). Arieianal, a prenylated benzoic acid from *Piper arieianum*. *Journal of Natural Products* 62:367–368.

Gupta, O. P., Nath, A., Gupta, S. C., and Srivastava, T. N. (1980). Preparation of semi-synthetic analogues of *Piper* amides and their antitubercular activity. *Bulletin of Medico-Ethno-Botanical Research* 1(1):99–106.

Gupta, S., Jha, A., Prasad, A. K., Rajwanshi, V. K., Jain, S. C., Olsen, C. E., Wengel, J., and Parmar, V. S. (1999). A new amide, N-cinnamoylpyrrole and other constituents from *Piper argyrophyllum*. *Indian Journal of Chemistry Section B—Organic Chemistry Including Medicinal Chemistry* 38:823–827.

Haller, H. L., McGovran, E. R., Goodhue, L. D., Sullivan, W. N. (1942). The synergistic action of sesamin with pyrethrum insecticides. *Journal of Organic Chemistry* 7:183–184.

Harborne, J. B. (1988). *Introduction to Ecological Biochemistry*. Academic Press, San Diego, California.

Heil, M., Delsinne, T., Hilpert, A., Schurkens, S., Andary, C., Linsenmair, K. E., Sousa, M. S., and McKey, D. (2002). Reduced chemical defence in ant-plants? A critical re-evaluation of a widely accepted hypothesis. *Oikos* 99:457–468.

Howard, J. J., Cazin, J., and Wiemer, D. F. (1988). Toxicity of terpenoid deterrents to the leafcutting ant *Atta cephalotes* and its mutualistic fungus. *Journal of Chemical Ecology* 14:59–69.

Hunter, M. D. (2001). Multiple approaches to estimating the relative importance of top–down and bottom–up forces on insect populations: Experiments, life tables, and time-series analysis. *Basic and Applied Ecology* 2:295–309.

Inatani, R., Nakatani, N., and Fuwa, H. (1981). Structure and synthesis of new phenolic amides from *Piper nigrum* L. *Agricultural and Biological Chemistry* 45(3):667–673.

Jacobs, H., Seeram, N. P., Nair, M. G., Reynolds, W. F., and McLean, S. (1999). Amides of *Piper amalago* var. *nigrinodum*. *Journal of the Indian Chemical Society* 76:713–717.

Jaramillo, M. A., and Manos, P. S. (2001). Phylogeny and patterns of floral diversity in the genus *Piper* (Piperaceae). *American Journal of Botany* 88:706–716.

Jenett-Siems, K., Mockenhaupt, F. P., Bienzle, U., Gupta, M. P., Eich, E. (1999). In vitro antiplasmodial activity of Central American medicinal plants. *Tropical Medicine & International Health* 4: 611–615.

Jones, D. G. (ed.). (1998). *Piperonyl Butoxide: The Insect Synergist*. Academic Press, London.

Joshi, B. S., Kamat, V. N., and Saksena, A. K. (1968). On the synthesis of piplartine and a synthesis of dihydropiplartine. *Tetrahedron Letters* 18(20):2395–2400.

Joshi, A. S., Li, X. C., Nimrod, A. C., ElSohly, H. N., Walker, L. A., and Clark, A. M. (2001). Dihydrochalcones from *Piper longicaudatum*. *Planta Medica* 67:186–188.

Kaga, H., Ahmed, Z., Gotoh, K., Orito, K. (1994). New access to conjugated dien- and eneamides. Synthesis of dehydropipernonaline, pipernonaline and related biologically active amides. *Synlett* 607–608.

Kang, I.-J., Wang, H.-M., Su, C.-H., and Chen, L.-L. (2001). Synthesis of dienamide natural products using a hypervalent iodine(III) reagent. *The Chinese Pharmaceutical Journal* 53:199–205.

Kiuchi, F., Nakamura, N., Saitoh, M., Komagome, K., Hiramatsu, H., Takimoto, N., Akao, N., Kondo, K., and Tsuda, Y. (1997). Synthesis and nematocidal activity of aralkyl- and aralkenylamides related to *Piper* amide in second-stage larvae of *Toxocara canis*. *Chemical and Pharmaceutical Bulletin* 45(4):685–696.

Ladenburg, A. and Scholtz, M. (1894). Synthese der Piperinsäure und des Piperins. *Berichte der Deutschen chemischen Gesellschaft*, 27: 2958.

Laird, S. A. (1999). The botanical medicine industry. In: Kate, K. and Laird, S. A. (eds.), *The Commercial Use of Biodiversity: Access to Genetic Resources and Benefit Sharing*. Earthscan, London.

Letourneau, D. K., and Dyer, L. A. (1998). Experimental test in lowland tropical forest shows top–down effects through four trophic levels. *Ecology* 79:1678–1687.

Likhitwitayawuid, K., Ruangrungsi, N., Lange, G. L., Decicco, C. P. (1987). Studies on Thai medicinal-plants. 5. Structural elucidation and synthesis of new components isolated from *Piper sarmentosum* (Piperaceae). *Tetrahedron* 43: 3689–3694.

Ma, D., and Lu, X. (1990). A convenient stereoselective synthesis of conjugated dienoic esters and amides. *Tetrahedron* 46(9):3189–3198.

Martins, A. P., Salgueiro, L., Vila, R., Tomi, F., Canigueral, S., Casanova, J., Da Cunha, A. P., and Adzet, T. (1998). Essential oils from four *Piper* species. *Phytochemistry* 49:2019–2023.

Masuoka, C., Ono, M., Ito, Y., Okawa, M., and Nohara, T. (2002). New megastigmane glycoside and aromadendrane derivative from the aerial part of *Piper elongatum*. *Chemical & Pharmaceutical Bulletin* 50:1413–1415.

McFerren, M. A., and Rodriguez, E. (1998). Piscicidal properties of piperovatine from *Piper piscatorum* (Piperaceae). *Journal of Ethnopharmacology* 60:183–187.

Menon, A. N., Padmakumari, K. P., and Jayalekshmy, A. J. (2002). Essential oil composition of four major cultivars of black pepper (*Piper nigrum* L.). *Journal of Essential Oil Research* 14:84–86.

Menon, A. N., Padmakumari, K. P., Jayalekshmy, A., Gopalakrishnan, M., and Narayanan, C. S. (2000). Essential oil composition of four popular Indian cultivars of black pepper (*Piper nigrum* L.). *Journal of Essential Oil Research* 12:431–434.

Miyakado, M., and Yoshioka, H. (1979). The Piperaceae amides. II: Synthesis of pipericide, a new insecticidal amide from *Piper nigrum* L. *Agricultural and Biological Chemistry* 43(11):2413–2415.

Miyakado, M., Nakayama, I., and Ohno, N. (1989). Insecticidal unsaturated isobutylamides from natural products to agrochemical leads. *ACS Symposium Series* 387:173–187.

Moen, J. H., Oksanen, L., Ericson, L., and Ekerholm, P. (1993). Grazing by food-limited microtine rodents on a productive experimental plant community: Does the "green desert" exist? *Oikos* 68:401–413.

Moreira, D. D., Guimaraes, E. F., and Kaplan, M. A. C. (1998). Non-polar constituents from leaves of *Piper lhotzkyanum*. *Phytochemistry* 49:1339–1342.

Moreira, D. D., Guimaraes, E. F., and Kaplan, M. A. C. (2000). A *C*-glucosylflavone from leaves of *Piper lhotzkyanum*. *Phytochemistry* 55:783–786.

Mundina, M., Vila, R., Tomi, F., Gupta, M. P., Adzet, T., Casanova, J., and Canigueral, S. (1998). Leaf essential oils of three Panamanian *Piper* species. *Phytochemistry* 47:1277–1282.

Nakatani, N., Inatani, R., and Fuwa, H. (1980). Structures and syntheses of two phenolic amides from *Piper nigrum* L. *Agricultural and Biological Chemistry* 44(12):2831–2836.

Navickiene, H. M. D., Alecio, A. C., Kato, M. J., Bolzani, V. D., Young, M. C. M., Cavalheiro, A. J., and Furlan, M. (2000). Antifungal amides from *Piper hispidum* and *Piper tuberculatum*. *Phytochemistry* 55:621–626.

Navickiene, H. M. D., Bolzani, V. D., Kato, M. J., Pereira, A. M. S., Bertoni, B. W., Franca, S. C., and Furlan, M. (2003). Quantitative determination of anti-fungal and insecticide amides in adult plants, plantlets and callus from *Piper tuberculatum* by reverse-phase high-performance liquid chromatography. *Phytochemical Analysis* 14:281–284.

Nelson, A. C., and Kursar, T. A. (1999). Interactions among plant defense compounds: A method for analysis. *Chemoecology* 9:81–92.

Novotny, V., Miller, S. E., Cizek, L., Leps, J., Janda, M., Basset, Y., Weiblen, G. D., and Darrow, K. (2003). Colonising aliens: Caterpillars (Lepidoptera) feeding on *Piper aduncum* and *P. umbellatum* in rainforests of Papua New Guinea. *Environmental Entomology* (in press).

Pande, A., Shukla, Y. N., Srivastava, R., and Verma, M. (1997). 3-Methyl-5-decanoylpyridine and amides from *Piper retrofractum*. *Indian Journal of Chemistry Section B—Organic Chemistry Including Medicinal Chemistry* 36:377–379.

Parmar, V. S., Jain, S. C., Bisht, K. S., Jain, R., Taneja, P., Jha, A., Tyagi, O. D., Prasad, A. K., Wengel, J., Olsen, C. E., and Boll, P. M. (1997). Phytochemistry of the genus *Piper*. *Phytochemistry* 46:597–673.

Parmar, V. S., Jain, S. C., Gupta, S., Talwar, S., Rajwanshi, V. K., Kumar, R., Azim, A., Malhotra, S., Kumar, N., Jain, R., Sharma, N. K., Tyagi, O. D., Lawrie, S. J., Errington, W., Howarth, O. W., Olsen, C. E., Singh, S. K., and Wengel, J. (1998). Polyphenols and alkaloids from *Piper* species. *Phytochemistry* 49:1069–1078.

Pring, B. G. (1982). Isolation and identification of amides from *Piper callosum*—Synthesis of pipercallosine and pipercallosidine. *Journal of the Chemical Society, Perkin Transactions 1* 1493–1498.

Richards, J. L., Myhre, S. M., and Jay, J. I. (2001). Total synthesis of piplartine, 13-desmethylpiplartine, and cenocladamide: Three compounds isolated from *Piper cenocladum*. *221st National Meeting of the American Chemical Society*. [Abstract]

Richards, J. L., Jay, J. I., and Pidcock, W. C., Agustsdottir, S. R. (2002). Improved synthesis of piplartine, 4'-desmethylpiplartine, and cenocladamide: Three compounds isolated from *Piper cenocladum*. *57th Northwest Regional Meeting of the American Chemical Society*. [Abstract]

Rocha, S. F. R., and Ming, L. C. (1999). *Piper hispidinervum*: A sustainable source of safrole. In: Janick, J. (ed.), *Perspectives on New Crops and New Uses*. ASHS Press, Alexandria, Virginia, pp. 479–481.

Rotherham, L. W., and Semple, J. E. (1998). A practical and efficient synthetic route to dihydropipercide and pipercide. *Journal of Organic Chemistry* 63:6667–6672.

Rügheimer, L. (1882). Künstliches Piperin. *Berichte der Deutschen chemischen Gesellschaft.* 15: 1390–1391.

Santos, B. V. D., and Chaves, M. C. D. (1999a). (E,E)-N-Isobutyl-2,4-octadienamide from *Piper marginatum*. *Biochemical Systematics and Ecology* 27:113–114.

Santos, B. V. D., and Chaves, M. C. D. (1999b). 2,4,5-Trimethoxypropiophenone from *Piper marginatum*. *Biochemical Systematics and Ecology* 27:539–541.

Santos, B. V. D., Da Cunha, E. V. L., Chaves, M. C. D., and Gray, A. I. (1998). Phenylalkanoids from *Piper marginatum*. *Phytochemistry* 49:1381–1384.

Schwarz, I., and Braun, M. (1999). Synthesis of naturally occurring dienamides by palladium-catalyzed carbonyl alkenylation. *Journal für Praktische Chemie* 341(1):72–74.

Scott, I. M., Puniani, E., Durst, T., Phelps, D., Merali, S., Assabgui, R. A., Sánchez-Vindas, P., Poveda, L., Philogène, B. J. R., and Arnason, J. T. (2002). Insecticidal activity of *Piper tuberculatum* Jacq. extracts: Synergistic interaction of piperamides. *Agricultural and Forest Entomology* 4:137–144.

Seeram, N. P., Lewis, P. A., Jacobs, H., McLean, S., Reynolds, W. F., Tay, L. L., and Yu, M. (1996). 3,4-Epoxy-8,9-dihydropiplartine. A new imide from *Piper verrucosum*. *Journal of Natural Products* 59:436–437.

Sengupta, S., Ray, A. B. (1987). The chemistry of Piper species: a review. *Fitoterapia* 58: 147–166.

Shoji, N., Umeyama, A., Saito, N. Takemoto, T., Kajiwara, A., and Ohizumi, Y. (1986). Dehydropipernonaline, an amide possessing coronary vasodilating activity, isolated from *Piper longum* L. *Journal of Pharmaceutical Sciences* 75:1188–1189.

Siddiqui, B. S., Gulzar, T., and Begum, S. (2002). Amides from the seeds of *Piper nigrum* Linn. and their insecticidal activity. *Heterocycles* 57:1653–1658.

Spino, C., Mayes, N., and Desfossés, H. (1996). Enantioselective synthesis of (+)- and (−)-dihydrokawain. *Tetrahedron Letters* 37(36):6503–6506.

Spring, F. S., and Stark, J. J. (1950). Piperettine from *Piper nigrum*: Its isolation, identification, and synthesis. *Journal of the Chemical Society* 1177–1180.

Srivastava, S., Gupta, M. M., Tripathi, A. K., and Kumar, S. (2000a). 1,3-Benzodioxole-5-(2,4,8-triene-methyl nonaoate) and 1,3-benzodioxole-5(2,4,8-triene-isobutyl nonaoate) from *Piper mullesua*. *Indian Journal of Chemistry Section B—Organic Chemistry Including Medicinal Chemistry* 39:946–949.

Srivastava, S., Verma, R. K., Gupta, M. M., and Kumar, S. (2000b). Chemical constituents of *Piper mullesua*. *Journal of the Indian Chemical Society* 77:305–306.

Stohr, J. R., Xiao, P. G., and Bauer, R. (1999). Isobutylamides and a new methylbutylamide from *Piper sarmentosum*. *Planta Medica* 65:175–177.

Stohr, J. R., Xiao, P. G., and Bauer, R. (2001). Constituents of Chinese *Piper* species and their inhibitory activity on prostaglandin and leukotriene biosynthesis in vitro. *Journal of Ethnopharmacology* 75:133–139.

Strunz, G. M., and Finlay, H. J. (1994). Concise, efficient new synthesis of pipercide, an insecticidal unsaturated amide from *Piper nigrum*, and related compounds. *Tetrahedron* 50(38):11113–11122.

Strunz, G. M., and Finlay, H. J. (1996). Expedient synthesis of unsaturated amide alkaloids from *Piper* spp.: Exploring the scope of recent methodology. *Canadian Journal of Chemistry* 74:419–432.

Synerholm, M. E., Hartzell, A., and Arthur, J. M. (1945). Derivatives of piperic acid and their toxicities toward houseflies. *Contributions from Boyce Thompson Institute* 13:433–442.

Terreaux, C., Gupta, M. P., and Hostettmann, K. (1998). Antifungal benzoic acid derivatives from *Piper dilatatum*. *Phytochemistry* 49:461–464.

Torquilho, H. S., Pinto, A. C., Godoy, R. L. D., and Guimaraes, E. F. (2000). Essential oil of *Piper cernum* Vell. var. *cernum* Yuncker from Rio de Janeiro, Brazil. *Journal of Essential Oil Research* 12:443–444.

van Genderen, M. H. P., Leclercq, P. A., Delgado, H. S., Kanjilal, P. B., and Singh, R. S. (1999). Compositional analysis of the leaf oils of *Piper callosum* Ruiz & Pav. from Peru and Michelia montana Blume from India. *Spectroscopy—An International Journal* 14:51–59.

Velpandian, T., Jasuja, R., Bhardwaj, R. K., Jaiswal, J., and Gupta, S. K. (2001). Piperine in food: Interference in the pharmacokinetics of phenytoin. *European Journal of Drug Metabolism and Pharmacokinetics* 26:241–247.

Vila, R., Milo, B., Tomi, F., Casanova, J., Ferro, E. A., and Canigueral, S. (2001). Chemical composition of the essential oil from the leaves of *Piper fulvescens*, a plant traditionally used in Paraguay. *Journal of Ethnopharmacology* 76:105–107.

Vila, R., Mundina, M., Tomi, F., Ciccio, J. F., Gupta, M. P., Iglesias, J., Casanova, J., and Canigueral, S. (2003). Constituents of the essential oils from *Piper friedrichsthalii* C.DC. and *P. pseudolindenii* C.DC. from Central America. *Flavour and Fragrance Journal* 18:198–201.

Viswanathan, N., Venkatachalam, B., Joshi, B. S., and von Philipsborn, W. (1975). Piperaceae alkaloids: Part III. Synthesis of N-isobutyl-11-(3,4-methylenedioxyphenyl)-undeca-2,4,6-*trans,trans,trans*,-trienoic amide and N-isobutyl-11-(3,4-methylenedioxyphenyl)-undeca-2,8,10-*trans,trans,trans*,-trienoic amide (Piperstachine). *Helvetica Chimica Acta* 58(7):2026–2035.

Wu, D., Nair, M. G., and DeWitt, D. L. (2002a). Novel compounds from *Piper methysticum* Forst (Kava Kava) roots and their effect on cyclooxygenase enzyme. *Journal of Agricultural and Food Chemistry* 50:701–705.

Wu, D., Yu, L., Nair, M. G., DeWitt, D. L., and Ramsewak, R. S. (2002b). Cyclooxygenase enzyme inhibitory compounds with antioxidant activities from *Piper methysticum* (Kava Kava) roots. *Phytomedicine* 9:41–47.

Wu, Q. L., Wang, S. P., Tu, G. Z., Feng, Y. X., and Yang, J. S. (1997). Alkaloids from *Piper puberullum*. *Phytochemistry* 44:727–730.

Yang, Y. C., Lee, S. G., Lee, H. K., Kim, M. K., Lee, S. H., and Lee, H. S. (2002). A piperidine amide extracted from *Piper longum* L. fruit shows activity against *Aedes aegypti* mosquito larvae. *Journal of Agricultural and Food Chemistry* 50:3765–3767.

Zeng, H. W., Jiang, Y. Y., Cai, D. G., Bian, J., Long, K., and Chen, Z. L. (1997). Piperbetol, methylpiperbetol, piperol A and piperol B: A new series of highly specific PAF receptor antagonists from *Piper betle*. *Planta Medica* 63:296–298.

8
Kava (*Piper methysticum*): Growth in Tissue Culture and *In Vitro* Production of Kavapyrones

Donald P. Briskin, Hideka Kobayashi, Mary Ann Lila, and Margaret Gawienowski
Department of Natural Resources and Environmental Sciences, University of Illinois at Urbana/Champaign

8.1. INTRODUCTION

As a phytomedicine, Kava (or "Kava Kava") is receiving considerable worldwide interest in its use as a treatment for anxiety, tension, agitation, and/or insomnia. This phytomedicine is produced from an extract of the root and rhizome of *Piper methysticum*. Although the term *Kava* actually refers to a Polynesian term for the intoxicating beverage produced from this tropical shrub, this word is now typically used in Western cultures as the common name for the plant as well.

Extensive clinical studies conducted primarily in Europe have shown an efficacy of the Kava phytomedicine that is comparable to benzodiazepines (e.g., valium) in the treatment of anxiety symptoms (Schulz *et al.* 1998, Volz and Kieser 1997, Lehmann *et al.* 1996). However, unlike benzodiazepines or other synthetic tranquilizers, no evidence has been found for the development of either a physical or psychological dependency on Kava arising from its use (Schulz *et al.* 1998 and references therein). Moreover, the use of Kava does not impair mental alertness and capability, and the frequency of problematic side effects also appears to be very low (<1.5%) (Schulz *et al.* 1998, Lehmann *et al.* 1996). In the United States, Kava is sold as an herbal "dietary supplement" in accordance with the 1994 Dietary Supplement Health and Education Act (DSHEA) and its use has increased dramatically over recent years (Brevoort 1998). Indeed, in 1998, Kava sales within the United States increased 470% over the previous year and all indications are that this medicinal plant may well be on its way to equal or even surpass St. John's wort in terms of popularity and use (Brevoort 1998).

8.2. ORIGINS OF KAVA USE AND DISCOVERY BY WESTERN CULTURES

The use of Kava as an intoxicating beverage originated in human cultures living in the Oceania island communities encompassed by Polynesia, Melanesia, and Micronesia (Singh 1992, Singh and Blumenthal 1997). The traditional method for producing the intoxicating beverage in native cultures typically involves harvesting of the rhizome and roots of *Piper methysticum*, chewing or grinding this plant material to produce a pulp, and the addition of water and coconut milk to generate the beverage (Singh and Blumenthal 1997, Schulz *et al*. 1998). When consumed, this Kava beverage produces a calming, relaxing effect without altering consciousness, and its use is associated with many ceremonies and social customs in the cultures of Oceania (Singh and Blumenthal 1997 and references therein). Europeans were first introduced to the use of Kava in the late 1700s during exploration of the South Pacific by Captain James Cook. A subsequent detailed botanical characterization of the Kava plant was conducted by Johann Georg Forster in 1777 (Singh 1992). It was Forster who noted the similarity of this plant to other members of the pepper family (Piperaceae) and named it *"Piper methysticum."* From a Latin transcription of the Greek term *methustikos* ("intoxicating drink"); this literally means "intoxicating pepper" (Singh and Blumenthal 1997).

8.3. DESCRIPTION OF KAVA (*Piper methysticum*) AND ITS GROWTH FOR USE IN KAVA PRODUCTION

The Kava plant is a highly branched, robust perennial shrub with large, heart-shaped leaves. It grows throughout the South Pacific Islands to as far east as Hawaii (Singh 1992, Singh and Blumenthal 1997). In addition to the South Pacific Islands, some small-scale production of Kava is found in regions of Australia (Singh 1992). Under high sunlight and in the warm, humid conditions of these regions, this plant can grow up to 6 m in height. However, when grown under commercial cultivation the plant is generally harvested after 3 to 5 years' growth when about 2 to 2.5 m tall (Singh and Blumenthal 1997). The fleshy rhizome and roots are harvested and dried for production of the phytomedicinal extract.

Kava plants are sterile and are propagated from stem cuttings (Lebot *et al*. 1991). When used for propagation, Kava stems are cut at the internodes to produce a series of small stem segments, each with a node. The segments are planted in the soil and new Kava plants are generated from the meristematic tissues present at the nodes. Work conducted by Lebot and Lévesque (1989) has shown that Kava (*Piper methysticum*) represents a sterile relative of *Piper wichmannii*, a plant geographically limited to the southwestern region of the South Pacific (i.e., New Guinea, The Solomon Islands, and Northern Vanuatu). With the growth and use of Kava occurring in South Pacific human cultures since prehistoric times, it is believed that this domesticated sterile plant became distributed across the South Pacific Islands with the geographical movements of these people (Lebot and Cabalion 1988 and references therein). A geographical survey and genetic analysis of Kava and its wild progenitor (*Piper wichmannii*) has suggested that Kava was originally domesticated by vegetative propagation of the wild progenitor and that it subsequently became sterile perhaps because of mutation. Although an array of Kava varieties with differences in appearence and patterns of kavapyrone production are recognized by growers (Lebot and Levesque 1996), analysis by Lebot *et al.* (1991, 1999) has shown that the genetic base of this plant is fairly narrow. These authors suggest that Kava varieties arose through human selection and preservation of somatic mutations from only a very small number of original clones.

8.4. ACTIVE PHYTOCHEMICALS PRESENT IN KAVA EXTRACTS

From intensive chemical and pharmacological studies conducted on Kava over the past 130 years, several key active constituents have been identified and characterized (Singh and Blumenthal 1997). Pharmacological activity of Kava appears to be primarily associated with a family of strylpyrones called "kavapyrones," which are present in active extracts of the rhizome and root. These compounds have been shown to have a variety of effects on the central nervous system, including local anesthesia, anticonvulsant activity, skeletal muscle relaxation, analgesia, and hypnosedation (Davies *et al.* 1992 and references therein). With regard to fundamental mechanisms of action, evidence has been presented for effects on several neurotransmitter systems including the glutamatergic (Schmitz *et al.* 1995, Gleitz *et al.* 1996), GABAergic (Jussofie *et al.* 1994), dopaminergic (Schelosky *et al.* 1995), and serotinergic (Walden *et al.* 1997) systems. Moreover, there is evidence that kavapyrones may act collectively or synergistically in inducing such effects upon the nervous system (Capasso and Calignano 1988, Davies *et al* 1992). Such evidence has involved the observation that purified, individual kavapyrones may show only small effects that are enhanced upon production of a mixture of kavapyrones. Indeed, this requirement for a mixture of different kavapyrones to yield potent psychotropic effects was observed early on in research on the phytochemical actions of this plant. Borche and Blount (1933) observed that individual kavapyrones did not yield the same potent effect as the crude kava root extract. Using mixtures of kavain, dihydrokavain, methysticin, dihydromethysticin, yangonin, and desmethoxyyangonin, Klohs *et al.* (1959) found evidence for synergism in the effects of these kavapyrones on blocking convulsion caused by strychnine.

Although 18 kavapyrones have been characterized at present, the 6 kavapyrones shown in Fig. 8.1 represent the major active consituents of Kava (He *et al.* 1997, Shao *et al.* 1998). The remaining 12 minor kavapyrones appear to be derivatives of either kawain, yangonin, or dihydromethysticin (Bruneton 1995, He *et al.* 1997). The 6 major kavapyrones can be conveniently separated and quantitated using high-performance liquid chromatographic methods (Gracza and Ruff 1980, Smith *et al.* 1984, Shao *et al.* 1998).

Although details regarding kavapyrone biosynthesis are still lacking, evidence from other systems such as *Equisetum arvense* gametophytes suggest that styrylpyrones may arise from a triketide produced by successive condensation of 2 malonyl-CoA molecules with a phenylpropanoid CoA-ester (Schröder 1997 and references therein). This is similar to reactions catalyzed by chalcone synthase except that two, rather than three, successive condensations involving malonyl CoA are involved. Recent studies have shown that kavapyrone levels in Kava roots are influenced by enviromental factors. In cultivated Kava plants, kavapyrone levels appear to increase with irrigation and mineral nutrient supplementation,

GROWTH IN TISSUE CULTURE AND PRODUCTION OF KAVAPYRONES

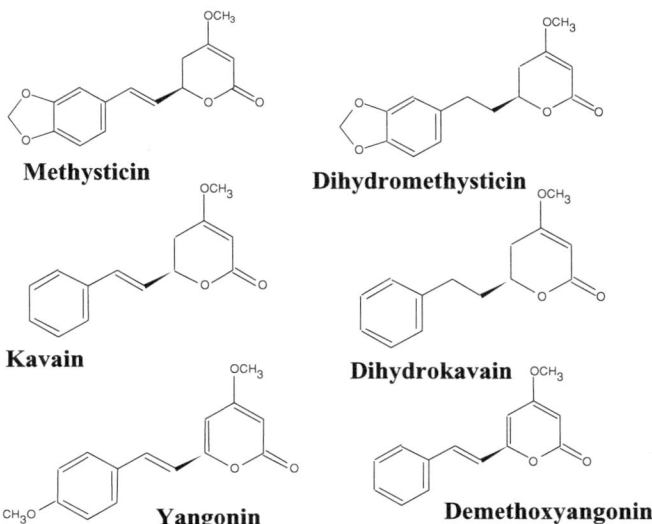

FIGURE 8.1. Chemical structures of the six major kavapyrones.

and decrease with shading (Lebot et al. 1999). Moreover, varietal differences in Kava also appear to have a role in determining the overall level of kavapyrone production (Lebot et al. 1999). For the production of modern phytomedicinal preparations of Kava, dried rhizomes/roots should contain about 3.5% kavapyrones, and typical extractions to produce the material for pills or capsules involve the use of either ethanol/water or acetone/water mixtures (Schulz et al. 1998).

Kavapyrones are also produced by the related *Piper* species *P. wichmannii* and *P. auritum* (syn. *P. sanctum*). With respect to the levels of individual kavapyrones found in extracts produced from *P. wichmannii*, there are a wide variety of chemotypes, with some showing a strong resemblence to patterns of kavapyrone production in Kava (Lebot and Levesque 1989). Although not well characterized phytochemically, *P. auritum* has been shown to produce the minor kavapyrone 5-methoxy-5,6-dihydromethysticin (Senguputa and Ray 1987).

8.5. ISSUES REGARDING THE POTENTIAL HEPATOTOXICITY OF KAVA EXTRACTS

There has been recent concern regarding a possible hepatotoxicity associated with the consumption of Kava-containing dietary supplements. In 2002, the United States Food and Drug Administration (FDA) issued a consumer advisory regarding use of Kava dietary supplements because of at least 25 reports of liver injury associated with Kava use (Center for Food Safety and Applied Nutrition—FDA 2002). Although there is one report of liver injury involving consumption of a synthetic kavapyrone (kavain), the role of kavapyrones in causing liver damage remains unclear (Denham *et al.* 2002). Although recent work by Johnson *et al.* (2003) has shown that hepatoxic quinoid metabolites of kavapyrones can be produced *in vitro* by liver microsome fractions, the metabolic pathway for production of these compounds represents only a minor route for detoxification and would only be relevant when other conjugation pathways become saturated or repressed. On the other hand, it has been shown that other phytochemicals potentially present in Kava extracts can be hepatotoxic. This includes piperidine alkaloids present in the aerial part of Kava as well as the bark of the plant (Dragull *et al.* 2003). This finding is of particular interest as bark peelings of Kava are sometimes used as an adulterant of Kava root preparations (Dragull *et al.* 2003).

8.6. SIGNIFICANCE OF TISSUE CULTURE GROWTH IN KAVA PRODUCTION AND PHYTOCHEMICAL RESEARCH

Plant tissue culture techniques could prove useful both from the practical standpoint of commerical Kava phytomedicinal production and in basic phytochemical research on this plant. Given the present and projected future world demand for Kava as a phytomedicine, there is serious concern as to how this demand will be met by conventional production methods (Brevoort 1998 and references therein). At present, Kava production is clearly limited by the growing time required prior to harvest (3–5 years at minimum) and the narrow habitat range in which the plant can be grown (tropical South Pacific). It is feared that such an inability to meet world supply needs for Kava could lead to overharvesting of the plant and possible endangerment of the species, or frequent adulteration of harvested plant materials by unscrupulous dealers. Although synthetic production of kavapyrones might appear as a viable alternative to use of the natural product, there is evidence that the beneficial effects of Kava may rely upon concerted actions and synergisms of active chemicals that could be difficult to replicate with a synthetic product (Capasso and Calignano 1988, Davies *et al.* 1992). In this respect, Kava tissue culture could provide a viable and cost-effective *in vitro* alternative for the production of kavapyrones. If cell cultures of kavaoyrone producing Kava can be successfully established, a Kava cell or organ culture could be utilized in a "bioreactor system" which could allow for continous kavapyrone production. There are a number of examples where use of plant tissue culture systems in bioreactors has yielded natural product amounts equal to or greater than that present in the intact plant (reviewed in Bhojwani and Razdan 1996). However, development of such a system with Kava would require identification of growth conditions and culture forms (e.g., suspension cell cultures, regenerated roots) that will promote a reasonable recovery of kavapyrones.

Tissue culture techniques would also be useful in support of conventional growth of Kava in cultivation. As the part of the Kava plant used for production of phytomedicinal products is the root and rhizome, conventional field production requires continual harvesting of plants and replanting. The traditional method for propagation of Kava involves planting stem cuttings containing nodes, and plants are generated from the meristematic tissue present at the nodes. Bacterial contamination is very problematic with this method and limits successful replanting. In our own experience, pathogenic bacterial contamination severely limited our attempts at growing Kava in laboratory pots. Using tissue culture methodologies such as shoot regeneration, root regeneration, or somatic cell embryogenesis, Kava plants could be propagated free of initial bacterial contamination (micropropagation). Such plants generated by tissue-culture-based micropropagation could be utilized with greater success in the replanting of Kava. As Kava plants are sterile, seeds are not produced and the only form of germplasm currently available are from intact plants. In this respect, tissue cultures could provide an alternative means for Kava germplasm preservation. Although continuous cell culture can result in genetic modifications unsuitable for germplasm storage, this problem could potentially be addressed by cryopreservation techniques (see Bhojwani and Razdan 1996 for review of methodology). The sterility of Kava plants also means that plant breeding cannot be used for cultivar development of Kava plants with modified or improved characteristics such as increased levels of kavapyrone production. Here, tissue culture methods such as somaclonal variation and perhaps even transformation could be utilized to generate plants with genetic variation and novel characteristics. This would be especially relevant for Kava because of its narrow genetic base (Lebot et al. 1991, Lebot and Levesque 1996).

Cell cultures would also be useful for studying the basic biochemistry of natural product production by this plant. The use of cell cultures could provide a more simple system for the use of radiotracers in the analysis and study of metabolic pathways. Although a number of active kavapyrones have been identified, their biosynthetic pathways and the interrelationships of pathway steps remain largely unknown in this plant.

The potential importance of applying tissue culture techniques in Kava production has been recognized, and to our knowledge several laboratory groups are currently attempting to establish Kava tissue cultures. However, the successful establishment of aseptic tissue cultures of Kava has been severely hampered by problems with endogenous bacterial, fungal, and viral contamination associated with explant material. Taylor and Taufa (1998) attempted to circumvent this problem by either treating Kava plants with fungicide and antibiotics prior to explant isolation or sterilizing Kava explant material and including antibiotics in the growth medium. The sterilization procedures utilized by these authors for the explant material included a single treatment with mercuric chloride and thermal treatment of the tissue. Nevertheless, the resultant Kava cultures either were initially contaminated or exhibited evidence of contamination after 3–5 weeks (Taylor and Taufa 1998).

It is interesting to note that problems with bacterial contamination have also occurred during attempts at culturing other *Piper* species (Bhat et al. 1995, Fitchet 1990, Philip et al. 1992). For example, Bhat et al. (1995) found that bacterial contamination occurred in 90% of initial cultures from explants of *Piper nigrum* (black pepper), *Piper longum* (pipli) and *Piper betle* (betel vine). These authors suggest that contamination in *Piper* species likely arises because of endogenous bacteria in plant material that are difficult to eliminate with the usual methods of explant preparation prior to culture.

8.7. ESTABLISHMENT OF KAVA CELL CULTURES AND THE DETERMINATION OF *IN VITRO* KAVAPYRONE PRODUCTION

In studies recently published (Briskin et al. 2001), we have succeeded in developing a methodology for establishment of *in vitro* Kava cell cultures using stem nodal tissue and leaf tissue as explant material. In our approach, explants of *Piper methysticum* have been introduced *in vitro* into a medium with Plant Preservative Mixture (PPM) (Plant Cell Technology, Washington, D.C.) at 2 ml/l, and to date, axenic cultures of calli, roots, and shoots have been successfully recovered with a success rate of 80% or more.

For generating cell cultures from stem nodal tissue explants, a two-step surface sterilization procedure was utilized in the preparation of explant tissue from the nodal region of Kava stems. The nodal region was excised from the stem and then soaked for 10 min in 0.1% sodium hypochlorite solution followed by four successive washes with sterile, distilled water. Explant tissue sections were removed from the nodal region using a cork borer and the tissue sections were surface sterilized and rinsed as above. As shown in Fig. 8.2(A) (left plate), this two-step surface-sterilization treatment of the explant material prior to transfer to a culture medium promoted the establishment of aseptic cultures displaying callus growth for the Kava cultivar "Makea." Aseptic callus cultures could also be generated in a similar way from the Kava cultivar "Awke."

In contrast, a single treatment of the Kava stem nodal section, followed by removal of nodal tissue, and then transfer to culture medium resulted in cultures that displayed contamination within 14 days (Fig. 8.2(A), right plate). Contamination was also observed when cultures were initiated from explants where a single 0.1% sodium hypochlorite treatment

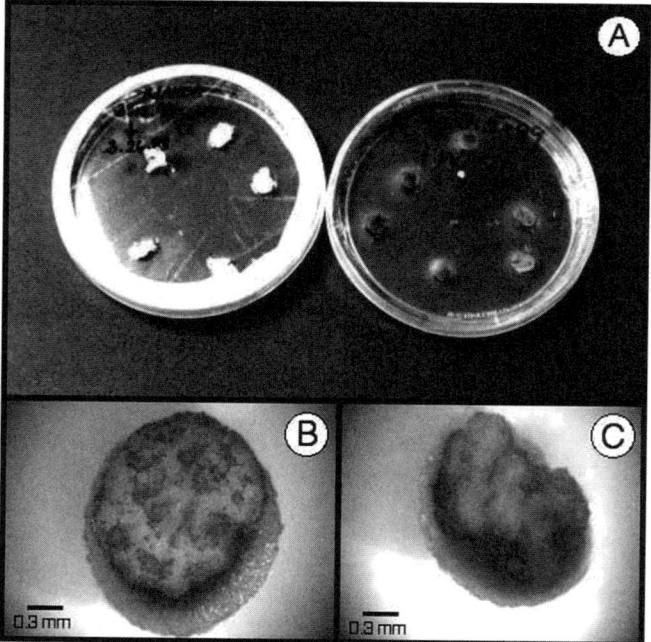

FIGURE 8.2. Kava callus cells generated from stem nodal tissue explants. (A) Effect of sequential 0.1% sodium hypochlorite washing of explants on culture contamination. Cultures on the left received two hypochlorite washes whereas the cultures on the right received a single wash. Cultures are in 100-mm petri plates. (B) Firm callus in a medium containing 2 mg/l 2,4-D. (C) Friable callus produced following transfer to a medium containing 1 mg/l 2,4-D and 1 mg/l NAA. (from Briskin *et al.* 2001)

was conducted on nodal tissue that was removed from a non–hypochlorite-treated stem section (data not shown). Hence, the inclusion of PPM as a biocide in the growth medium was observed to be crucial for establishment of axenic Kava cultures. When this biocide was omitted from the medium, aseptic Kava tissue cultures could not be established. In subsequent culture steps, PPM was also present in the growth medium to prevent the recurrence of contamination.

With a growth medium containing 2 mg/l 2,4-D, callus cells were generated from the stem node explant material which could be subcultured on a similar medium (Fig. 8.2(B)). Friable callus could also be generated with the transfer of the firm Kava callus to a growth medium containing both 1 mg/l 2,4-D and 1 mg/l naphthaleneacetic acid (NAA) (Fig. 8.2(C)).

For the successful establishment of Kava cell cultures from leaf explant material, the following protocol was utilized. Only very young, new expanding leaves were collected as explants. The vegetative material was gently rubbed with running tap water and a drop of commercial detergent, followed by 10 min of agitation in soapy water. Leaves were then thoroughly surface-sterilized by agitation in 10% commercial bleach (0.525% sodium hypochlorite) for 5 min. Subsequently, explants were triple-rinsed with sterile, double-distilled water, prior to explanting, and PPM was incorporated into the culture medium with half-strength Murashige and Skoog basal salts (Murashige and Skoog 1962) to combat persistent microbial infestation. Although contamination remained problematic, this combination of tactics permitted generation of Kava cultures that could be rescued from the predominantly contaminated stock, and established and maintained indefinitely as independent cultures.

Callus cells developed from the explant material after 21 days in this medium (Fig. 8.3(A)) and were viable following subculture in the same medium. With continued culture of callus generated from leaf material on this medium containing 2 mg/l NAA, root organogenesis began after 1–2 months (Fig. 8.3(B)). A close-up of the roots emerging from Kava callus showed the presence of root hairs (Fig. 8.3(C)), and roots were evident following continual culture in this same medium for an additional 2 months (Fig. 8.4). On the other hand, transfer of callus cells from leaves to a medium containing 1 mg/l 2,4-D and either 1 mg/l benzyladenine (BA) or 2 mg/l NAA resulted in the generation of friable callus after 10 days that could be maintained by subculture on either an agar-solidified or liquid medium (data not shown).

To determine if the cultures produced kavapyrones, extracts were prepared from intact plant tissues and callus cells, and the kavapyrone content was determined by high-performance liquid chromatography (HPLC). From analysis of the extracts using HPLC, it was evident that the cultured cells did produce kavapyrones but at a level that was much lower than that found for root tissues from which Kava phytomedicinal chemicals are typically isolated (Table 8.1). Relative to root tissue, recovery in the total amount of the six major kavapyrones in extracts produced from callus ranged from 1.5 to 2.2% for the two Kava cultivars examined in this study (Table 8.1). Although a lower total kavapyrone recovery was found for callus cells, a comparison of HPLC chromatograms for extracts from roots and callus from the Kava cultivar "Makea" showed that the six major kavapyrones present in root extracts were also generated in callus cells (Fig. 8.5). Moreover, there are some similarities in the relative proportions of these individual chemicals that are produced by roots and callus (Fig. 8.5).

TABLE 8.1
Total Recovery of Kavapyrones in Extracts from Kava Plants and Kava Callus Cells (from Briskin *et al.* 2001)

Sample	% Kavalactones (g/100 g)	% Recovery
Makea root	11.2	100.0
Makea callus	0.242	2.2
Awke root	8.92	100.0
Awke callus	0.130	1.5

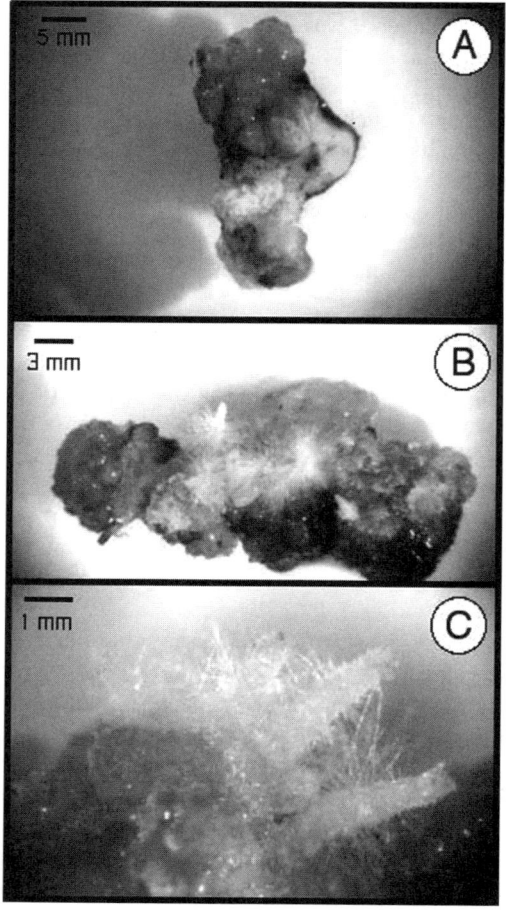

FIGURE 8.3. Kava callus cell cultures generated from leaf tissue explants. (A) Firm callus generated following 21 days of growth on a medium containing 2 mg/l 2,4-D. (B) Root organogenesis following continued culture on a medium containing 2 mg/l 2,4-D. (C) Close-up of roots emerging from Kava callus showing root hairs. (from Briskin *et al.* 2001)

TABLE 8.2
Distribution of Kavapyrones in Kava Plant Samples and Kava Callus Cells from Two Kava Cultivars. Values Represent the Percentage of Each Kavapyrone Relative to the Total (from Briskin et al. 2001)

Sample	Methysticin	Dihydromethysticin	Kavain	Dihydrokavain	Yangonin	Desmethoxyyangonin
"Makea" Cultivar						
Root	21.1	4.9	53.3	8.9	5.6	6.2
Stem	12.0	20.9	27.5	29.1	5.9	4.5
Callus	20.5	12.4	49.9	8.9	7.6	0.4
"Awke" Cultivar						
Root	12.9	7.6	48.9	13.7	6.3	1.6
Stem	23.7	11.9	18.5	30.6	7.4	7.9
Callus	19.3	4.4	50.0	6.5	10.1	9.7

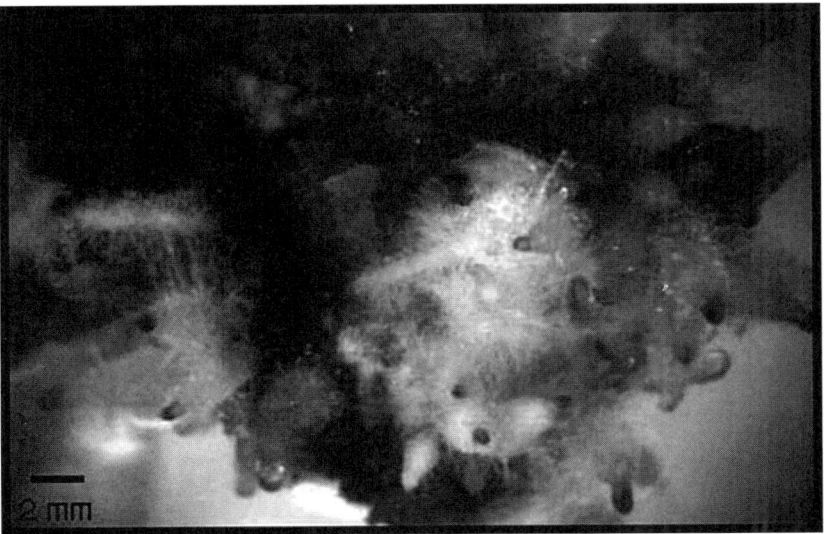

FIGURE 8.4. Root development by Kava (Piper methysticum) following growth over 4 months in a medium containing 2 mg/l 2,4-D. (from Briskin et al. 2001)

The recovery of the six major kavapyrones for extracts prepared from root, stem, and callus cells from both the "Makea" and "Awke" Kava cultivars is presented in Table 8.2. The level of individual kavapyrones could not be determined for leaf extracts because of an interference with the assay. For both Kava cultivars, kavain and methysticin were the predominate kavapyrones present in root extracts, and the reduced forms of these phytochemicals (dihydrokavain, dihydromethysticin) were present at a lower level. On the other hand, dihydrokavain and dihydromethysticin were present at a higher level relative to kavain and methysticin in stem extracts. This elevation in dihydromethysticin and dihydrokawain in stem extracts as compared to root extracts has been previously observed for several cultivars

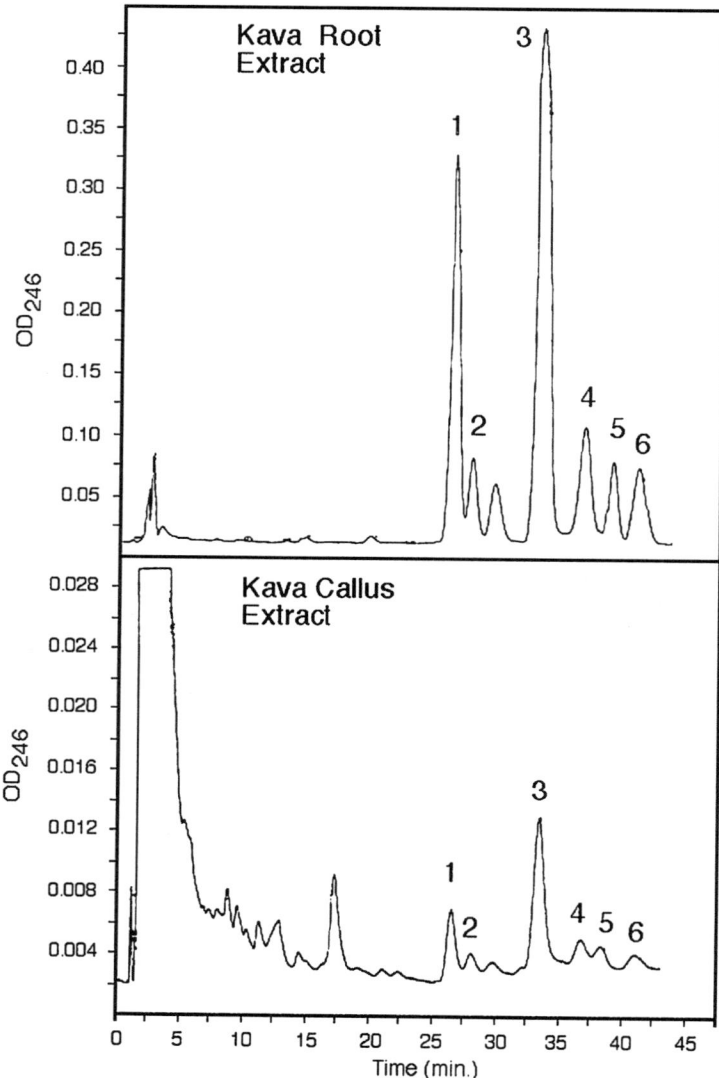

FIGURE 8.5. Profiles for kavapyrone production by Kava root tissue and culture cells generated from stem nodal sections. Extracts from roots and callus cells were prepared and subjected to HPLC as described by Briskin et al. (2001). The numbered peaks correspond to 1 – methysticin, 2 – dihydromethysticin, 3 – kavain, 4 – dihydrokavain, 5 – yangonin, 6 – desmethoxyyangonin.

of Kava (Lebot et al. 1999). Although callus cultures for the "Makea" Kava cultivar were produced from shoot bud explant material, the pattern of kavapyrone production appeared to be more similar to that observed for root extracts with kavain and methysticin representing the predominate kavapyrones. Likewise, whereas callus cultures for the "Awke" Kava cultivar were produced from leaf explant material, the pattern of kavapyrone production again involved a predominance of kavain and methysticin similar to roots. This would also be in contrast to what has been reported for leaf extracts (using a different assay) where dihydrokavain and dihydromethysticin represent the main kavapyrones (Lebot et al. 1999). Hence, the generation of callus cells would appear to alter the relative pattern of kavapyrone production from that observed for the original explant tissue to a pattern similar to that produced in roots, though the amount of kavapyrone produced is lower relative to roots from which the phytomedicine and intoxicating beverage are produced. A low recovery of phytochemicals in cell cultures relative to the original plant material has been a common limitation to *in vitro* production of secondary metabolites (Roja and Rao 2000, Verpoorte et al. 1998, and references therein). As a potential route to achieve an elevated level of *in vitro* kavapyrone production we are currently examining phytochemical recovery from Kava root cultures and Kava cultures transformed using *Agrobacterium rhizogenes*. If successful, this could prove useful for large-scale production of Kava medicinal compounds using bioreactor systems.

8.8. REGENERATION OF VIABLE KAVA PLANTS FROM KAVA CELL CULTURES

From the callus cell cultures of Kava established by the above methods, it was possible to regenerate viable Kava plants by first promoting shoot organogenesis from the callus cells and then allowing these microshoots to establish roots in culture. Shoot morphogenesis was achieved with the Kava growth medium modified to contain 1.0 mg/l benzyladenine (BA). The cytokinin and microshoots were then transferred to a medium free of plant growth regulator (Fig. 8.6(A, B)). In contrast, we found that a medium with a higher auxin ratio with respect to cytokinin caused spontaneous rhizogenesis of callus cultures (data not shown). From the shoot cultures it was then possible to obtain intact regenerated Kava plants in culture that remained viable following acclimatization and then transfer to a commercial soil-free potting medium (Fig. 8.6(C)). We have allowed the regenerated plants to grow in the greenhouse, and currently our first set of regenerated Kava plants are over 2 feet tall and demonstrate a growth habit similar to Kava plants propagated via cuttings (Fig. 8.7). Moreover, these plants produce kavapyrones in a similar manner to the *in vivo* plants from which explants were initially generated (Kobayashi et al. 2003).

8.9. SUMMARY AND PERSPECTIVE

The work presented in this chapter demonstrates the applicability of tissue culture approaches in Kava production and fundamental research on this plant. Although fungal and bacterial contamination had previously prevented successful establishment of Kava cell cultures, an approach of careful tissue disinfecting followed by culture in a biocide-containing medium allowed for production of aseptic cell cultures. Through modification of growth-regulator levels in this medium, it was then possible to regenerate intact plants from the cell cultures. Kavapyrones as the medicinally active phytochemicals produced by the

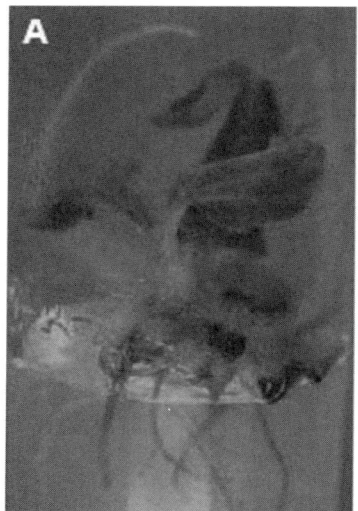

FIGURE 8.6. Shoot culture and regeneration of Kava in tissue culture. (A) A single *in vitro* kava shoot regenerated from callus. (B) Kava shoot multiplication and subsequent development. (C) Acclimatized kava plants in the greenhouse.

plant are present in extracts produced from cultured cells and regenerated plants, although in the case of cultured cells, phytochemical levels are relatively low when compared to the original plant material.

With further refinement of the current Kava plant regeneration system, our long-range goals are to elucidate biological mechanisms of kavapyrone production, and to improve existing germplasm material through means of mutagenesis and genetic engineering.

FIGURE 8.7. Kava plants regenerated from tissue culture cells growing in commercial potting mixture.

The successful culture of Kava could also lead to new approaches for large-scale propagation of this plant. Moreover, kavapyrone production systems based on bioreactors may be developed as an alternative source or a substitution of Kava for phytomedicine production if levels of kavapyrone production in cell cultures can be enhanced.

REFERENCES

Bhat, S. R., Chandel, P. S., and Malik, S. K. (1995). Plant regeneration from various explants of cultured *Piper* species. *Plant Cell Reports* 14:398–402.

Bhojwani, S. S., and Razdan, M. K. (1996). *Plant Tissue Culture: Theory And Practice, A Revised Edition*. Elsevier, Amsterdam.

Borche, W., and Blount, B. K. (1933). Undersuchungen über die bestandteile der Kawawurzel XIII. Mitteil, über einige neue stoffe aus technischem. *Chemische Berichte* 66:803–806.

Brevoort, P. (1998). The booming U.S. botanical market. A new overview. *HerbalGram* 44:33–46.

Briskin, D., Kobayashi, H., Metha, A., Gawienowski, M., Ainsworth, L., and Smith, M. A. L. (2001). Production of kavapyrones by Kava (*Piper methysticum*) tissue cultures. *Plant Cell Reports* 20:556–561.

Bruneton, J. (1995). *Pharmacognosy, Phytochemistry, Medicinal Plants*. Lavisior Ltd., Paris.

Capasso, A., and Calignano, A. (1988). Synergism between the sedative action of kava extract and D,L-kawain. *Acta Therapeutica* 14:249–256.

Center for Food Safety and Applied Nutrition. U.S. Food and Drug Administration. (March 25, 2002). Kava-containing dietary supplements may be associated with liver injury. Available from: http://www.cfsan.fda.gov/~dms/addskava.html.

Davies, L. P., Drew, C. A., Duffield, P., Johnston, G. A. R., and Jamieson, D. D. (1992). Kavapyrones and resin: Studies on $GABA_A$, $GABA_B$, and benzodiazepine binding sites in rodent brain. *Pharmacology and Toxicology* 71:120–126.

Denham, A., McIntyre, M., and Whitehouse, J. (2002). Kava—the unfolding story: Report on a work-in-progress. *The Journal of Alternative and Complementary Medicine* 8:237–263.

Dragull, K., Yoshida, Y. A., and Tang, C.-S. (2003). Piperidine alkaloids from *Piper methysticum*. *Phytochemistry* 63:193–198.

Fitchet, M. (1990). Tissue culture propagation of black pepper (*Piper nigrum* L.). *Acta Horticultura* 275:285–291.

Gleitz, J., Beile, A., and Peters, T. (1996). The action of kavain, a kavapyrone prepared from the psychotropic remedy *Piper methysticum*, on presynaptic neurotransmission. *European Neuropsychopharmacology* 6:S26–S27.

Gracza, L., and Ruff, P. (1980). Einfache methode zur trennung und quantitativen bestimmung von kawa-laktonen durch hochleistungs-flüssigkeits-chromatographie. *Journal of Chromatograhy* 93:486–490.

He, X.-G., Longze, L., and Lian, L.-Z. (1997). Electrospray high performance liquid chromatography–mass spectrometry in phytochemical analysis of kava (*Piper methysticum*) extract. *Planta Medica* 63:70–74.

Johnson, B. M., Qiu, S.-X., Zhang, S., Zhang, F., Burdette, J. E., Yu, L., Bolton, J. D., and van Breemen, R. B. (2003). Identification of novel electrophilic metabolites of *Piper methysticum* Forst. (Kava). *Chemical Research in Toxicology* 16:733–740.

Jussofie, A., Schmiz, A., and Hiemke, C. (1994). Kavapyrone enriched extract from *Piper methysticum* as modulator of the GABA binding site in different regions of rat brain. *Psychopharmacology* 116:469–474.

Klohs, M. W., Keller, F., Williams, R. E., Toekes, M. I., and Cronheim, E. G. (1959). A chemical and pharmacological investigation of *Piper methysticum* Forst. *Journal of Medicinal and Pharmaceutical Chemistry* 1:95–99.

Kobayashi, H., Gawienowski, M. C., Lila, M. A., and Briskin, D. P. (in press). Kavapyrone production during tissue culture and regeneration of Kava (*Piper methysticum*). *Phytochemistry*.

Lebot, V., and Cabalion, P. (1988). *Kavas of Vanuatu. Cultivars of* Piper methysticum. Technical Paper No. 195, South Pacific Commission, Noumea.

Lebot, V., and Lévesque, J. (1989). The origin and distribution of kava (*Piper methysticum* Forst. f., Piperaceae): A phytochemical approach. *Allertonia* 5:223–280.

Lebot, V., and Lévesque, J. (1996). Genetic control of kavalactone chemotypes in *Piper methysticum* cultivars. *Phytochemistry* 43:397–403.

Lebot, V., Aradhya, M. K., and Manshardt, R. M. (1991). Geographical survey of genetic variation in Kava (*Piper methysticum* Forst. f.) and *P. wichmannii*. *Pacific Science* 45:169–185.

Lebot, V., Johnson, E., Zheng, Q.-Y., McKern, D., and McKenna, D. J. (1999). Morphological, phytochemical and genetic variation in Hawaiian cultivars of 'awa (Kava, *Piper methysticum*, Piperaceae). *Economic Botany* 53:407–418.

Lehmann, E., Kinzler, E., and Friedemann, J. (1996). Efficacy of a special Kava extract (*Piper methysticum*) in patients with states of anxiety, tension and excitedness of non-mental origin. A double-blind placebo-controlled study of four weeks treatment. *Phytomedicine* 3:113–119.

Murashige, T., and Skoog, F. (1962). A revised medium for rapid growth and bioassays with tobacco tissue cultures. *Plant Physiology* 15:473–497.

Philip, V. J., Dominic, J., Triggs, G. S., and Dickinson, N. M. (1992). Micropropagation of black pepper (*Piper nigrum* Linn.) through shoot tip cultures. *Plant Cell Reports* 12:41–44.

Roja, G., and Rao, P. S. (2000). Anticancer compounds from tissue cultures of medicinal plants. *Journal of Herb, Spice, and Medicinal Plants* 7:71–102.

Schelosky, L., Raffauf, C., Jendroska, K., and Poewe, W. (1995). Kava and dopamine antagonism. *Journal of Neurology* 58:639–640.

Schmitz, D., Zhang, C. L., Chatterjee, S. S., and Heinemann, U. (1995). Effects of methysticin on three different models of seizure like events studied in rat hippocampal and entorhinal cortex slices. *Archives of Pharmacology* 351:348–355.

Schröder, J. (1997). A family of plant-specific polyketide synthases: Facts and predictions. *Trends in Plant Science* 2:373–378.

Schulz, V., Hänsel, R., and Tyler, V. E. (1998). *Rational Phytotherapy*. Springer-Verlag, Berlin.

Sengupta, S., and Ray, A. B. (1987). The chemistry of *Piper* species: A review. *Fitoterapia* 58:147–166.

Shao, Y., He, K., Zheng, B. L., and Zheng, Q. Y. (1998). Reversed-phase high-performance liquid chromatographic method for quantitative analysis of the six major kavalactones in *Piper methysticum*. *Journal of Chromatography* 825:1–8.

Singh, Y. N. (1992). Kava. An overview. *Journal of Ethnopharmacology* 37:13–45.

Singh, Y. N., and Blumenthal, M. (1997). Kava: An overview. *HerbalGram* 39:33–55.

Smith, R. M., Thakrar, H., Arowolo, T. A., and Shafi, A. A. (1984). High performance liquid chromatography of kava lactones from *Piper methysticum*. *Journal of Chromatography* 283:303–308.

Taylor, M., and Taufa, L. (1998). Decontamination of kava (*Piper methysticum*) for *in vitro* propagation. *Acta Hort.* 461:267–274.

Verpoorte, R., van der Heijden, R., ten Hoopen, H. J. G., and Memlink, J. (1998). Metabolite engineering for the improvement of plant secondary metabolite production. *Plant Tissue Culture and Biotechnology* 4:3–20.

Volz, H. P., and Kieser, M. (1997). Kava-Kava extract WS 1490 versus placebo in anxiety disorders. A randomized placebo-controlled 25-week outpatient trial. *Pharmacopsychiatry* 30:1–5.

Walden, J., von Wegerer, J., Winter, U., and Berger, M. (1997). Actions of kavain and dihydromethysticin on ipsapirone-induced field potential changes in the hippocampus. *Human Psychopharmacology* 12:265–270.

9

Phylogenetic Patterns, Evolutionary Trends, and the Origin of Ant–Plant Associations in *Piper* section *Macrostachys*: Burger's Hypotheses Revisited

Eric. J. Tepe, Michael A. Vincent, and Linda E. Watson
Department of Botany, Miami University, Oxford, Ohio

9.1. INTRODUCTION

In *Flora Costaricensis*, William Burger of the Field Museum in Chicago considered eight of the *Piper* species that occur in Costa Rica to make up the *P. obliquum* complex, and six additional species to be closely allied to it (Fig. 9.1; Burger 1971). *Piper calcariformis*, an additional species assignable to the *P. obliquum* complex, was later described by Tebbs (1989). Although Burger did not apply a system of infrageneric classification in *Flora Costaricensis* (1971), these 15 species of the *P. obliquum* complex and its allies represent all of the Costa Rican members of *Piper* sect. *Macrostachys* (Miq.) C.DC. Besides discussing the relationships between species, Burger (1972) also made hypotheses concerning evolutionary trends among the Costa Rican *Piper* species. In addition, he was among the first to note that certain *Piper* species have a tendency to support associations with ants. All known obligate ant-plants among New World *Piper* are found within the *P. obliquum* complex and its allies. In this chapter, we revisit Burger's hypotheses regarding systematic relationships and evolutionary hypotheses, and explore the evolution of ant–plant associations more fully in the broader context of *Piper* sect. *Macrostachys* using a molecular phylogenetic approach.

Molecular phylogenies are exceptionally useful as independent frameworks to study the evolution of ecological associations and the morphological traits related to them (Armbruster 1992, 1993, 1994, Dodd *et al*. 1999). *Piper* sect. *Macrostachys*, for many

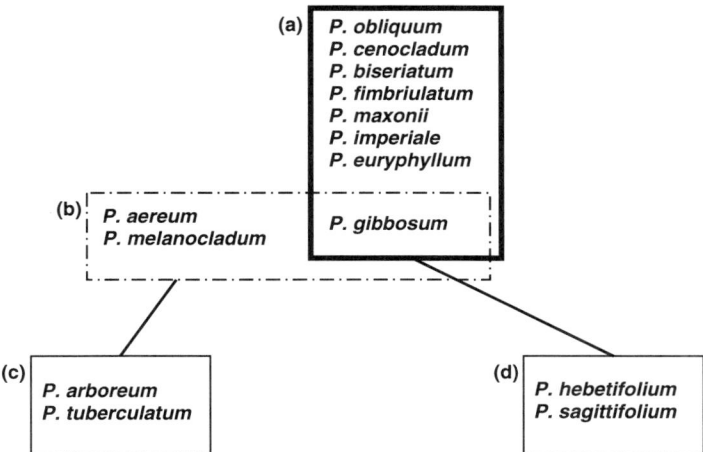

FIGURE 9.1. Systematic relationships among the Costa Rican members of *Piper* subg. *Macrostachys* (i.e., the "*P. obliquum* complex") as proposed in *Flora Costaricensis* (Burger, 1971). The species in box (a) constitute the "*P. obliquum* complex." Burger proposed a close relationship between *P. euryphyllum* and *P. gibbosum*, but he also suggested that *P. gibbosum* was closely related to the species in box (b). In addition, he suggested that *P. gibbosum* was a potential link, along with its allies in (b), between the species in (c) and (a), and by itself between (d) and (a). With the exception of *P. calcariformis*, which was described later, the species in this figure represent all of the Costa Rican species of *Piper* subg. *Macrostachys*.

reasons, presents an ideal system for studying the evolution of ant–plant associations using phylogeny. For example, a range of ant associations, from obligate to facultative to none, provides the potential for studying evolutionary patterns and trajectories that may result in obligate mutualisms. The number of origins of obligate associations is of interest since several of the obligate myrmecophytes are morphologically divergent for many traits (e.g., *P. sagittifolium* vs. *P. fimbriulatum*), whereas the characters associated with ants are remarkably constant in all five obligate species.

In this chapter, we examine interspecific relationships within the *P. obliquum* complex and allied species *sensu* Burger (1971; i.e., the Costa Rican members of sect. *Macrostachys*), as well as representative species of sect. *Macrostachys* that occur outside of Costa Rica (Table 9.1). We utilize the phylogeny as an independent framework to gain insight into systematic relationships, and evolutionary patterns and trends within the section, including those that led to obligate ant–plant associations. This approach will allow us to address the following questions:

i) Is the *P. obliquum* species complex *sensu* Burger monophyletic within sect. *Macrostachys*? Do the species of the *P. obliquum* complex, along with allied species, also form a monophyletic group?
ii) Are Burger's hypotheses of relationships within the *P. obliquum* complex and its affinities to allied species supported?
iii) Are the evolutionary trends within the *P. obliquum* complex and allied species supported, as proposed by Burger?
iv) What are the patterns of morphological change that led to obligate ant–plant interactions in *Piper* sect. *Macrostachys*?

TABLE 9.1
Species Included in This study. Asterisks Denote Species That We Have Observed and Collected in the Field

Species	Collector/Accession #	Source
INGROUP		
P. aereum Trel.	I.A. Chacon 2213 (MO)	Costa Rica
P. arboreum Aublet*	E. Tepe 377 (MU)	Costa Rica
P. archeri T. & Y.	M.A. Jaramillo 87 (DUKE), GenBank AF275178	Colombia
P. begoniicolor T. & Y.	T. Croat 69629 (MO)	Colombia
P. biseriatum C.DC.*	E. Tepe 141 (MU)	Costa Rica
P. calcariformis Tebbs*	A. Estrada 2397 (CR)	Costa Rica
P. cenocladum C.DC.*	E. Tepe 393 (MU)	Costa Rica
P. cernuum Vell.	G. Hatschbach 46665 (MU)	Brazil
P. cogolloi Callejas	R. Callejas 6431 (MO)	Colombia
P. euryphyllum C.DC.*	E. Tepe 410 (MU)	Costa Rica
P. fimbriulatum C.DC.*	E. Tepe 115 (MU)	Costa Rica
P. gibbosum C.DC.*	E. Tepe 411 (MU)	Costa Rica
P. gigantifolium C.DC.	S. Mori 12866 (NY)	Brazil
P. hebetifolium Burger*	E. Tepe 448 (MU)	Costa Rica
P. imperiale-a (Miq.)C.DC.*	E. Tepe 402 (MU)	Costa Rica
P. imperiale-b (Miq.)C.DC.*	E. Tepe 473 (MU)	Costa Rica
P. marsupiatum T. & Y.*	MBG #931716	Cultivated, Missouri Botanical Garden
P. maxonii C.DC.*	E. Tepe 370 (MU)	Costa Rica
P. melanocladum C.DC.*	E. Tepe 134 (MU)	Costa Rica
P. nobile C.DC.	T. Croat 60527 (MO)	Venezuela
P. obliquum R. & P.*	E. Tepe 114 (MU)	Costa Rica
P. obtusilimbum C.DC.*	MBG #930887	Cultivated, Missouri Botanical Garden
P. perareolatum C.DC.	A. Gentry 74657 (MO)	Peru
P. pseudonobile C.DC.	Boyle 3521 (MO)	Ecuador
P. sagittifolium C.DC.*	E. Tepe 116 (MU)	Costa Rica
P. subglabribracteatum C.DC.	A. Cogollo 7784 (MO)	Colombia
P. sp. nov.*	E. Tepe 94 (MU)	Costa Rica
OUTGROUP		
P. auritum Kunth*	M.A. Jaramillo 63 (DUKE), GenBank AF 275175	Colombia
P. friedrischtalii C.DC.*	E. Tepe 131 (MU)	Costa Rica
P. garagaranum C.DC.*	M.A. Jaramillo 73 (DUKE), GenBank Af275162	Colombia
P. lacunosum Kunth*	E. Tepe 443 (MU)	Costa Rica
P. multiplinervium C.DC.*	M.A. Jaramillo 139 (DUKE), GanBank AF 275168	Colombia

9.2. TAXONOMIC HISTORY OF *Piper* sect. *Macrostachys* (MIQ.) C.DC.

The taxonomy of *Piper* L. has been problematic since tropical exploration increased in the 18th and 19th Centuries. Despite the attention of many notable botanists (Kunth 1839, Miquel 1843–1844, de Candolle 1869, 1923), the taxonomy of *Piper* remains difficult and enigmatic (Yuncker 1958, Bornstein 1989). Among factors contributing to a difficult taxonomy are that many species have been described within the past 100 years on the basis of sterile and/or fragmentary material (e.g., Trelease 1929, Trelease and Yuncker 1950), and that several recent authors have applied drastically different species concepts (Trelease 1929, Trelease and Yuncker 1950, Tebbs 1989, 1990, 1993).

More recently, a number of Floras have aided in clarifying the taxonomy and reducing the number of accepted names in *Piper* (Standley and Steyermark 1952, Burger 1971, Yuncker 1972, 1973, 1974, Howard 1988, Steyermark 1984, Callejas 2001), as have a smaller number of revisionary studies (Steyermark 1971, Smith 1975, Callejas 1986, Bornstein 1989). In addition, the circumscription of *Piper* has been debated, with varying numbers of segregate genera recognized by different authors (Kunth 1839, Miquel 1843–1844, de Candolle 1923, Trelease and Yuncker 1950). Great strides have been made in recent years toward a more robust system of infrageneric classification based on cladistic analyses of morphological and molecular evidence (Callejas 1986, Jaramillo and Manos 2001, E. Tepe *et al.*, unpubl. data; Chapter 10). These phylogenetic studies support *Piper* L. in a broad sense as well as many of the traditional sections recognized by earlier authors (i.e., Kunth 1839, Miquel 1843–1844, de Candolle 1869, 1923).

Kunth (1839) was the first author to attempt to provide a system of infrageneric classification for *Piper*, and thus was the first to group all of the known species that would come to constitute *Piper* sect. *Macrostachys* in a section denoted *Species Piperi obliquo... propinqua.* Miquel (1843–1844) transferred these species to *Artanthe* Miq. (now synonymous with *Piper*) sect. *Macrostachys* Miq. In the most recent genuswide classification, de Candolle (text: 1869, key: 1923) submerged *Artanthe* into an expanded *Piper* L., but separated the species of *Piper* sect. *Macrostachys* into several different sections on the basis of differing leaf venation patterns and bract characters. However, Callejas (1986) and Tebbs (1989) regrouped these into *Piper* sect. *Macrostachys* (Miq.) C.DC. on the basis of inflorescence, fruit, leaf base, petiole, and bract characters. *Piper* sect. *Macrostachys* is one of seven currently recognized Neotropical sections (Tebbs 1989; however, see Jaramillo and Manos 2001 and Chapter 10 for a different view), and contains approximately 50 species.

9.3. NATURAL HISTORY OF *Piper* sect. *Macrostachys*

Most species of *Piper* sect. *Macrostachys* (Figs. 9.2–9.4; Fig. 9.4 is typical *Macrostachys*) are large-leaved shrubs and small trees with pendulous inflorescences that are typically longer than the leaves. They are characteristic of deep-shade areas in the moist understory of wet tropical forests ranging from southern Mexico to northern Argentina. Several species of sect. *Macrostachys* represent stages along a continuum of ant associations, ranging from casual and periodic to highly specialized and obligatory (Risch *et al.* 1977, E. Tepe, unpubl. data). The obligate ant-plants in sect. *Macrostachys*—reported heretofore only from Costa Rica—are characterized by a number of adaptations (Fig. 9.2). In young plants, ants nest inside petiolar cavities that are tightly folded into a tube (Fig. 9.2; Burger 1972, Risch *et al.* 1977, Letourneau 1998). As the plant increases in size, the stem becomes hollow (Fig. 9.2; Risch *et al.* 1977). A pore between the petiolar and stem cavities allows for movement of ants between the two domatia. The primary source of nutrients for the ants appears to be plant-produced pearl bodies. The pearl bodies are located on the

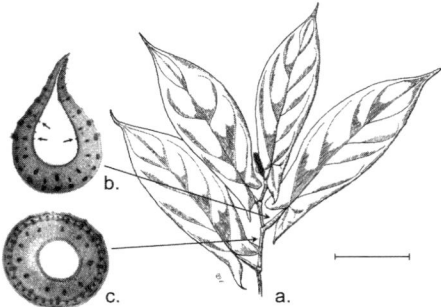

FIGURE 9.2. Obligate myrmecophyte. (a) *Piper sagittifolium* (a morphologically atypical member of *Piper* sect. *Macrostachys*). (b) Petiole cross section of *P. sagittifolium*, arrows indicate pearl bodies. (c) Stem cross section of *P. sagittifolium* from third internode. Scale bar = 5 mm for sections and 10 cm for illustration.

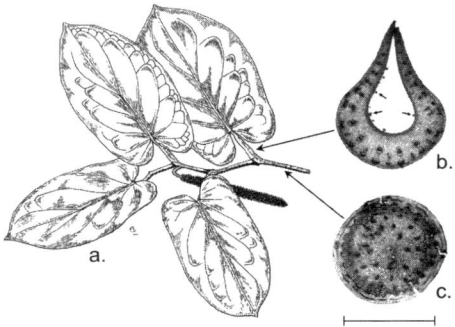

FIGURE 9.3. Obligate myrmecophyte. (a) *Piper calcariformis*. (b) Petiole cross section of *P. calcariformis*, arrows indicate pearl bodies. (c) Stem cross section of *P. sagittifolium* from third internode. Scale bar = 5 mm for sections and 15 cm for illustration.

adaxial side of the petioles (i.e., inside the tube, Fig. 9.2(B), arrows) such that they are only available to organisms inside the petiolar cavity, and are produced in large numbers only in the presence of the ant mutualist *Pheidole bicornis* Forel (Risch and Rickson 1981). Separate studies have produced different estimates of the nutritive composition of the pearl bodies with values ranging from 22 to 48% (dry weight) for lipids, negligible amounts of up to 2.1% for carbohydrates, and 10 to 24% for proteins (Risch *et al.* 1977, Fischer *et al.* 2002). *Piper calcariformis*, an obligate myrmecophyte with closed petioles and pearl bodies (Fig. 9.3), is unique among the obligate ant-plants in that it consistently has solid stems (Fig. 9.3).

The mutualism between *Ph. bicornis* and the myrmecophytic species of *Piper* is appropriately referred to as obligate because the fitness and survival of the plants are compromised when uninhabited by ants (Risch 1982, Letourneau 1983, 1998), and because *Ph. bicornis* appears to be specific to these *Piper* species; further, *Ph. bicornis* seemingly never forages off of its host plants (Risch *et al.* 1977, Letourneau 1998). The plants benefit from the obligate association with ants primarily through reduced damage by stem borers and a lower incidence of fungal infections of the inflorescences and infructescences (Letourneau 1998).

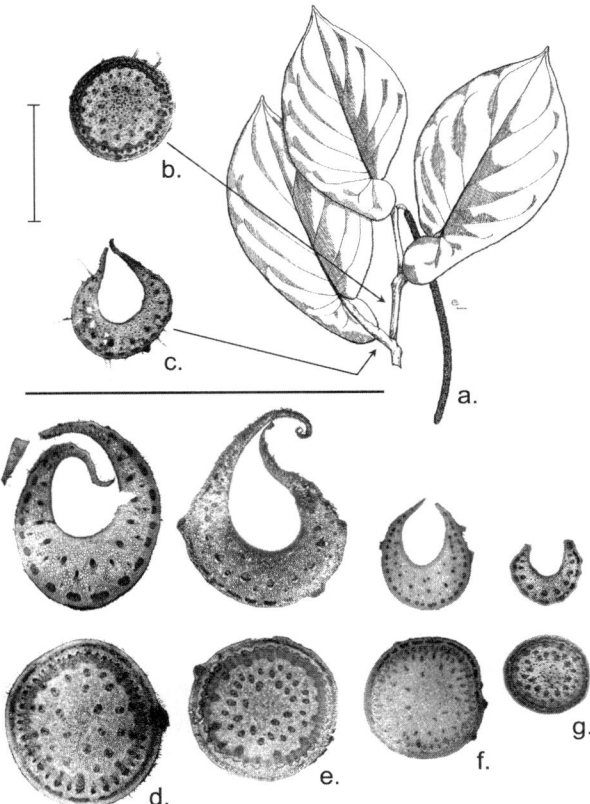

FIGURE 9.4. Facultative and nonmyrmecophytes. (a) *Piper biseriatum* (typical *Macrostachys*). (b) Stem cross section of *P. biseriatum*. (c) Petiole cross section of *P. biseriatum*. (d) Petiole and stem cross sections of *P. imperiale*, individual a facultative myrmecophyte. (e) *P. euryphyllum*, predicted, but not observed facultative myrmecophyte. (f) *P. imperiale*, individual a nonmyrmecophyte, compare with (d) for intraspecific variation in petiole closure. (g) *P. gibbosum*, nonmyrmecophyte. Scale bar = 5 mm for sections and 15 cm for illustration.

Myrmecophytic species of *Piper* have characteristic sheathing petioles that become tightly folded into a tube, whereas nonobligate myrmecophytes have solid stems and petioles that are closed to varying degrees (Fig. 9.4). In some facultative ant-plant species, colonies of ants are occasionally found nesting inside petiolar cavities (E. Tepe, unpubl. data); however, the ant residents are members of a suite of arboreal ant species that nest opportunistically in a variety of available, nonterrestrial cavities (Ward 1991, Byrne 1994, Alonso 1998, Orivel and Dejean 1999). Both the ant and plant species involved in facultative associations are capable of surviving without the other, and it is likely that these ant–plant associations are ephemeral (Alonso 1998). Ants apparently excavate the pith of the obligate myrmecophytes (Risch *et al*. 1977, Letourneau 1998); however, medulary vascular bundles are often scattered throughout the pith in species with stems that do not become hollow (e.g., Fig. 9.3) and may preclude excavation by ants. Tough tissue appears to deter pith excavation in some species of *Macaranga* (Euphorbiaceae; Fiala and Maschwitz 1992a).

Finally, in all remaining species of sect. *Macrostachys* encountered in the field, the petioles are either very small or the margins do not come together to form a tube (Fig. 9.4(F–G)), and no ants have been observed living on or in these plants (E. Tepe, pers. obs.). In summary, five species of *Piper* have obligate associations, two or more species have facultative associations, and numerous species have no associations with ants. All of these species are members of *Piper* sect. *Macrostachys*. It is within this continuum of ant–plant interactions that clues may be found to better understand the evolution of these mutually beneficial interactions.

9.4. PHYLOGENETIC RELATIONSHIPS IN *Piper* sect. *Macrostachys*

Our phylogenetic analyses of nucleotide sequences of the internal transcribed spacers (ITS) of nuclear ribosomal DNA for 32 taxa (27 ingroup and 5 outgroup species) converged upon a single Maximum Likelihood (ML) tree (Fig. 9.5) and an identical Bayesian tree. Trees with similar topologies were also recovered in four equally most parsimonious trees (MPT) in an analysis of ITS in which gaps were treated as missing data, and 27 MPTs when gaps were coded as separate presence/absence characters following Simmons and Ochoterena (2000). The topology recovered in the ML and Bayesian analyses is not in conflict with any of the MPTs with or without gap coding (Fig. 9.6), and none of the topologies of the MPTs contradict the hypotheses discussed here. The only differences among MPT and ML/Bayesian topologies is the level of support for clades C and D when gaps are treated as missing data, and support for clade C when gaps are coded as presence/absence characters. (For methods see Appendix 9.1)

Our results are concordant with several recent studies that support the monophyly of *Piper* sect. *Macrostachys* (Miq.) C.DC. (Callejas 1986, Jaramillo and Manos 2001, Tepe *et al.*, unpubl. data). Its monophyly is supported in studies with broad taxonomic sampling (Jaramillo and Manos 2001; Chapter 10) and in this study with the selection of closely related sister species. Section *Macrostachys s.l.* (Fig. 9.5, clade A), including sect. *Hemipodium* (Miq.) C.DC., represented here by *P. arboreum* (Tebbs 1989, Jaramillo and Manos 2001; Chapter 10), is strongly supported as monophyletic, with a Bayesian posterior probability (PP) of 1.0 and a bootstrap (BS) value of 94%. Section *Hemipodium* is sister to the remainder of sect. *Macrostachys* (Fig. 9.5, clade B), which is likewise strongly supported (1.0 PP, 76% BS). Posterior probabilities and bootstrap values have been considered the upper and lower limits of support, respectively (Douady *et al.* 2003), with bootstrap representing an overly conservative estimate of support in many cases (Wilcox *et al.* 2002). Short branch lengths and unresolved clades corroborate the results of Jaramillo (Chapter 10) that sect. *Macrostachys* appears to be recently derived through rapid diversification.

The sect. *Macrostachys* clade (Fig. 9.5, clade A) is somewhat unresolved, particularly in the parsimony analysis, with one major clade weakly supported within it (clade C). Within clade C, two subclades (D and E) are resolved, but weakly supported. Within the unresolved region of the *Macrostachys* clade, a small subclade that includes *P. melanocladum* and *P. aereum* represents a subgroup of *Macrostachys* species characterized by relatively small leaves lacking basal lobes, and with distributions restricted to

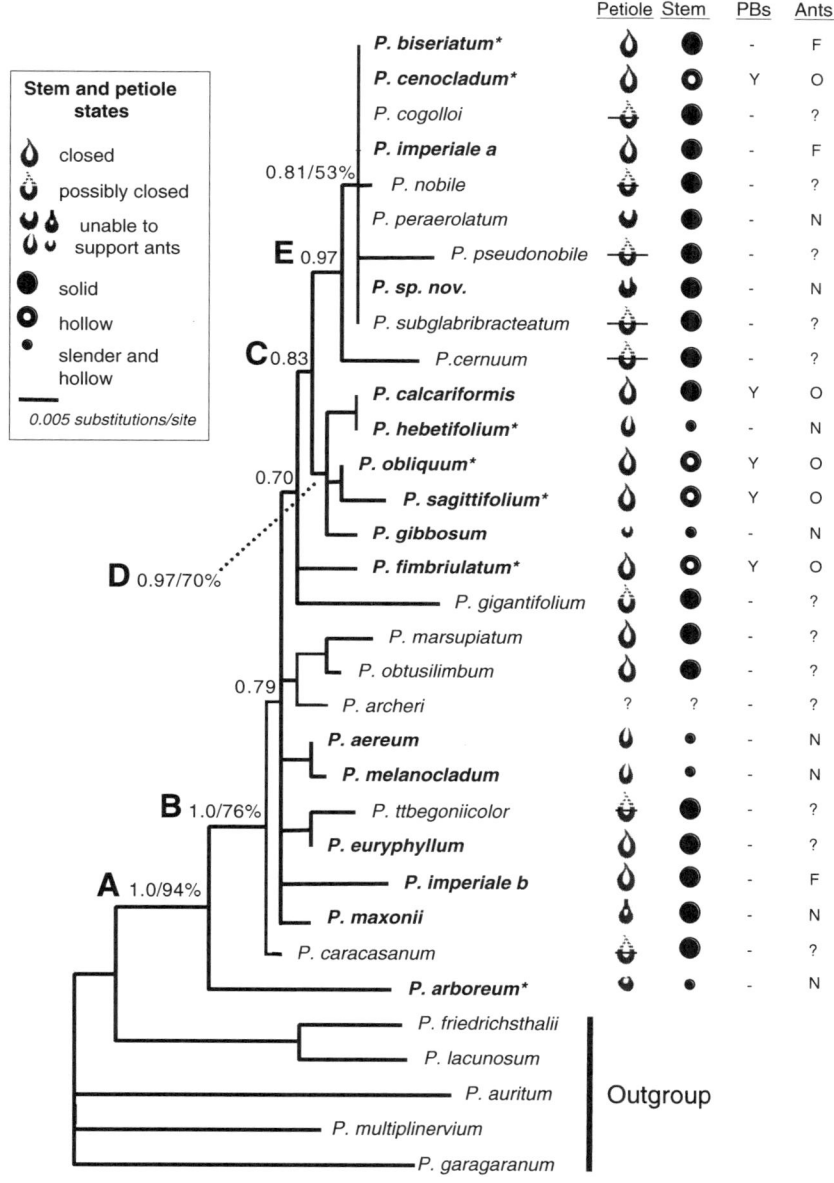

FIGURE 9.5. The single tree for *Piper* subg. *Macrostachys* from ML analysis of ITS sequence data (CI = 0.82, RI = 0.78). Numbers indicate Posterior Probabilities/ Bootstrap values above 50%. Support for branches not indicated on tree are included in text. Species names in bold are found in Costa Rica, and asterisks (*) denote species in which more than one collection from different locations were included in the analysis and were resolved in the same position on the tree. Stem and Petiole columns represent cross sections. PBs are pearl bodies found inside the petiole chambers. The ant column indicates whether obligate "O," facultative "F," or no "N," associations with ants have been observed. Species that have not been observed, but that are predicted to have facultative associations with ants are indicated by "?" See Appendix 9.1 for methods.

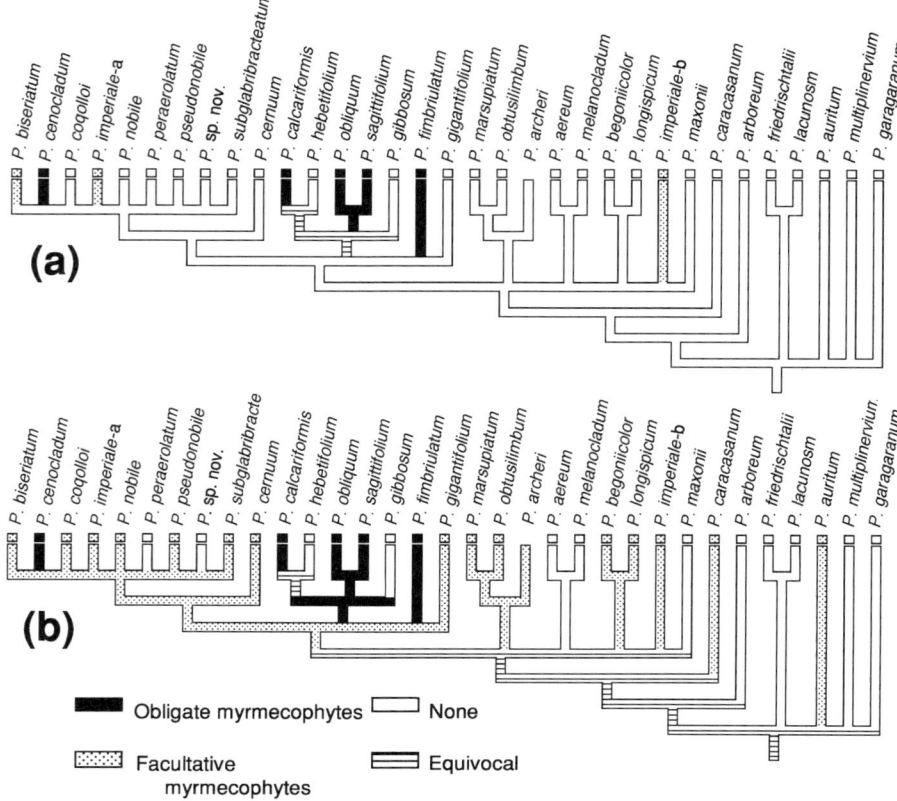

FIGURE 9.6. Reconstruction of the evolutionary history of myrmecophytism in *Piper* subg. *Macrostachys* based on (a) observed associations and (b) predicted associations.

southern Central America and adjacent northwestern South America (1.0 PP, 88% BS). Another clade includes *P. euryphyllum* (= *P. longispicum* C.DC.) and *P. begoniicolor* (1.0 PP, 98% BS). These two species, along with several others (including at least one species that occurs at high altitudes in Costa Rica), have been informally recognized as an Andean lineage within sect. *Macrostachys* by R. Callejas (pers. comm.). This Andean subclade is morphologically distinct and is resolved by our data; thus it appears, pending the inclusion of additional species in future analyses, that Callejas' informal rank is supported. However, the small clade that includes *P. archeri*, *P. obtusilimbum*, and *P. marsupiatum*, which are all South American species, is not strongly supported (0.74 PP, 20% BS).

The major clade within sect. *Macrostachys* (Fig. 9.5, clade C), although not present in all MP trees, is present in both ML and Bayesian analyses. Within clade C, two sister subclades (D and E) are present. The unresolved placements of *P. fimbriulatum* and *P. gigantifolium* are sister to clade C. In several of the MP trees, *P. fimbriulatum* and *P. gigantifolium* are sister to each other, but are placed in a large, unresolved portion of the tree. Subclade D includes rare species (except *P. obliquum*, which is widespread) that have distributions restricted to Costa Rica and adjacent Panama. Subclade E is internally unresolved by the ITS, with the exception of *P. cernuum*, which is resolved as basal to the remainder of sect. *Macrostachys s.s.* It includes species of Costa Rican endemics, as well as South American species.

Many of the species included in our analyses were represented by more than one individual from more than one location (asterisks, Fig. 9.5). In all cases except for *P. imperiale*, all replicates of a given species are resolved as closest relatives and were therefore excluded in final analyses. In contrast, different collections of *P. imperiale* are placed distant to each other: in clade E, and at the base of the tree as unresolved (Fig. 9.5). Taxonomic determinations for some collections from Costa Rica are difficult because they tend to have traits that are characteristic of both *P. imperiale* and *P. biseriatum*. Furthermore, these ambiguous specimens are unequivocally resolved among individuals of *P. biseriatum* and *P. imperiale* in our ITS trees. It is possible that these problematic specimens represent hybrids, or species not yet identified. Although the collections included in this study were unambiguously identified, the placement of *P. imperiale* collections in two distant parts of the tree may indicate possible gene flow between *P. biseriatum* and *P. imperiale*. If hybridization is in fact implicated, this would represent the first report of hybridization within *Piper* in the wild. Thus far, the only report of hybridization in *Piper* is of artificial crosses between the cultivated pepper, *P. nigrum*, and several of its close relatives (Sasikumar et al. 1999).

9.5. BURGER'S HYPOTHESES REVISITED

9.5.1. *Systematic Relationships*

The *P. obliquum* complex (*sensu* Burger 1971) comprises eight Costa Rican species that Burger recognized as being closely related (Fig. 9.1). In addition to this complex, he posited that six other Costa Rican species were more closely allied to this complex than to any other *Piper* species that occur in Costa Rica (Fig. 9.1). The *P. obliquum* complex along with the allied species constitute all of the Costa Rican members of sect. *Macrostachys*, and, although these 14 species are each other's closest relatives among Costa Rican *Piper* species (the geographic extent of Burger's study), based on the ITS phylogeny, neither the *P. obliquum* complex nor the allied species constitute a monophyletic group when analyzed with additional species of sect. *Macrostachys* that occur outside of Costa Rica (Figs. 9.5 and 9.6). To further test the nonmonophyly, we constrained our ML analysis to force monophyly of the *P. obliquum* complex. The resulting likelihood scores of the trees were significantly higher than that of the most likely tree ($P = 0.001$, Shimodaira–Hasegawa test; Shimodaira and Hasegawa 1999). This result is consistent with all analyses of the ITS. In addition, it is clear that Burger did not believe the *P. obliquum* complex to be monophyletic because he proposed several close relationships between species in the *P. obliquum* complex with members of sect. *Macrostachys* not present in Costa Rica.

Although Burger (1971) proposed few systematic relationships within the *P. obliquum* complex itself, he did suggest that several groups of species were closely related to it (Fig. 9.1). He had doubts regarding the affinities of *P. gibbosum*. In some instances, he included it in the *P. obliquum* complex, most closely related to *P. euryphyllum*, but later in the text he included it in a group with *P. aereum* and *P. melanocladum* (Fig. 9.1). Indeed, *P. gibbosum* has characters in common with both groups. It shares leaf shape, habitat, and the presence of styles (rather than sessile stigmas) with *P. euryphyllum*, and smaller leaf size, leaf texture, and venation patterns with *P. aereum* and *P. melanocladum*. Our analyses do not support a close relationship between *P. gibbosum* and either one of these two species groups but, instead, place *P. gibbosum* in the clade containing *P. obliquum* and several other species (Fig. 9.5, clade D). Although the positions of *P. euryphyllum* and *P. aereum*/*P. melanocladum* are not fully resolved, they do not appear to be closely related to *P. gibbosum*.

Secondly, Burger (1971) was the first to suggest that *P. arboreum* and *P. tuberculatum* were closely related to species of sect. *Macrostachys* (Fig. 9.1) on the basis of anther form and the lack of an apically developed prophyll. This conclusion has been strongly supported by morphological (Callejas 1986) and molecular data (Jaramillo and Manos 2001; Chapter 10). Again, Burger believed that *P. aereum* and *P. melanocladum* are closely related to each other (Fig. 9.1), and that these two species, along with *P. gibbosum*, form a link between *P. arboreum* and *P. tuberculatum* and the remainder of the *P. obliquum* complex. This conclusion is logical since these five species (i.e., Fig. 9.1(B–C)) are relatively slender and small-leaved plants when compared to the rest of sect. *Macrostachys*. Indeed, *P. arboreum* does represent a group of species in our analyses that includes *P. tuberculatum* and possibly *P. cordulatum* (a Panama endemic, not sampled) that is placed outside of the core *Macrostachys* s.s. clade (i.e., Fig. 9.5, clade B), and *P. aereum* and *P. melanocladum* are supported as closely related, as Burger suggested. However, the affinities of *P. aereum* and *P. melanocladum* to the remainder of the *Macrostachys* species cannot be determined since the placement of this clade is unresolved in the ITS trees. *Piper caracasanum* is placed between *P. arboreum* and the rest of *Macrostachys* (clade B), but its position is not well supported. Increased sampling and the addition of more variable markers may resolve these nodes, and *P. aereum* and *P. melanocladum* may indeed turn out to represent the link that Burger suggested (Fig. 9.1). Again, *P. gibbosum* is resolved in clade D (Fig. 9.5) and thus is apparently not closely related to these species.

Lastly, Burger suggested that, although *P. sagittifolium* is a morphologically unusual species, unlike any other in Costa Rica (Fig. 9.2), it is allied to the *P. obliquum* complex through *P. hebetifolium* and *P. gibbosum* (Fig. 9.1). Our data strongly support a close relationship for these three species, with the addition of *P. obliquum*, and the more recently described *P. calcariformis* (Fig. 9.5, clade D). However, rather than being distantly related to the *P. obliquum* complex, these species form a clade that is nested fully within the core of sect. *Macrostachys*. The placement of *P. sagittifolium* relative to the remainder of the section is the most apparent inconsistency between our analyses and Burger's hypotheses. Burger believed *P. sagittifolium* to be isolated and primitive among the Costa Rican *Macrostachys* species. However, a number of morphological characters appear to support clade D (Fig. 9.5), including flowers with styles (most members of sect. *Macrostachys* outside of clade D have sessile stigmas), long and recurved stigma lobes, apiculate anthers, and a number of vegetative traits that support obligate associations with ants. In fact, with the exceptions of *P. hebetifolium* and *P. gibbosum*, all members of this clade are involved in obligate associations with *Pheidole bicornis*, the obligate plant ant. The most recent circumscription of *Piper obliquum* R&P (Tebbs 1989) suggests that it is an extremely variable and widespread species. However, this broad circumscription includes a number of well-defined taxa that warrant specific status (Callejas 2001). The form of *P. obliquum* represented in this clade (Fig. 9.6, Clade D) appears to be a unique form of the species that is restricted to southern Central America. Of the species included in our analysis, *P. nobile*, *P. pseudonobile*, *P. archeri*, and *P. caracasanum* have all been considered synonymous with *P. obliquum* at one time or another. This topological distribution does not support the broad concept of *P. obliquum* adopted by some authors.

9.5.2. Evolutionary Trends

In addition to proposing systematic relationships among species of Costa Rican *Piper*, Burger (1972) postulated a number of evolutionary trends. This discussion, however, is limited to those trends related to *Macrostachys* specifically. Burger considered the flowers of *P. sagittifolium* to represent the primitive condition in *Piper*. This conclusion was based on the idea that Piperaceae is derived from an ancestor similar to *Saururus* Mill. (Saururaceae; Burger 1972), the well-supported sister group to the Piperaceae (Tucker *et al.* 1993, Savolainen *et al.* 2000). It follows then, that the flowers of primitive *Piper* species should be plesiomorphic and resemble those of *Saururus*. Indeed, similarities in the flowers of *P. sagittifolium* and *Saururus* include relatively large parts, long styles with long, divergent stigmas, and unusually large anthers (~1 mm!) on long filaments (to ~2 mm). Furthermore, the inflorescences of *P. sagittifolium* are not as tightly packed as is typical of many *Piper* species but, instead, are more loosely associated, similar to those of *Saururus*. Our analyses strongly support the placement of *P. sagittifolium* within sect. *Macrostachys*. If our phylogeny is accurate, parsimony then suggests that any resemblance between the flowers of *P. sagittifolium* and *Saururus* are homoplasies rather than plesiomorphies.

Anthers of the majority of *Piper* have latrorse dehiscence through longitudinal slits. A tendency toward upward dehiscence of anthers is found in several groups of Neotropical *Piper* species, including sect. *Macrostachys*. This shift from lateral to upward dehiscence is presumably due to the tight, cylindrical packing of flowers in several species groups (Burger 1972, Jaramillo and Manos 2001). Upward dehiscence among Neotropical *Piper* species is achieved in two distinct ways. Anthers with parallel thecae and apical dehiscence are found exclusively in a small group of scandent and climbing species (e.g., *P. xanthostachyum* C.DC., sect. *Churumayu*; Burger 1972). In contrast, upward dehiscence has also been attained through broadening of the lower part of the anther connective in species of sections *Radula* and *Macrostachys* (Burger 1972). This results, in the most extreme cases, in the anther thecae being oriented end to end, 180° to each other. Dehiscence is technically lateral, but because of the expanded connective and the altered orientation of the thecae, pollen is effectively released apically. Sections *Radula* (e.g., *P. aduncum* and *P. hispidum*) and *Macrostachys* are exceedingly dissimilar morphologically; thus it is not surprising that Burger (1972) proposed that upward dehiscence via expanded connective evolved independently in these two lines. Molecular evidence, however, strongly supports a sister group relationship between *Radula* and *Macrostachys* (Jaramillo and Manos 2001; Chapter 10) suggesting that a tendency toward upward anther dehiscence through expanded connectives may be a synapomorphy that unites these two sections.

9.6. ANT–PLANT ASSOCIATIONS IN *Piper* sect. *Macrostachys*

Several recent studies have used phylogenetic analyses to address the evolution of ant–plant mutualisms (Michelangeli 2000, Blattner *et al*. 2001, Brouat *et al*. 2001, Davies *et al*. 2001). Interestingly, each study has revealed a unique pattern of evolution of ant–plant associations including numbers of origins and losses of ant associations, trends and correlations in associated plant morphologies, and whether or not plants with intermediate morphologies and facultative associations with ants represent plant species that are ancestrally intermediate between plant species with obligate associations and those with none. Our use of the term *intermediate* throughout this paper is not with regard to a stage in the evolutionary process, but rather to describe species that are morphologically intermediate and that have facultative associations with ants, i.e., the associations are neither obligate nor random and ephemeral.

Numerous origins of myrmecophytes have been reported in the Melastomataceae, with representatives in nine genera from several tribes (Gleason 1931, Whiffin 1972, Vasconcelos 1991, Morawetz *et al*. 1992, Michelangeli 2000). A cladistic analysis of morphology for *Tococa* revealed at least two origins of associations with ants with a minimum of one loss (Michelangeli 2000). Furthermore, the presence, location, and morphology of the domatia appear to be rather plastic at the species level and above (Michelangeli 2000). In *Leonardoxa africana* (Fabaceae), a single origin of ant interactions among four subspecies is supported, each representing a different degree of association with ants (McKey 1984, 1991, 2000, Chenuil and McKey 1996, Brouat *et al*. 2001). Brouat *et al*. (2001) sequenced several chloroplast markers for the four subspecies of *L. africana*. The subspecies are not resolved as monophyletic groups, but rather are intermixed. Although evolutionary hypotheses could not be determined for *L. africana*, the study elegantly illustrates how apparent gene flow or insufficient molecular divergence between taxa can obscure evolutionary interpretations (Brouat *et al*. 2001). In the Asian genus *Macaranga* (Euphorbiaceae), up to four independent origins of ant–plant associations and numerous reversals are supported through a molecular phylogenetic analysis (Blattner *et al*. 2001, Davies *et al*. 2001). In addition, several species of *Macaranga* exhibit traits previously thought to be intermediate in the evolution of ant mutualisms (Fiala and Maschwitz 1992a). However, these species appear to represent a distinct clade, separate from other myrmecophytes, with an independent origin of myrmecophytism rather than evolutionary intermediates (Blattner *et al*. 2001, Davies *et al*. 2001). Phylogenetic studies of ants have also been used to study ant–plant associations with similarly diverse results (Ward 1991, 1999, Ayala *et al*. 1996, Chenuil and McKey 1996).

9.6.1. Origins and Evolutionary Trends

The somewhat unresolved nature of portions of the ITS tree (Figs. 9.5 and 9.6), and, to some degree, the absence of natural history information for a number of South American species do not permit us to make unambiguous hypotheses regarding the number of origins of obligate and facultative myrmecophytism in species of *Piper* sect. *Macrostachys*. Mapping known ant–plant associations onto the ITS trees, however, suggests that obligate associations evolved independently at least twice, but not more than four times (Fig. 9.6). Again, although the ML (= Bayesian) tree is presented in this paper (Fig. 9.6), none of the MP trees contradict the hypotheses presented here (see Fig. 9.6). It is possible to detect reliably the presence of obligate ant–plant associations from herbarium specimens owing to the presence of persistent, conspicuous pearl bodies in the petiole chambers, and, in most specimens, the remains of numerous ants or ant parts. Examination of herbarium specimens does not suggest the presence of any obligately myrmecophytic species in sect. *Macrostachys* additional to those included in this study.

The obligate myrmecophyte *P. cenocladum*, placed in clade E (Fig. 9.5) is isolated from the other four species of obligate myrmecophytes and, thus, presumably represents an independent origin. Three of the remaining myrmecophytes are placed in clade D. Two equally parsimonious reconstructions of obligate myrmecophytism are possible in this clade (Fig. 9.6). One reconstruction has a single gain of obligate associations with two independent and complete losses. Alternately, obligate associations could have evolved twice independently within this clade: once in the *P. obliquum/P. sagittifolium* clade (0.96 PP, 96% BS) and again in *P. calcariformis*.

The position of the obligate myrmecophyte, *P. fimbriulatum*, is unresolved. Thus it is unclear whether this species represents an additional origin of obligate myrmecophytism. *Piper fimbriulatum* resembles the rest of the species in clade D (Fig. 9.5) in having long, recurved stigma lobes, but differs markedly in leaf texture and absence of a style (i.e., sessile stigmas). Additional species and more informative markers are required to further evaluate the hypotheses proposed here, and to resolve the ambiguities in our analyses.

Facultative associations are more ephemeral and inconspicuous than obligate associations. Often only one to several petioles on a given plant are occupied by ants (E. Tepe, pers. obs.). On the basis of species that we have actually observed in the field with resident ants, our analyses suggest that facultative associations evolved independently two to three times (Fig. 9.6). However, large sheathing petioles with persistent margins are typical of a majority of *Macrostachys* species, and it is likely that the petioles of many more species than we have observed in the field form closed shelters and are therefore possibly inhabited by ants. Detection of facultative myrmecophytes in herbarium specimens is more ambiguous than it is for obligate associations. The degree of petiole closure is not preserved in dried specimens, but petiole morphology can suggest that facultative associations are likely to occur in a given species. Examination of herbarium specimens has led us to predict that the phenomenon of facultative associations is probably much more common and widespread than we have observed in the field. When predicted facultative associations are mapped onto the tree, a single origin is supported within the *Macrostachys* clade (Fig. 9.6). Consequently, the hypothesis of the number of origins of facultative associations based only on plants observed *in situ* with resident ants is the most conservative, and it is undoubtedly an underestimation.

9.6.2. *Evolution of the Mutualism*

9.6.2a. *Obligate associations and hollow stems*

All obligate myrmecophytes in sect. *Macrostachys* are characterized by tightly closed petioles with pearl body production localized on the inner surface of the petiolar tube. Accordingly, the distribution of these two characters parallels obligate myrmecophytism (Fig. 9.6). Similarly, hollow stems are restricted to the plants with obligate associations. But because the stems never become hollow in *P. calcariformis*, a minimum of two and a maximum of three independent origins are required to explain the distribution of hollow stems, or stems that the ants are capable of or inclined to excavate, as the case may be. None of the facultative ant-plants studied thus far have hollow stems; however nonhomologous cavities are occasionally formed by stem-boring insects (E. Tepe, pers. obs.).

According to our current phylogenetic hypothesis based on the ITS data, the most parsimonious reconstructions of the evolution of obligate associations with ants is three to four independent gains, with up to two losses (Fig. 9.6). The number of gains depends on the placement of *P. fimbriulatum* and the resolution of the clade containing *P. obliquum*; the number of losses depends entirely upon the resolution of the latter. It seems almost certain, however, that the origin of myrmecophytism in *P. cenocladum* is independent of that for the *P. obliquum* clade (Fig. 9.5, clade E).

The hypothesis that obligate associations evolved twice in Clade D (Fig. 9.5), once for *P. obliquum/P. sagittifolium*, and independently in *P. calcariformis*, is intriguing because *P. calcariformis* is unique among the obligate ant-plants in that it consistently has solid stems (Fig. 9.3). The stem anatomy of *P. calcariformis* is different from the other four obligate myrmecophytes in that the medulary vascular bundles are scattered throughout the pith (which is the most common arrangement among species of sect. *Macrostachys*). In contrast, the medulary bundles of the obligate species with hollow stems are arranged in a second ring just interior to the primary ring (Fig. 9.2(C)). In these plants, the pith is reportedly excavated by the resident ants (Risch *et al.* 1977, Letourneau 1998, Dyer and Letourneau 1999). It is unknown whether bundles throughout the pith of *P. calcariformis* and the rest of the large-stemmed species of *Macrostachys* (Fig. 9.5) would preclude excavation by ants, but the bundles of the other four obligate myrmecophyte species with hollow stems do not extend into the area of pith that becomes hollow (Fig. 9.2). Alternatively, if the obligate myrmecophytes in the *P. obliquum* clade (Fig. 9.5, clade D) are the result of a single origin, the loss of the associations in *P. gibbosum* and *P. hebetifolium* is not surprising, since neither of these species is morphologically suited to support ant residents; both of these species have small petioles that do not close tightly, and stems that are more slender than any of its close relatives (Fig. 9.5). Furthermore, the petiole margins of *P. gibbosum* are caducous, leaving behind a broadly U-shaped petiole (Fig. 9.4(G)). It appears that the slender stem morphology is derived in these two species, perhaps in response to their mid- to high elevational habitats. It is possible that the mutualism between *Piper* and *Pheidole bicornis* is not as stable at higher elevations as at lower elevations. Myrmecophytic species of *Cecropia* are more frequently found with ant inhabitants at lower elevations, and increasingly less so as altitude increases (Wheeler 1942, Janzen 1973). It is possible that the same phenomenon is responsible for the lack or loss of associations in *P. gibbosum* and *P. hebetifolium*.

Multiple origins and losses of obligate myrmecophytism appear to be common in the ant–plant associations that have been studied in other genera thus far. For example, two to four origins, with numerous losses, are supported in *Macaranga* (Euphorbiaceae; Blattner *et al.* 2001, Davies *et al.* 2001), and a minimum of two gains and one loss are supported in *Tococa* (Melastomataceae; Michelangeli 2000, also see Davidson and McKey 1993). Davies *et al.* (2001) proposed that multiple origins of myrmecophytism in *Macaranga* may be the result of the combination of certain morphological traits and a specific ecological and biogeographical setting. For example, the ancestors of the Malesian species of *Macaranga* that have given rise to myrmecophytes appear to have had large-diameter stems with soft pith and food bodies, and they occurred in a climate that allowed uninterrupted food body production (Davies *et al.* 2001). In other words, it appears that myrmecophytism has most likely evolved in plants that had a morphological predisposition for supporting such associations, and that were located in constant, tropical environments. According to this hypothesis, the combination of such characters as large, sheathing petioles, large-diameter stems, possibly widespread pearl body production, and a climatologically constant habitat that also includes *Pheidole bicornis* have contributed to the development of ant–plant associations in *Piper*.

A number of studies have demonstrated that ant partners of obligate associations are rarely species-specific, and that no ant–plant association studied thus far is the result of parallel cladogenesis (Mitter and Brooks 1983, Ward 1991, Ayala *et al.* 1996, Chenuil and McKey 1996). The fact that a single ant species, *Ph. bicornis*, is associated with all five obligate species of sect. *Macrostachys*, excluding the possibility of cryptic species of ants, precludes the possibility that species-for-species coevolution between ant and plant species has taken place. In fact, that one ant species is obligately associated with several plant species is not surprising. Frequent host switching of an ant species among host plant species appears to be common among many groups of plant ants, namely *Azteca* (Ayala *et al.* 1996), *Crematogaster* (Blattner *et al.* 2001), and *Pseudomyrmex* (Ward 1991). Thus, it appears that multiple origins of obligate ant-plants, and host switching by obligate plant ants, is common and that *Ph. bicornis* is capable of switching between the different obligate myrmecophyte species in *Piper* as well.

The mutualism with ants in obligate myrmecophytes is not maintained by a single plant character, but rather a suite of characters that is implicated in the associations. As the number of concurrent characters increases among species (i.e., tightly closed petiole sheaths, obligate-type pearl bodies, restricted areas of pearl body production, hollow stems in four of the five species), it becomes increasingly unlikely that these characters evolve in parallel or through convergence. In order to test the possibility that this suite of characters, and thus obligate myrmecophytism, evolved only once, we constrained our analysis to force the monophyly of the obligate myrmecophytes. The MP analysis produced the shortest trees, which were 11 steps longer than the shortest tree in our unconstrained analysis (125 vs. 114 steps). The ML analysis resulted in trees with likelihood values significantly higher than the most likely tree's from the unconstrained analysis when they were compared with the Shimodaira–Hasegawa test ($P=0.001$). When the ML analysis was constrained to force the obligates into a monophyletic clade with the facultative myrmecophytes as a paraphyletic clade basal to them, so as to suggest that the facultatives are evolutionary transitions between species without associations and with obligate associations, the resulting trees were also longer (MP: 129 vs. 114 steps, ML: $P = 0.002$). Thus, according to our current analyses, neither the obligate nor the facultative myrmecophytes in *Macrostachys* can be explained by a single evolutionary origin, and the facultative myrmecophytes are not transitional between obligate myrmecophytes and species that lack associations with ants.

9.6.2b. Petiolar domatia and facultative associations

Given that petioles form the primary domatia in *Piper* ant-plants, evolution of facultative myrmecophytism cannot be discussed separately from petiolar morphology. In fact, no close, species-specific relationship has developed in any ant–plant system studied thus far that only provides food for the ant partners, but no shelter (Fiala and Maschwitz 1992b). The petiole cavity of the obligate species is tightly closed throughout its length, and the petiole margins are pressed tightly to the stem such that little, if any, water running down the stem enters the cavity. This morphology is exceptionally constant among the five species of obligate ant-plants in *Piper*. The degree of petiole closure of the facultative myrmecophytes is more variable, but several petioles on a given plant are often closed enough so as to provide sufficient shelter for ant colonies.

The ITS phylogeny supports two to three independent origins of facultative myrmecophytism, based solely on the species that we have observed in the field with ants nesting in the petioles. However, we have only had the opportunity to study 14 species of *Macrostachys* in the field (Table 9.1). On the basis of the examination of herbarium specimens of species that we have not studied *in vivo*, we have observed that most species of sect. *Macrostachys* have large, sheathing petioles; consistent with the observed correlation of plant morphology and ant occupancy of species that we have observed in the field, we believe that facultative mutualisms are much more taxonomically and geographically widespread than we have reported here. If these potentially facultative ant-plants are mapped onto the phylogeny, then a single origin of facultative associations near the base of sect. *Macrostachys* is supported, with several independent losses, and with the obligate myrmecophytes derived from the facultative species (Fig. 9.6). Under this scenario, facultative myrmecophytes may represent evolutionary precursors to the obligate myrmecophytes. In fact, Risch *et al.* (1977) suggested that the mutualism between ants and *Piper* might have originated with the evolution of large, sheathing petioles. This suggestion and our predictive tree corroborate the findings of Fiala and Maschwitz (1992b) that domatia are the most important plant trait for the development of myrmecophytism. However, additional data and taxa are needed before this hypothesis can be more fully tested.

The ant genera that have been found nesting in the petioles of *P. biseriatum* and *P. imperiale* (e.g., *Crematogaster* spp., *Solenopsis* spp., *Wassmannia* spp., and other species of *Pheidole*) are opportunistically nesting, arboreal ants (Hölldobler and Wilson 1990, Orivel and Dejean 1999). These ants apparently nest in petioles of *Piper* species whenever they encounter one that provides sufficient protection from the environment.

Arboreal ants occasionally nest in cavities formed by other stem-boring insects in stems, petioles, and even the leaf midveins (E. Tepe, pers. obs.). Ward (1991) noted that a number of arboreal pseudomyrmecine ants have a tendency to nest in cavities in living plant parts, as opposed to dead, hollow twigs, as is typical of most opportunistic, arboreal pseudomyrmecines. He suggested that ant species that nest in living plant parts might lend insights into the evolution of obligate ant–plant associations (Ward 1991). However, phylogenetic studies of *Pheidole* are currently unavailable that would allow us to determine the nesting habits of species related to *Ph. bicornis*. Fiala *et al.* (1994) found that the presence of facultative ants in *Macaranga* can dramatically reduce damage by herbivores, and therefore may be important in driving plants toward more complex and mutually beneficial associations. This is likely the case in *Piper* as well.

9.6.2c. Pearl Bodies

Fischer *et al.* (2002) studied the chemical composition of pearl bodies of the four hollow-stemmed species of obligate myrmecophytes in *Piper* and found that, with the exception of slightly different levels of soluble carbohydrates and proteinaceous nitrogen in *P. sagittifolium*, pearl body composition did not vary significantly between species. This similarity in pearl body composition could be explained by common ancestry, but may also be explained by selective pressures exerted by the nutritional requirements of the ants. The ants derive the majority of their sustenance from the pearl bodies (Fischer *et al.* 2002). Food bodies are undeniably important, but in *Piper*, as in *Macaranga*, they appear to be second to domatia as the most important factor in the development of obligate ant–plant associations (Fiala and Maschwitz 1992b).

9.6.2d. Origin of ant-associated plant structures

In all plant genera that have obligate mutualisms with ants, very few, if any, of the plant parts implicated in the associations evolved completely *de novo*. With the possible exceptions of the Beltian bodies in *Acacia* and the collagen-containing Müllerian bodies of *Cecropia*, all ant-associated plant traits are modifications of preexisting structures (Janzen 1966, Rickson 1973). In *Piper*, the petioles of the obligate species are not fundamentally different from those of many species of sect. *Macrostachys*, except that they are more tightly folded and more consistently closed. In fact, Fiala and Maschwitz (1992b) noted that only *Macaranga* species with a predisposition for domatia developed into obligate myrmecophytes, and this appears to hold true for *Piper* as well. Stem anatomy of hollow-stemmed species appears to differ in that the medulary vascular bundles do not extend as far into the pith as in species with solid stems (Fig. 9.2). The arrangement of vascular bundles is novel, but again, are no more than modifications of preexisting structures.

Although no pearl bodies were found in the petioles of the facultative ant-plant species, we have frequently observed structures resembling pearl bodies on leaf and young shoot surfaces of *P. aduncum*, *P. nigrum*, and *P. tuberculatum* growing in greenhouses and even on leaves of *P. auritum* for sale in a Mexican market in Nashville, TN (E. Tepe, pers. obs.). Furthermore, they have been observed on young shoots and leaves on a number of *Piper* species in the field (L. Dyer, pers. comm.). It is possible that these structures function as generalized ant attractants in some species of *Piper*, in lieu of extrafloral nectaries, but may not have been reported from plants in the field because they are removed continuously by ants and leave no macroscopically visible trace. The difficulty of detecting food bodies on exposed plant surfaces has also been reported in *Macaranga* (Fiala and Maschwitz 1992b); however, microscopically visible traces of food bodies have been observed (Hatada *et al.* 2001). These extrapetiolar bodies have a somewhat different appearance than those found in the petioles of obligate myrmecophytes in that they are recognizably larger and more translucent. However, they also contain lipids and proteins as do the pearl bodies of obligate myrmecophytes (Sudan IV and Bromphenol Blue spot staining respectively; methods from Baker and Baker 1975), and around 2% carbohydrates when freshly extracted contents were measured with a refractometer (E. Tepe, unpubl. data; method from Kearns and Inouye 1993). If, in fact, these bodies are homologous with the pearl bodies found in the petioles of the obligate species, then pearl bodies, which are so important to the maintenance of ant–plant mutualisms (Fiala and Maschwitz 1992b), are also preexisting structures modified only in size and location.

9.7. CONCLUSIONS

The ITS phylogeny presented here further supports the monophyly of *Piper* sect. *Macrostachys*. Several ambiguities remain, however, due to a lack of variation in ITS. Further examination of the systematic relationships is currently under way, with more variable molecular markers and the addition of the remaining species of sect. *Macrostachys*. Examination of William Burger's hypotheses regarding evolutionary trends and relationships in sect. *Macrostachys* using our phylogenetic analyses reveal that his ideas are astonishingly insightful and are, for the most part, supported by our data. For example, although the *P. obliquum* complex and its allies are not monophyletic, they do represent all of the Costa Rican species of *Macrostachys*. Furthermore, the affinities that he suggested between many species are supported. Burger appeared to be misled by several homoplasious morphological characters, however, in his proposed affinities of *P. sagittifolium* to the rest of the *P. obliquum* complex. He believed that *P. sagittifolium* was rather isolated and "primitive" among Costa Rican *Piper* species. Instead, it appears to be nested well within *Macrostachys* and has retained plesiomorphic characters. Obligate myrmecophytism appears to have evolved independently two to four times. Observed facultative myrmecophytes have evolved one to three times, but the association is predicted to be much more widespread and common than we have observed. The introduction of DNA sequencing and phylogenetic analysis has provided additional data and new insights into relationships and evolutionary patterns, and allows us to revisit previously proposed hypotheses with a renewed perspective.

APPENDIX 9.1

DNA was isolated from silica gel–dried and herbarium leaf material for species of sect. *Macrostachys* and representative outgroup species from other subgenera, using a modified mini-prep CTAB procedure (Doyle and Doyle 1987). DNA was PCR-amplified for the internal transcribed spacers (ITS) of nuclear ribosomal DNA using the published primers of Blattner (1999). PCR products for ITS were sequenced on an ABI 310 or ABI 3100 automated DNA sequencer using ET terminator chemistry, or downloaded from GenBank for four species (Table 9.1). Sequences analyzed included sequences of ITS 1 and 2 and the 5.8s nrDNA. The extreme ends of ITS 1 and 2 were excluded because of dubious sequence quality in some accessions. The aligned matrix was 551 base pairs (bp) long with raw sequences ranging from 525bp in *P. spoliatum* to 538bp in *P. pseudonobile*. Within *Piper* sect. *Macrostachys*, sequences ranged from 93.1% similar between *P. arboreum* and *P. pseudonobile* to 100% similar between *P. cenocladum* and *P. subglabribracteatum*, with a mean of 97.5% similarity. In the aligned matrix, 134 characters were variable and 54 were parsimony informative. Thirteen gaps were coded (see below), seven of which were parsimony informative.

Sequence data were aligned using Clustall W (Thompson *et al.* 1996), and analyzed with PAUP* 4.0 (Swofford 2001) for Parsimony and Maximum Likelihood (ML) analyses and MrBayes (Huelsenbeck and Ronquist 2001) for Bayesian analyses to construct phylogenies of sect. *Macrostachys*. For the ML and Bayesian analyses, the appropriate model parameters were selected using a Hierarchical Likelihood Ratio Test (HLRT) in Modeltest (Posada and Crandall 1988). The model that best fit the data corresponded to the K80 (K2P) +G model. Gaps were treated as missing data and coded as separate presence/absence characters following the guidelines of Simmons and Ochoterena (2000) and analyzed with Maximum Parsimony. Trees were rooted with outgroups from outside Sect. *Macrostachys* using Fitch Parsimony. Maximum Likelihood and Maximum Parsimony analyses were run employing the following options: 100 random addition sequences, tree bisection reconnection (TBR), MULTREES in effect (Swofford 2001). Bayesian trees were generated using MrBayes (Huelsenbeck and Ronquist 2001), applying the same model used for the ML analysis. The analysis was run for 1,000,000 generations using four Markov Chain Monte Carlo chains and randomly generated starting trees. Trees were sampled every 100 generations, resulting in 10,000 saved trees. To avoid artifacts from the nature of the analysis, the first 2,500 trees were discarded to account for burn-in, i.e., the generations required for the analysis to reach optimality. To retrieve a single tree and the posterior probability values, the remaining 7,500 trees were used to construct a majority rule consensus tree. Branch support was evaluated using bootstrap values (Felsenstein 1985) and Bayesian posterior probabilities (Larget and Simon 1999, Huelsenbeck and Ronquist 2001). Bootstrap support for recovered nodes was estimated from analysis of 1,000 pseudoreplicate data sets using maximum parsimony. All polytomies are considered soft polytomies.

Alternate topologies were evaluated by constraining analyses to correspond to several hypotheses (see text). Because the null hypothesis of the Shimodaira–Hasegawa test is that all trees are equally good explanations of the data, this test is most appropriate for comparing alternate tree topologies (Shimodaira and Hasegawa 1999, Goldman *et al.* 2000). The log-likelihood scores of these constraint analyses were compared to the score of the unconstrained tree using the Shimodaira–Hasegawa test option in PAUP* 4.0 using RELL bootstrap with 1,000 pseudoreplicates and the same model parameters as above. MacClade 4.0 (Maddison and Maddison 2000) was used to map the ant-related traits onto the molecular phylogenies (Cunningham 2001, Omland 2001).

The morphological and anatomical data are based on observations and collections of 79 populations of all of the 15 Costa Rican species of *Piper* sect. *Macrostachys*. In addition, 1,067 herbarium specimens from throughout the range of sect. *Macrostachys* were examined for the presence of hollow stems, ants, pearl bodies, and petiole morphology. Micrographs are hand sections of FAA-fixed stems and petioles collected from living specimens in the field. Unless otherwise indicated, all stem sections are taken from the third internode, and the petiole sections are taken from the midpoint between the stem and the end of the sheathing margins. Sections were stained with Toluidine Blue and photographed with a SPOT High Resolution Digital Camera through a Nikon SMZ-2T dissecting microscope.

REFERENCES

Alonso, L. E. (1998). Spatial and temporal variation in the ant occupants of a facultative ant plant. *Biotropica* 30:201–213.
Armbruster, W. S. (1992). Phylogeny and the evolution of plant–animal interactions. *Bioscience* 42:12–20.
Armbruster, W. S. (1993). Evolution of plant pollination systems: Hypotheses and tests with the Neotropical vine *Dalechampia*. *Evolution* 47:1480–1505.

Armbruster, W. S. (1994). Early evolution of *Dalechampia* (Euphorbiaceae): Insights from phylogeny, biogeography, and comparative ecology. *Annals of the Missouri Botanical Garden* 81:302–316.

Ayala, F. J., Wetterer, J. K., Longino, J. T., and Hartl, D. L. (1996). Molecular phylogeny of *Azteca* ants (Hymenopetra: Formicidae) and the colonization of *Cecropia* trees. *Molecular Phylogenetics and Evolution* 5:423–428.

Baker, H. G., and Baker, I. (1975). Studies of nectar constitution and pollinator–plant coevolution. In: Gilbert, L. E., and Raven, P. H. (eds.), *Animal and Plant Coevolution*. University of Texas Press, Austin, pp. 100–140.

Blattner, F. R. (1999). Direct amplification of the entire ITS region from poorly preserved plant material using recombinant PCR. *BioTechniques* 27:1180–1186.

Blattner, F. R., Weising, K., Bänfer, G., Maschwitz, U., and Fiala, B. (2001). Molecular analysis of phylogenetic relationships among myrmecophytic *Macaranga* species (Euphorbiaceae). *Molecular Phylogenetics and Evolution* 19:331–344.

Bornstein, A. J. (1989). Taxonomic studies in the Piperaceae–1. The pedicellate pipers of Mexico and Central America (*Piper* subgen. *Arctottonia*). *Journal of the Arnold Arboretum* 70:1–55.

Brouat, C., Gielly, L., and McKey, D. (2001). Phylogenetic relationships in the genus *Leonardoxa* (Leguminosae: Caesalpinioideae) inferred from chloroplast *trnL* intron and *trnL–trnF* intergenic spacer sequences. *American Journal of Botany* 88:143–149.

Burger, W. C. (1971). Piperaceae. In: Burger, W. C. (ed.), *Flora Costaricensis. Fieldiana, Bot.* 35:1–227.

Burger, W. C. (1972). Evolutionary trends in the Central American species of *Piper* (Piperaceae). *Brittonia* 24:356–362.

Byrne, M. M. (1994). Ecology of twig-dwelling ants in a wet lowland tropical forest. *Biotropica* 26:61–72.

Callejas, R. (1986). *Taxonomic Revision of* Piper *subgenus* Ottonia *(Piperaceae)*. Ph.D. dissertation, City University of New York.

Callejas, R. (2001). Piperaceae. In: Stevens, W. D., Ulloa, C., Pool, A., and Montiel, O. M. (eds.), *Flora de Nicaragua*, t. 1. Missouri Botanical Garden Press, St. Louis, Missouri, pp. 1928–1984.

Chenuil, A., and McKey, D. (1996). Molecular phylogenetic study of a myrmecophyte symbiosis: Did *Leonardoxa*/ant associations diversify via cospeciation? *Molecular Phylogenetics and Evolution* 6:270–286.

Cunningham, C. W. (2001). Some limitations of ancestral character–state reconstruction when testing evolutionary hypotheses. *Systematic Biology* 48:665–674.

Davies, S. J., Lum, S. K. Y., Chan, R., and Wang, L. K. (2001). Evolution of myrmecophytism in western Malesian *Macaranga* (Euphorbiaceae). *Evolution* 55:1542–1559.

Davidson, D. W., and McKey, D. (1993). The evolutionary ecology of symbiotic ant–plant relationships. *J. Hymen. Res.* 2:13–83.

de Candolle, C. (1869). Piperaceae. In: de Candolle, A.(ed.), *Prodromus* 16:235–471.

de Candolle, C. (1923). Piperacearum clavis analytica. *Candollea* 1:65–415.

Dodd, M. E., Silvertown, J., and Chase, M. W. (1999). Phylogenetic analysis of trait evolution and species diversity variation among angiosperm families. *Evolution* 53:732–744.

Douady, C. J., Delsuc, F., Boucher, Y., Doolittle, W. F., and Douzery, E. J. P. (2003). Comparison of Bayesian and maximum likelihood bootstrap measures of phylogenetic reliability. *Molecular Biology and Evolution* 20:248–254.

Doyle, J. J., and Doyle, J. L. (1987). A rapid DNA isolation procedure for small amounts of fresh leaf tissue. *Phytochemical Bulletin* 19:11–15.

Dyer, L. A., and Letourneau, D. K. (1999). Relative strengths of top–down and bottom–up forces in a tropical forest community. *Oecologia* 119:265–274.

Felsenstein, J. (1985). Confidence limits on phylogenies: An approach using the bootstrap. *Evolution* 39:783–791.

Fiala, B., and Maschwitz, U. (1992a). Domatia as most important adaptations in the evolution of myrmecophytes in the paleotropical tree genus *Macaranga* (Euphorbiaceae). *Plant Systematics and Evolution* 180:53–64.

Fiala, B., and Maschwitz, U. (1992b). Food bodies and their significance for obligate ant-association in the tree genus *Macaranga* (Euphorbiaceae). *Botanical Journal of the Linnean Society* 110:61–75.

Fiala, B., Grunsky, H., Maschwitz, U., and Linsenmair, K.E. (1994). Diversity of ant–plant interactions: Protective efficacy in *Macaranga* species with different degrees of ant association. *Oecologia* 97:186–192.

Fischer, R. C., Richter, A., Wanek, W., and Mayer, V. (2002). Plants feed ants: Food bodies of myrmecophytic *Piper* and their significance for the interaction with *Pheidole bicornis* ants. *Oecologia* 133:186–192.

Gleason, H. A. (1931). The relationships of certain myrmecophilous melastomes. *Bulletin of the Torrey Botanical Club* 58:73–85.

Goldman, N., Anderson, J. P., and Rodrigo, A. G. (2000). Likelihood-based tests of topologies in phylogenetics. *Systematic Biology* 49:652–670.

Hatada, A., Ishiguro, S., Itioka, T, and Kawano, S. (2001). Myrmecosymbiosis in the Bornean *Macaranga* species with special reference to food bodies (Beccarian bodies) and extrafloral nectaries. *Plant Species Biology* 16:241–246.

Hölldobler, B., and Wilson, E. O. (1990). *The Ants*. Springer-Verlag, Berlin.

Howard, R. A. (1988). Piperaceae. In: Howard, R. A. (ed.), *Flora of the Lesser Antilles: Leeward and Windward Islands*. Arnold Arboretum, Harvard University, Jamaica Plains, MA, pp. 10–32.

Huelsenbeck, J., and Ronquist, F. (2001). MrBayes: Bayesian inference of phylogeny. *Bioinformatics* 17:754–755.

Janzen, D. H. (1966). Coevolution of mutualism between ants and acacias in Central America. *Evolution* 20:249–275.

Janzen, D. H. (1973). Dissolution of mutualism between *Cecropia* and its *Azteca* ants. *Biotropica* 5:15–28.

Jaramillo, M. A., and Manos, P. S. (2001). Phylogeny and patterns of floral diversity in the genus *Piper* (Piperaceae). *American Journal of Botany* 88:706–716.

Kearns, C. A., and Inouye, D. W. (1993). *Techniques for Pollination Biologists*. University Press of Colorado, Niwot.

Kunth, K. S. (1839). Bemerkungen über die familie der Piperaceen. *Linnaea* 13:562–726.

Larget, B., and Simon, D. (1999). Markov chain Monte Carlo algorithms for the Bayesian analysis of phylogenetic trees. *Molecular Biology and Evolution* 16:750–759.

Letourneau, D. K. (1983). Passive aggression: An alternative hypothesis for the *Piper–Pheidole* association. *Oecologia* 60:122–126.

Letourneau, D. K. (1998). Ants, stem-borers, and fungal pathogens: Experimental tests of a fitness advantage in *Piper* ant-plants. *Ecology* 79:593–603.

Maddison, W., and Maddison, D. (2000). *MacClade Version 4.0: Analysis of Phylogeny and Character Evolution*. Sinauer, Sunderland, Massachusetts.

McKey, D. (1984). Interaction of the ant-plant *Leonardoxa africana* (Caesalpiniaceae) with its obligate inhabitants in a rainforest in Cameroon. *Biotropica* 16:81–99.

McKey, D. (1991). Phylogenetic analysis of the evolution of a mutualism: *Leonardoxa* (Caesalpinaceae) and its ant associates. In: Huxley, C. R., and Cutler, D. F. (eds.), *Ant–Plant Interactions*. Oxford University Press, Oxford.

McKey, D. (2000). *Leonardoxa africana* (Leguminosae: Caesalpinioideae): A complex of mostly allopatric subspecies. *Adansonia, sér* 3, 22:71–109.

Michelangeli, F. A. (2000). A cladistic analysis of the genus *Tococa* (Melastomataceae) based on morphological data. *Systematic Botany* 25:211–234.

Miquel, F. A. (1843–1844). *Systema Piperacearum*. H. A. Kramer, Rotterdam.

Mitter, C., and Brooks, D. R. (1983). Phylogenetic aspects of coevolution. In: Futuyma, D. J., and Slatkin, M. (eds.), *Coevolution*. Sinauer, Sunderland, Massachusetts, pp. 65–98.

Morawetz, W., Henzl, M., and Wallnöfer, B. (1992). Tree killing by herbicide producing ants for the establishment of pure *Tococa occidentalis* populations in the Peruvian Amazon. *Biodiversity and Conservation* 1:19–33.

Omland, K. E. (2001). Assumptions and challenges of ancestral state reconstructions. *Systematic Biology* 48:604–611.

Orivel, J., and Dejean, A. (1999). L'adaptation à la vie arboricole chez les fourmis. *Année Biól.* 38:131–148.

Posada, D., and Crandall, K. (1988). Model test: Testing the model of DNA substitution. *Bioinformatics* 14:817.

Rickson, F. R. (1973). Review of glycogen plastid differentiation in Mullerian body cells of *Cecropia peltata*. *Annals of the New York Academy of Sciences* 210:104–114.

Risch, S. J. (1982). How *Pheidole* ants help *Piper* plants. *Brenesia* 19/20:545–548.

Risch, S., McClure, M., Vandermeer, J., and Waltz, S. (1977). Mutualism between three species of tropical *Piper* (Piperaceae) and their ant inhabitants. *American Midlands Naturalist* 98:433–444.

Risch, S. J., and Rickson, F. R. (1981). Mutualism in which ants must be present before plants produce food bodies. *Nature* 291:149–150.

Sasikumar, B., Chempakam, B., George, J. K., Remashree, A. B., Devasahayam, S., Dhamayanthi, K. P. M., Ravindran, P. N., and Peter, K. V. (1999). Characterization of two interspecific hybrids of *Piper*. *Journal of Horticultural Science and Biotechnology* 74:125–131.

Savolainen, V., Chase, M. W., Hoot, S. B., Morton, C. M., Soltis, D. E., Bayer, C., Fay, M. F., deBruin, A. Y., Sullivan, S., and Qiu, Y. (2000). Phylogenetics of flowering plants based on combined analysis of plastid *atpB* and *rbcL* gene sequences. *Systematic Biology* 49:306–362.

Shimodaira, H., and Hasegawa, M. (1999). Multiple comparisons of log-likelihoods with applications for phylogenetic inference. *Molecular and Biological Evolution* 16:1114–1116.

Simmons, M. P., and Ochoterena, H. (2000). Gaps as characters in sequence-based phylogenetic analyses. *Systematic Biology* 49:369–381.

Smith, A.C. (1975). The genus *Macropiper* (Piperaceae). *Journal of the Linnean Society of Botany* 71:1–38.

Standley, P. C., and Steyermark, J. A. (1952). Piperaceae. In: Standley, P. C., and Steyermark, J. A. (eds.), *Flora of Guatemala. Fieldiana, Botany* 24:228–37.

Steyermark, J. A. (1971). Notes on the genus *Sarcorhachis* Trel. (Piperaceae). *Pittieria* 3:29–38.

Steyermark, J. A. (1984). Flora of Venezuela. In: Steyermark, J. A. (ed.), *Flora of Venezuela*, vol. 2, part 2. Instituto Nacional de Parques, Ediciones Fundación, Caracas, Venezuela.

Swofford, D. L. (2001). *PAUP*. Phylogenetic Analysis Using Parsimony (*and Other Methods). Version 4.* Sinauer Associates, Sunderland, Massachusetts.

Tebbs, M. C. (1989). Revision of *Piper* (Piperaceae) in the New World: 1. Review of characters and taxonomy of *Piper* section *Macrostachys*. *Bulletin of the British Museum of Natural History (Botany)* 19:117–158.

Tebbs, M. C. (1990). Revision of *Piper* (Piperaceae) in the New World. 2. The taxonomy of *Piper* section *Churumayu*. *Bulletin of the British Museum of Natural History (Botany)* 20:193–236.

Tebbs, M. C. (1993). Revision of *Piper* (Piperaceae) in the New World. 3. The taxonomy of *Piper* section *Lepianthes* and *Radula*. *Bulletin of the British Museum of Natural History (Botany)* 23:1–50.

Thompson, J. D., Higgins, D. G., and Gibson, T. J. (1996). *Clustal W: Sequence Alignment Program, Version 1.6*. European Molecular Biology Laboratory, Heidelberg, Germany.

Trelease, W. (1929). The Piperaceae of Costa Rica. *Contributions of the U.S. National Herbarium* 26:115–226.

Trelease, W., and Yuncker, T. G. (1950). *The Piperaceae of Northern South America*. University of Illinois Press, Urbana.

Tucker, S. C., Douglas, A. W., and Liang, H. X. (1993). Utility of ontogenetic and conventional characters in determining phylogenetic relationships of Saururaceae and Piperaceae (Piperales). *Systematic Botany* 18:614–641.

Vasconcelos, H. L. (1991). Mutualism between *Miaeta guianensis* Aubl., a myrmecophytic melastome, and one of its ant inhabitants: Ant protection against insect herbivores. *Oecologia* 87:295–298.

Ward, P. S. (1991). Phylogenetic analysis of Pseudomyrmecine ants associated with domatia-bearing plants. In: Huxley, C. R., and Cutler, D.F. (eds.), *Ant–plant interactions*. Oxford University Press, Oxford, pp. 335–349.

Ward, P. S. (1999). Systematics, biogeography and host plant associations of the *Pseudomyrmex viduus* group (Hymenoptera: Formicidae), *Triplaris*- and *Tachigali*-inhabiting ants. *Zoological Journal of the Linnean Society* 126:451–540.

Wheeler, W. M. (1942). Studies of Neotropical ant plants and their ants. *Bulletin of the Museum of Comparative Zoology Harvard University* 90:1–262.

Whiffin, T. (1972). Observations on some upper Amazon formicarial Melastomataceae. *Sida* 5:33–41.

Wilcox, T. P., Zwickl, D. J., Heath, T. A., and Hillis, D. M. (2002). Phylogenetic relationships of the dwarf boas and a comparison of Bayesian and bootstrap measures of phylogenetic support. *Molecular and Phylogenetic Evolution* 25:361–371.

Yuncker, T. G. (1958). The Piperaceae—a family profile. *Brittonia* 10:1–7.

Yuncker, T. G. (1972). The Piperaceae of Brazil—I: *Piper*, groups I, II, III, IV. *Hoehnea* 2:19–366.

Yuncker, T. G. (1973). The Piperaceae of Brazil—II: *Piper*, group V; *Ottonia*; *Pothomorphe*; *Sarchrhachis*. *Hoehnea* 3:29–284.

Yuncker, T. G. (1974). The Piperaceae of Brazil—III: *Peperomia*; taxa of uncertain status. *Hoehnea* 4:71–413.

10
Current Perspectives on the Classification and Phylogenetics of the Genus *Piper* L.

M. Alejandra Jaramillo[1] *and Ricardo Callejas*[2]

[1]*Departamento de Bioquímica Médica, Universidade Federal do Rio de Janeiro, Rio de Janeiro, RJ, Brazil*

[2]*Departamento de Biología, Universidad de Antioquia, Apartado Aereo 1226, Medellin, Colombia*

10.1. INTRODUCTION

Piper is one of the most diverse genera among the basal lineages of angiosperms, and one of the most diverse genera in tropical wet forests around the world. *Piper* species are shrubs, climbers, and herbs abundant in the understory of tropical wet forests. Ecologically, *Piper* species are important structural components of the forest understory, especially in the Neotropics (Gentry 1990); they also make up a significant portion of the diet of frugivorous bats in the subfamilies Caroliinae and Sternodermatinae (Fleming 1981, 1985). A few species of *Piper* are known for their economic value. The most important of these is black pepper (*Piper nigrum* L.), the best known agricultural product in the genus. In addition, there are other species exploited by indigenous peoples in different parts of the geographic range of *Piper*: *P. betle* L., the "betle leaf" is chewed with the betle nut in Asia, whereas *P. methysticum* Forst.f., Kava Kava, is a traditional drink from the South Pacific that has been introduced to Europe and North America as an herbal medicine. Both of these species are used in traditional practices for their narcotic properties. Other species of *Piper* are used as condiments: *P. auritum* HBK, acuyo, is used in Mexico and *P. lolot* C.DC., lolot leaves, is used in Viet Nam. Many species of *Piper* are also used by traditional societies in tropical countries around the world for their antiinflammatory and analgesic properties; as such, they may have immense potential for the pharmaceutical industry (Ehringhaus 1997).

Piper is more diverse in the American tropics than in the Asian tropics (ca. 700 vs. 300 species, respectively); fewer numbers of species are found in the islands of the South Pacific (ca. 40 spp.) and in Africa (only two native species). They are abundant in low- and mid-elevation forests, rarely reaching up to 2,500 m in elevation in the tropical Andes. The centers of *Piper* species diversification include southeast Asia, southern México, the Andes, the Chocó, Amazonia and the Atlantic Forest of Brazil. Historically speaking, the geographic areas best studied for *Piper* taxonomy are Central America and the northern Andes. Most of the knowledge of the ecology of the genus has been derived from studies conducted in Central America for the past 30 years. Here we will present a short history of the classification of *Piper* and insights on the current knowledge of the phylogenetic relationships within Piperales and also within *Piper*.

10.2. CLASSIFICATION

The first classification systems proposed for the family Piperaceae were published almost simultaneously in the mid-19th century (Kunth 1839, Miquel 1843–1844). These treatments divided the family into 8 and 20 different genera, respectively, including species from the entire distribution of Piperaceae. Later, de Candolle proposed a new classification in which most of the genera recognized by Kunth and Miquel were included in two large genera: *Piper* and *Peperomia*, each with ~1,000 species and several sections (deCandolle 1866, 1923). DeCandolle subdivided *Piper* into 11 sections, many of them corresponding to groupings recognized by Miquel and Kunth at the genus level. These early treatments had the advantage of including species from the whole geographic range of the family, thus presenting a more integrated picture of the classification. The authors also provided infrageneric classifications for *Piper* (and also for *Peperomia*) using a combination of vegetative and reproductive characters.

10.2.1. Getting Cluttered

During the 20th century, the classification of *Piper* became rather chaotic as a result of three important developments: (1) treatments became fragmentary; (2) new genera were segregated from *Piper* on the basis of gross morphology; and (3) a large number of new species were described without clear subgeneric affinities. After the pioneering studies by Kunth, Miquel, and deCandolle, treatments of *Piper* became restricted to specific countries and regions. During the 1900s, monographs were produced for the following countries and/or regions: the Malay Peninsula, Panama, the Philippines, northern South America, New Guinea, the South Pacific Islands, Java, Costa Rica, Brazil, Taiwan, and Venezuela (Ridley 1924, Trelease 1927, Quisumbing 1930, Trelease and Yuncker 1950, Chew 1953, Backer and van den Brink 1963, Burger 1971, Yuncker 1972, 1973, Liu and Wang 1975, Smith 1975, Steyemark 1984). Despite the extensive geographical coverage, these treatments presented very fragmentary accounts of the taxonomy of the genus. In addition, only two of these localized treatments used the subgeneric arrangement of de Candolle (Ridley 1924, Quisumbing 1930); the majority of monographs did not use any infrageneric classification.

Beginning in 1927, several new genera were identified in the American tropics; these included *Trianaeopiper*, *Sarcorhachis*, *Arctottonia*, *Lindeniopiper*, and

Pleiostachyopiper (Trelease 1927, 1928, 1929, 1930, 1934, 1940). One new subgenus, *Penninervia*, was described in the Philippines (Quisumbing 1930). Diagnostic characters for these new genera included gross (and labile) morphological characters such as leaf venation or inflorescence shape and position (e.g., *Arctottonia, Pleyostachyopiper*, and *Trianaeopiper*). Several new species were described without specifying subgeneric affiliation. Two of the great specialists in Piperaceae, W. Trelease and T.G. Yuncker, suggested that determining the floral characters of *Piper* species (especially in herbarium specimens) was "difficult or impossible"; therefore, it did not seem to them to be practical to maintain the subdivisions on the basis of these characters (Trelease and Yuncker 1950). The small size of the flowers in *Piper* lead taxonomists to rely on gross morphological characters (as Trelease and Yuncker did), assuming uniformity in the reproductive characters or simply ignoring the variation present in flower, fruit, and seed morphology. Other vegetative characters, such as indumentum, prophyll morphology, and plant architecture, were also overlooked. Last and often not considered, the poor sampling of many areas, the very few and fragmentary collections available for many species, and the lack of knowledge on vegetative variation due to plant architecture contributed to a meaningless and poorly resolved taxonomy.

10.2.2. Getting Articulated

Thankfully, in the last 30 years, many efforts have been made toward developing a more comprehensive infrageneric classification for *Piper*. Botanists have revived the old segregates suggested by Kunth and Miquel at the genus level, and considered them as subgenera or sections within *Piper* (Callejas 1986, Bornstein 1989, Tebbs 1989, 1993). Early efforts to characterize the infrageneric relationships within *Piper* using morphological cladistics revealed extensive homoplasy; thus, it was not possible to provide resolution at the subgenus level (Callejas 1986). Given the challenge that the use of morphological characters has presented in *Piper*, molecular sequence data have been particularly useful in providing a larger number of characters for the classification of the genus and its segregates (Jaramillo and Manos 2001). The use of molecular tools (Jaramillo and Manos 2001; Chapter 9) paired with a close examination of morphological characters (Callejas 1986) has allowed for the reconstruction of phylogenetic relationships within the genus and a better understanding of the evolutionary history of the group.

10.3. PHYLOGENY

10.3.1. Phylogenetic Relationships of the Piperales: A Test of Piper's Monophyly

The circumscription of Piperales has changed in the last few decades. The traditional diagnosis of Piperales emphasized the lack of perianth and included the families Piperaceae, Saururaceae, and Chloranthaceae (Cronquist 1981). Molecular phylogenetic analyses of the basal lineages of angiosperms have suggested a new circumscription for the order (APG 1998, 2003). On the one hand, the perianth-less family Chloranthaceae has been removed from the Piperales on the basis of morphological (Endress 1987) and molecular evidence (Chase *et al.* 1993, Soltis *et al.* 2000, Nickrent *et al.* 2002). On the other hand, based on recent phylogenetic analyses, the order has been extended to include some families

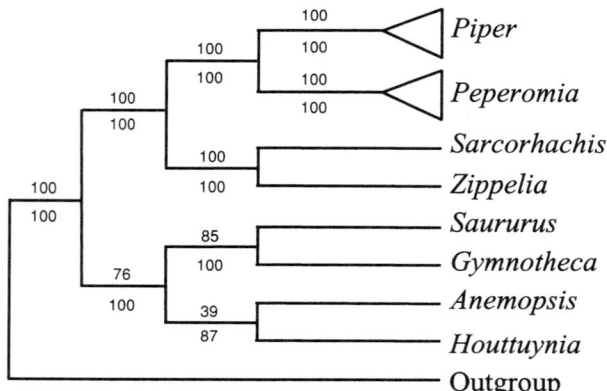

FIGURE 10.1. Summary of the phylogenetic relationships of the genera of perianth-less Piperales. Based on four genes phylogenetic analyses from two independent studies (rbcL, ATPB and 18S, Jaramillo et al. (submitted); and matK, Wanke, 2003). Numbers above branches are maximum likelihood bootstrap (BS) values from the Jaramillo et al. study (Jaramillo et al., in revision). Numbers below branches are posterior probabilities from Wanke's bayesian phylogenetic analysis (Wanke, 2003).

that exhibit a perianth, i.e., Aristolochiaceae, Hydnoraceae, and Lactoridaceae (Nickrent et al. 2002).

Two independent analyses of the perianth-less Piperales using chloroplast and nuclear DNA sequence data have helped to clarify the relationships among the genera in the order further (rbcL, 18S, and ATPB in Jaramillo et al., 2004; matK in Wanke 2003). These analyses included at least one species from each genus and several taxa for most genera. The resulting phylogenies had two well-supported clades, corresponding to the families Piperaceae and Saururaceae (Fig. 10.1). The pantropical family Piperaceae has over 2,000 species, whereas its temperate counterpart, Saururaceae, has only six species. Within the Piperaceae, the monophyly of *Piper* has never been questioned, and is well supported (see above), but ignored. The circumscription of the genus still appears unclear given that a large number of groups are continuously included or segregated. The phylogeny presented here clarifies, to a certain extent, the circumscription of *Piper*. It suggests that there are two well-supported clades in Piperaceae: /Pipereae, consisting of *Piper* and *Peperomia*, and /Zippeliae, consisting of *Zippelia* and *Sarcorhachis*. *Piper* and *Peperomia* are large genera with over 1,000 species each and pantropical distributions; *Zippelia* and *Sarcorhachis* include less than 10 taxa, and are less known than the more diverse genera. In the past, *Zippelia* and *Sarcorhachis* had been included in a broader circumscription of *Piper*. However, our molecular data substantiate their status as distinct clades (genera, if you wish) of the Piperaceae.

Zippelia is a monotypic genus distributed in Southern Asia. *Zippelia* plants are very different from Asian pipers in that they are small herbs with hermaphroditic flowers whereas most *Piper* species in the Paleotropics are dioecious climbers. Recently, a group of Chinese botanists led by Dr. Liang (Kunming Academy of Science, Beijing) have provided a major contribution to our knowledge of *Zippelia* with detailed morphological studies on flower development (Liang and Tucker 1995) and female embryogenesis (Lei et al. 2002).

Currently, *Zippelia* may be one of the best-known taxa in Piperaceae in these aspects, although we still need to learn more about the ecology of this group.

Sarcorhachis is distinguished by its climbing habit and by the way in which its fruits are embedded in the inflorescence rachis. The genus has about nine species of disjunct distribution in the Caribbean and Central and South America. It is a puzzling group that does not exhibit a clear relationship with any species of *Piper* in particular, and its current distribution suggests a relictual geography. Compared with the detailed knowledge on *Zippelia* morphology, little is known about the morphology and development of *Sarcorhachis*. Despite the fact that *Sarcorhachis* is common in perturbed forests, the plants only flower when they reach the highest portion of the canopy (10 m tall) making it difficult to collect the developing inflorescences for morphology studies. *Sarcorhachis* is being studied by Tatiana Arias (under the direction of R. Callejas, Universidad de Antioquia, Colombia). Hopefully this study will shed light into the evolution of its unique inflorescence type and its close relationship to *Zippelia*.

10.3.2. Infrageneric Relationships of Piper

Using molecular sequences of the fast-evolving ITS region (the internal transcribed spacer of the ribosomal nDNA) and a Bayesian approach for the phylogenetic analysis, a genus level phylogeny was produced to examine the relationships within *Piper* (see Appendix 10.1 for details on the methodology). An exemplar sampling (91 species in total) collected by the authors in tropical America and Asia, and a few taxa provided by collaborators (R. O. Gardner, C. Castro, and A. F. Oliveira) were included in this study. The resulting phylogeny (Fig. 10.2) suggests that within *Piper* an old vicariant event created three major clades of distinct geographic distribution: (a) American tropics, (b) Asian tropics, and (c) South Pacific Islands. There are several well-supported subclades (posterior probabilities $\geq 85\%$), including *Enckea, Ottonia, Radula, Macrostachys, Pothomorphe*, and *Macropiper*. These clades coincide with segregates that have been recognized since the earliest classifications of the genus (Kunth 1839, Miquel 1843–1844), and also with clades described in earlier papers by the authors (Callejas 1986, Jaramillo and Manos 2001). The above-mentioned monophyletic groups can be recognized by a mixture of reproductive and vegetative characters. Here we provide a characterization of the strongly supported clades.

10.3.2a. Neotropical Taxa

Enckea Kunth. /Enckea includes shrubs and treelets, with palmately veined leaves and hermaphroditic flowers loosely arranged in the inflorescence. Flowers can be pedicellate (previously recognized as the genus or subgenus *Arctottonia*) or not, with 4 (–6) stamens and 3–4 carpels. /Enckea is a group of approximately 40 species, with a geographic range that extends from Central America to southern South America. The clade has mainly diversified in Central America, where it is predominantly found in drier forests. The South American components of this group, i.e., *P. reticulatum, P. nicoyanum, P. laevigatum*, and *P. nudilimbum* (= *Pleiostachyopiper* Trel.) (with sessile flowers) are plants from mesic conditions. This clade was identified by Yuncker as a group of plants with palmately

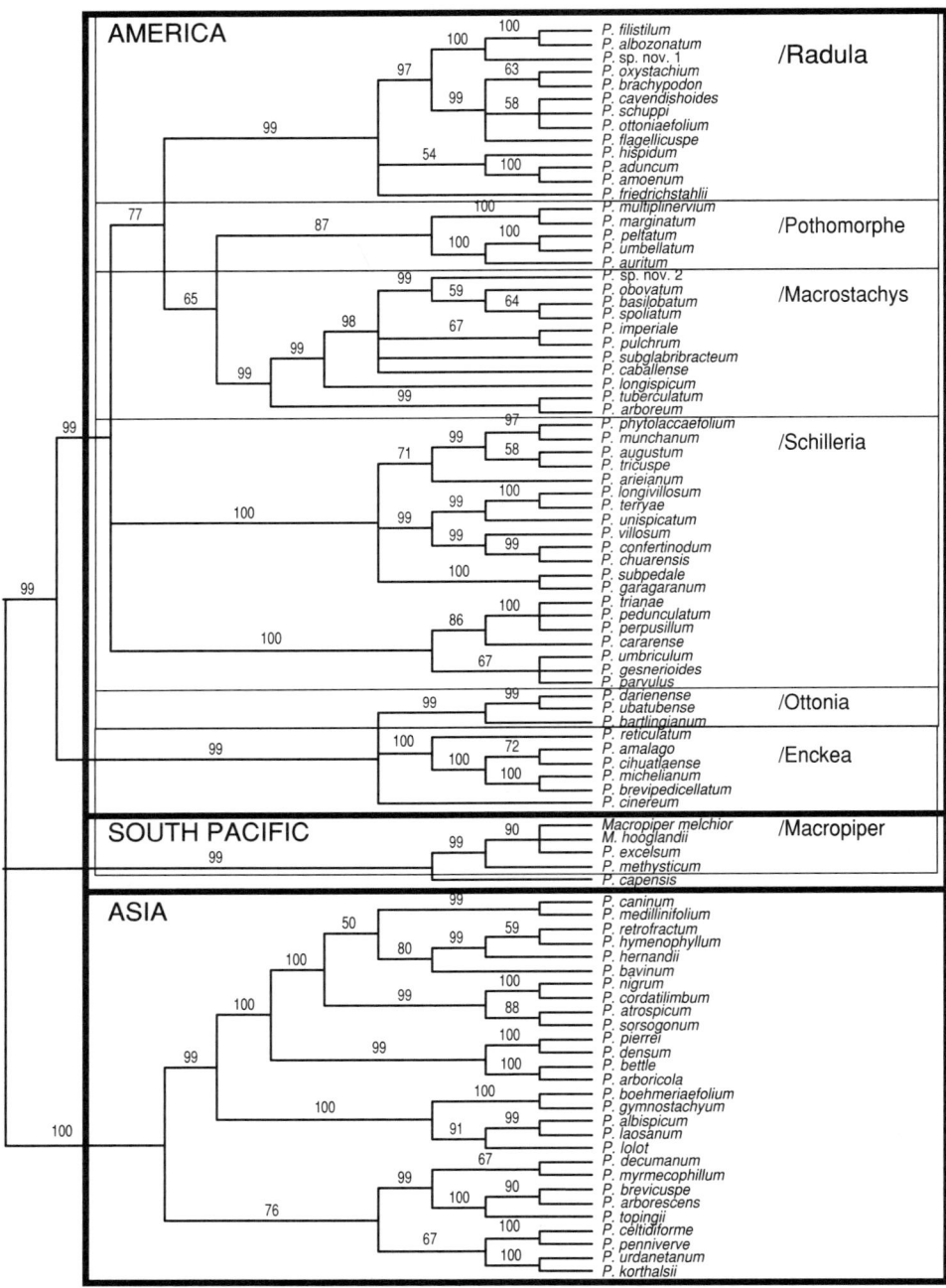

FIGURE 10.2. Phylogenetic relationships within the genus *Piper* based on ITS sequence data (topology is a 50% majority rule consensus). Numbers above branches are posterior probabilities derived from a bayesian phylogenetic analysis.

nerved leaves (Trelease and Yuncker 1950, Yuncker 1972), excluding *P. marginatum*. The pedicellate members of /Enckea, recognized as the genus or subgenus *Arctottonia*, were revised by A. Bornstein (Bornstein 1989); he is currently working on a revision of the whole /Enckea clade. *Arctottonia* has 14 species distributed west of the Sierra Madre Occidental and east of the Sierra Madre Oriental in Mexico. All of the species in *Arctottonia* exhibit pedicel elongation after fertilization; other than this character, they are morphologically very similar to all other members of /Enckea. *Arctottonia* species grow intermixed with deciduous trees and shrubs in the drier regions of Mexico and Central America. Little is known about their biology or their phylogenetic relationships. Bornstein suggested that there are two species assemblages within *Arctottonia*, i.e., one on western and the other or eastern Central America (Bornstein 1989). Rigorous testing of the nature of these assemblages will provide insights on the evolution of the genus *Piper* in the northernmost range of its distribution in the western hemisphere.

Ottonia Sprengel. /Ottonia includes small- and medium-sized shrubs, with pinnately veined leaves, hermaphroditic flowers loosely arranged in the inflorescence, flowers pedicellate or not (pedicel elongation occurs after fertilization) with 4 stamens and 4 carpels, and elliptic–globose fruit. /Ottonia was recently revised by one of us (Callejas 1986) as a subgenus of a broadly circumscribed *Piper*. The treatment included 21 species, most of them occurring in the Amazon and the Atlantic Forest, with only one species west of the Andes (i.e., *P. darienense*). We reanalyzed Callejas' data set using the current version of PAUP* (see details in Appendix 10.1). Among the 55 characters used, only 28 were parsimony informative; the analysis produced 1,175 most parsimonious trees of 50 steps length, CI = 0.56 and RI = 0.74 (Fig. 10.3). The resulting phylogeny is very similar to the tree obtained by Callejas (1986). Although there is little resolution in general, there are two well-supported clades. Clade A has three Amazonian species: *P. alatabaccum*, *P. francovilleanum*, and *P. bartlinguianum*. Clade B includes *P. ubatubense* and *P. scutifolium*, both species from the Atlantic Forest. The strong support for these clades suggests that geographically differentiated assemblages form monophyletic groups. However, a pair of Amazonia–Atlantic Forest species, *P. ottonoides* and *P. klotzschianum*, form a clade with marginal bootstrap support as well. The support for this clade challenges the conclusion derived from clades A and B. Furthermore, given the poor resolution of the present phylogeny, there is no evidence to test the number of origins of pedicellate flowers, a character used as diagnostic of /Ottonia in the past. A molecular phylogeny will certainly provide more evidence to test the relationships within the group and, thus, hypotheses about the character evolution further. Special attention needs to be given to the diversification of /Ottonia in the Atlantic Forest, where 75% of the species (almost restricted to the south of Bahia) occur, and the group's relationship to the Amazonian taxa. Members of /Ottonia are of particular interest to the pharmaceutical industry; more than half of the species in the group are known as jaborandi, a Tupi–Guarani name that refers to its use for treating poisonous snake bites. /Ottonia is a very interesting group given its geographic distribution, flower morphology, and great potential for pharmacology; however, its basic biology is still poorly known.

Radula Miquel. /Radula species are, in general, medium-size shrubs, with pinnately veined, membranaceous leaves and bisexual flowers that are tightly arranged in erect or arched inflorescences. Flowers have four stamens and three carpels. The /Radula clade includes ~100 species distributed throughout the American tropics. Species in this clade are common in open areas, as they are some of the most aggressive colonizers in the wet

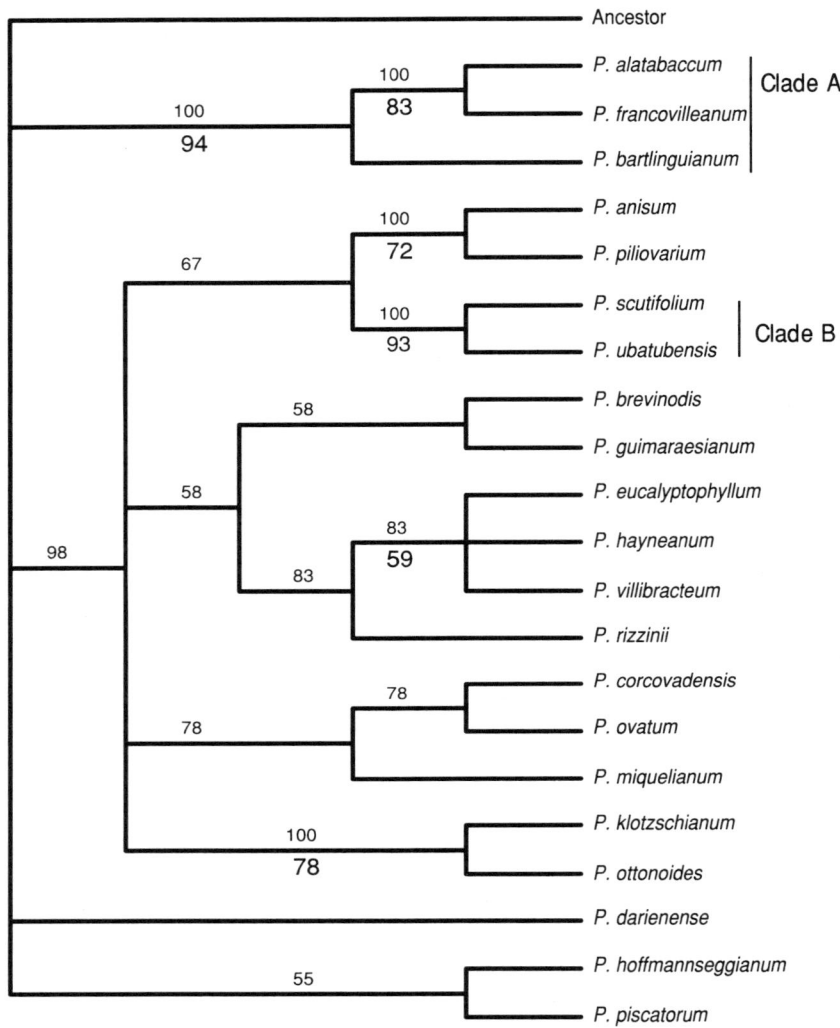

FIGURE 10.3. Phylogenetic relationships within the clade /Ottonia. Topology shown is a 50% majority rule consensus. Numbers above branches are percentages on the majority rule consensus. Numbers below the branches are BS values.

tropics; several species are widely distributed along roadsides. This colonizing ability makes /Radula easy to find in large numbers; consequently, it is one of the best-studied groups ecologically. *P. aduncum*, a member of this clade, has been introduced in Asia and the South Pacific, where it has become a very aggressive competitor with the local flora (Rogers and Hartemink 2000). Margaret Tebbs recently revised this group (Tebbs 1993) focusing on Central American species. A new revision with extensive sampling from South America is also needed.

Macrostachys Miquel. /Macrostachys species are shrubs reaching up to 6 m in height; because branching occurs much later than in species of the /Radula clade, these plants have the appearance of single-stemmed shrubs. Leaves are pinnately nerved, commonly with prominent basal lobes. Plants often have very long and pendulous inflorescences. Flowers are hermaphroditic, rarely arranged in a banded pattern (typical also of /Radula species), with four stamens and three carpels; the seeds are oblong. This group is most diversified in South America, occurring in the Andes as well as in the wet lowland forests of the Amazon and the Atlantic Forest. Some of the Central American members of this group have close relations with ants of the genus *Pheidole* and have been the subject of intensive studies (Chapter 2). *Macrostachys* was revised by Margaret Tebbs .(1989). She focused on collections from Central America and reduced most South American species to synonymy with *P. obliquum*. Tebbs' excessive synonymy manifests lack of knowledge of the growing habits and field characteristics of the plants, which are very important in the taxonomy of the group. Close examination of collections from South America suggests that *P. obliquum* is a taxon restricted to the mid-elevations of the Andean region in Peru and Bolivia (R. Callejas, pers. obs.) in Peru and Bolivia, where it is a rather uncommon plant and quite distinct from all the taxa lumped under this name. Recently, Eric Tepe engaged in a phylogenetic study of the Central American species of /Macrostachys to study the evolution of plant–ant mutualisms in *Piper*. The resulting phylogeny from his study suggests several origins of this interesting interaction (Chapter 9). Tepe plans to expand the sampling to South America and test the multiple-origins hypothesis. This more extensive study focusing on South American taxa is crucial to understanding the ecology and evolution of the group.

/Radula and /Macrostachys are very large groups for which our sampling should be considered very preliminary. Despite the strong support for these clades based on the molecular data collected to date, we believe that a much larger sampling of these groups, especially the addition of more taxa from the Andes and the Amazon, may bring to light relationships still obscured by the limited sampling available. For example, our study includes several species of /Macrostachys with long and pendulous inflorescences suggesting that this is a synapomorphy for the clade; however, this condition is not generalized, and its evolution needs to be examined with a larger sampling of the group.

Pothomorphe –Miquel. /Pothomorphe (= *Lepianthes* Raffinesque) includes shrubs and herbs with palmately and pinnately veined leaves that thrive in open areas. All parts of the plants have a strong spicy smell due to the high concentration of safrole. Flowers are hermaphroditic, tightly arranged in the inflorescences, each with two or four stamens and three carpels. The flowering branches are unbranched, often with internodes reduced; they appear congested, with umbellate axillary inflorescences (Burger 1971). A group traditionally including only two species easily recognized by "axillary umbels," *P. umbellatum* and *P. peltatum*, also includes other taxa without umbels such as *P. auritum*, *P. marginatum*, and *P. heydei*. /Pothomorphe belongs to the tropical American clade. The pantropical distribution of the umbellate inflorescence taxa is a consequence of its secondary dispersal by humans. *P. umbellatum* is used by indigenous people as a muscle relaxant during child birth (Ehringhaus 1997).

It is difficult to assign a clade name to the remaining species since they include divergent characters and taxa from several groups recognized earlier. For now, this clade has been called /Schilleria, a group difficult to characterize because it has a lot of generalized, plesiomorphic characters. Species of /Schilleria are medium-size shrubs, mostly highly branched, with pinnately veined leaves, and hermaphroditic flowers with four stamens and three carpels. Most of the clades found in this phylogeny of *Piper* were either named as distinct genera by Miquel, or as sections of his genus *Artanthe* (including Kunth's genera *Steffensia* and *Schilleria*). *Artanthe* is a large and diverse group that includes ten different sections according to Miquel (1843–1844). It still needs a lot of detailed analyses and more sampling in the Amazon, Southern Andes, and Atlantic Forest to allow us to understand the relationships among the remaining taxa of *Piper* in the American tropics.

10.3.2b. South Pacific and Asian taxa

The other two major clades, the South Pacific and the Asian clades, are not as well characterized. The South Pacific clade comprises mainly the members of /Macropiper. /Macropiper includes shrubs with palmately veined leaves. Flowers are unisexual, loosely arranged in colorful inflorescences, umbellate or individual; male flowers have three stamens and female flowers have 3–4 carpels. /Macropiper is restricted to the South Pacific Islands, where there are ca. 10 species. A recent taxonomic revision suggests that the center of diversity of the /Macropiper is the Fijian archipelago, with six species total, three of them endemic to the region (Smith 1975). Many species of /Macropiper are used by indigenous people as remedies and the wide distribution in the South Pacific Islands may be a result of human introduction; however, no data are available to verify this hypothesis. The relationship of *P. methysticum*, Kava Kava, to /Macropiper is not clear. Some authors emphasize the inflorescence differences (solitary vs. umbellate) between /Macropiper and *P. methysticum* (Lebot and Lèvesque 1989). However, the phylogeny presented here shows a close relationship between *P. methysticum* and /Macropiper. On the basis of the phylogenetic data available, we can conclude that *P. methysticum* is more closely related to /Macropiper than to any other clade of tropical Asian or American *Piper*. However, a broader sampling may show that there are other species from the South Pacific that are more closely related to *P. methysticum* than /Macropiper. Fortunately, Vicente Garcia (a doctoral student in the University of California at Berkeley) has started phylogenetic and ethnobotanical research in this interesting group of *Piper*. Preliminary data suggest that /Macropiper diversification is the result of an adaptive radiation in the islands. Despite the large variation in morphology, /Macropiper species seem to be genetically very similar.

The Asian clade has approximately 300 species. There are several well-supported clades, but they are more difficult to interpret because they do not correspond to previously recognized groupings. The phylogenetic data presented here suggest that morphological characters used to differentiate subgeneric taxa in the past are homoplasious. Old treatments of the genus *Piper* used characters such as peltalte versus adnate bracts, or sessile versus pedicellate fruits to differentiate subgeneric taxa within the Asian species (Miquel 1843–1844, deCandolle 1923); however, these characters appear to have multiple origins. A conscientious examination of classical morphological characters should be made in order to ascertain their real significance for the taxonomy of the group. With the restricted sampling available, it is difficult to identify definitive clades and their diagnostic characters.

Some clades seem to be characterized by a distinct growing habit, e.g., small subshrubs, i.e., *P. boehmeriaefolium*, *P. gymnostachyum*, *P. albispicum*, *P. laosanum*, and *P. lolot*. Others seem to be characterized by their association with ants: *P. decumanum*, *P. myrmecophilum*, *P. brevicuspe*, *P. arborescens*, and *P. toppingii*. The species sampled here, from the section *Penninervia*, form a monophyletic group, suggesting that this section is a natural group. Section *Penninervia* (Quisumbing 1930) is a small group consisting of only four species, with pinnately nerved leaves, very elongated anther connectives, and erect, orange female inflorescences; they are endemic to the Philippine Islands. There are several well-supported and distinct clades among the Asian *Piper* species, suggesting that the genus is highly differentiated in this region. Because our knowledge of *Piper* in its Asian distribution is very limited, a larger sampling is needed to better understand its phylogenetic relationships and improve its classification. In addition, special efforts should be made to study their reproductive biology, and their population structure at the demographic and genetic levels.

The origin of *Piper* species in Africa is puzzling since one of the species, *P. capensis*, is a shrub with bisexual flowers, similar to the tropical American taxa. The other African species, *P. guieenense*, is a climber with unisexual flowers, like Asian taxa. The only species included in the phylogeny presented here is *P. capensis*; it appears to belong to the South Pacific clade. This relationship suggests that *Piper* may have been dispersed to Africa instead of being separated by an old vicariant event during the continental drift. This result should be considered preliminary; further analyses using more molecular markers are needed to confirm this relationship.

We acknowledge that our sampling, less than 10% of the species in the genus, is still very scarce. However, we believe this is a good starting point to continue analyzing the evolution of *Piper* species within a phylogenetic perspective.

10.4. EVOLUTIONARY ASPECTS

10.4.1. Flower Morphology

Piper flowers are very reduced, however they exhibit definite patterns of variation that have been important for the classification of the genus. The major difference is that American species have bisexual flowers whereas Asian species are dioecious. Dioecy occurs by abortion of stamens in female flowers or carpels in male flowers .(Lei and Liang 1998). Besides this major difference, most variation occurs in the number of stamens. *Piper* flowers may have anywhere between 1 and 10 stamens. Nevertheless, most American species have 4 stamens and most Asian taxa have 3 stamens. The number of stamens seems to be correlated with inflorescence structure, at least in American species (Fig. 10.4). Reduction in stamen number appears to be associated with tight packaging of flowers in some inflorescences (Jaramillo and Manos 2001). This hypothesis was first presented by Shirley Tucker after studying flower development in Central American species. Stamens in Piperaceae are stereotypically initiated with a pair in the lateral plane, followed by a primordium in the anterior position, and another single stamen in the posterior position (Tucker 1982). In looser inflorescences, such as in *P. amalago*, another pair of primordia is originated in the lateral plane of the flower (Tucker 1982); in tighter inflorescences like in *P. umbellatum*,

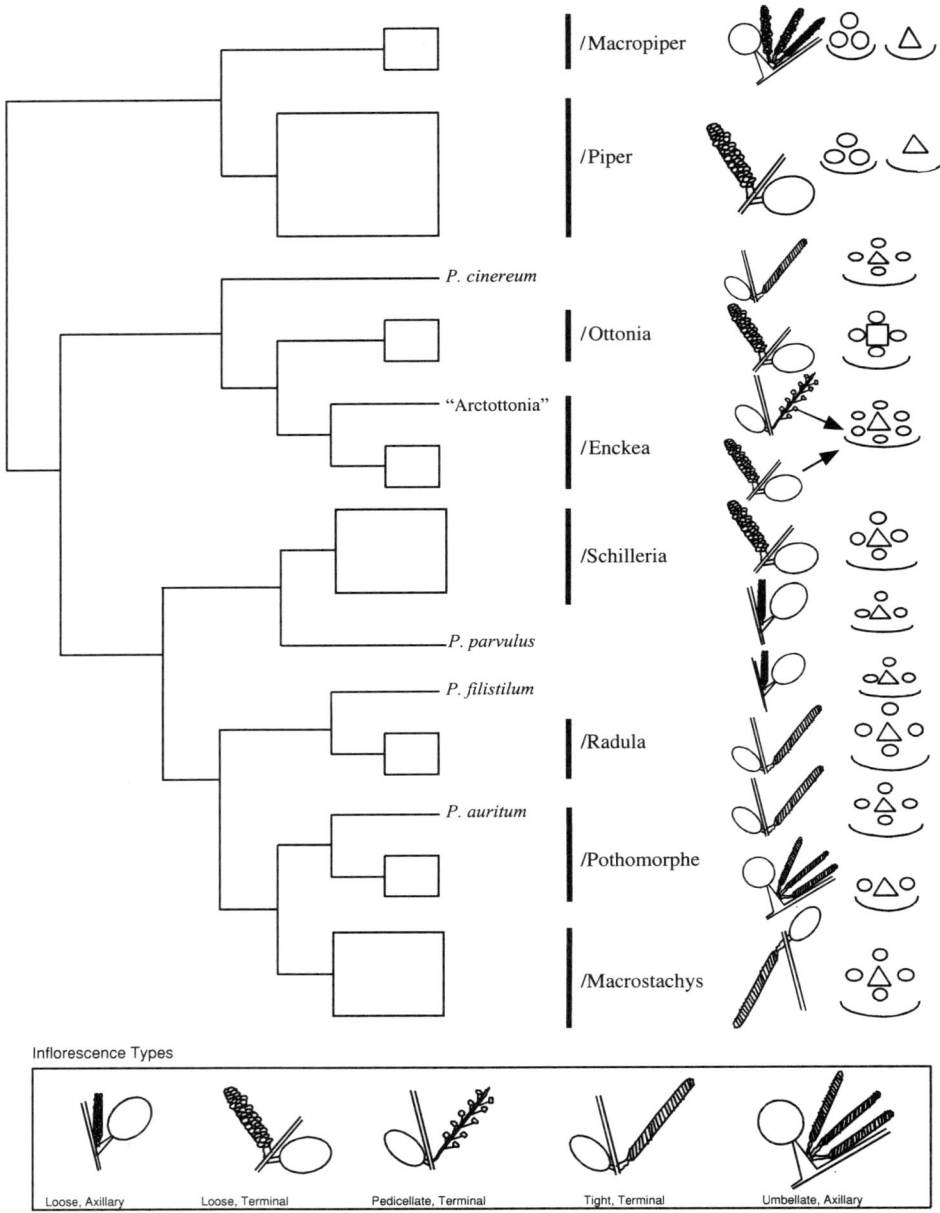

FIGURE 10.4. Evolutionary history of flowers and inflorescences in the genus *Piper*. Simplified phylogeny from Fig. 10.2. In each floral diagram, circles are stamens, triangles and squares are ovaries, arching lines are floral bracts.

only the first pair of stamens originates (Tucker *et al.* 1993). However, this explanation does not seem to be universal. The evolution of single staminate flowers occurred both in the American and Asian tropics. The single staminate flowers of *P. khorthalsii* (Philippines) and *P. kegelii* (Amazon Basin) cannot be explained by packaging of the inflorescences since, in both species, flowers are loosely arranged in the inflorescences. Some clues to understanding the evolution of these unique flowers may be found by studying their flower development and reproductive biology.

Besides differences in reproductive system (dioecy vs. hermaphroditism) and stamen number, other categories of variation can be found in *Piper* flowers. Some of this variation is exhibited in the relative size of anthers and filaments, the orientation of the anther dehiscence, and the length of the style and stigma lobes. However, it is difficult to talk about the patterns of evolution of these characters; they are usually used as differentiating characters and are not useful to group and classify taxa. More studies of structure development as well as investigations linking morphologies with function may shed light on other flower-related evolutionary trends in the genus.

10.4.2. Plant Architecture

In addition to flower variability, most variation among *Piper* taxa is in their architecture. *Piper* species occur as shrubs of different sizes (50 cm to 6 m) and shapes (highly ramified or single-stemmed) and as lianas; despite these differences, *Piper* species are very homogeneous in their construction. Their architecture can be characterized by Petit's model (Hallé *et al.* 1978), which consists of a monopodial, orthotropic trunk that grows continuously, and gives rise to lateral, plagiotropic, sympodial branches (Fig. 10.5). In *Piper*, the branches are composed of monophylls, modules constituted by an internode, an inflorescence, a prophyll, and a leaf (Blanc and Andraos 1983). Differences in appearance result from modifications of the ramification pattern and are related to the light environment in which the plant grows and the clade affiliation. An upright appearance, characteristic of the /Radula and /Macrostachys clade, has an unbranched main axis with lateral branches that ramify occasionally (Fig. 10.5). A more rounded appearance is exhibited by taxa of /Ottonia, /Schilleria, and /Enckea, where the main axis branches and has reiterations along its whole length, and the lateral branches are often ramified (Fig. 10.5). Species that have an upright look are found mostly in forest gaps and edges. Species with a ramified principal axis are favored in the understory, where the plants receive very faint light (Blanc and Andraos 1983).

A particular case of modification in the architecture as an adaptation to the environment is that of the dwarf pipers of the Chocó Region (northwestern slope of the Andes). These plants were all included in *Trianaeopiper* Trelease, an endemic genus from the Chocó; they are very small herbs with axillary inflorescences. An initial phylogenetic analysis of the genus *Piper* showed that species recognized as *Trianaeopiper* are embedded within *Piper* and do not form a clade (Jaramillo and Manos 2001). Closer examination of these species showed that they are part of three (or two) distinct clades (Fig. 10.6) and the apparent axillary position of the inflorescences is a byproduct of the general reduction of axis that produced a dwarf appearance (Jaramillo and Callejas 2004). A dwarf habit is achieved by the reduction of some or all of the following architectural elements:

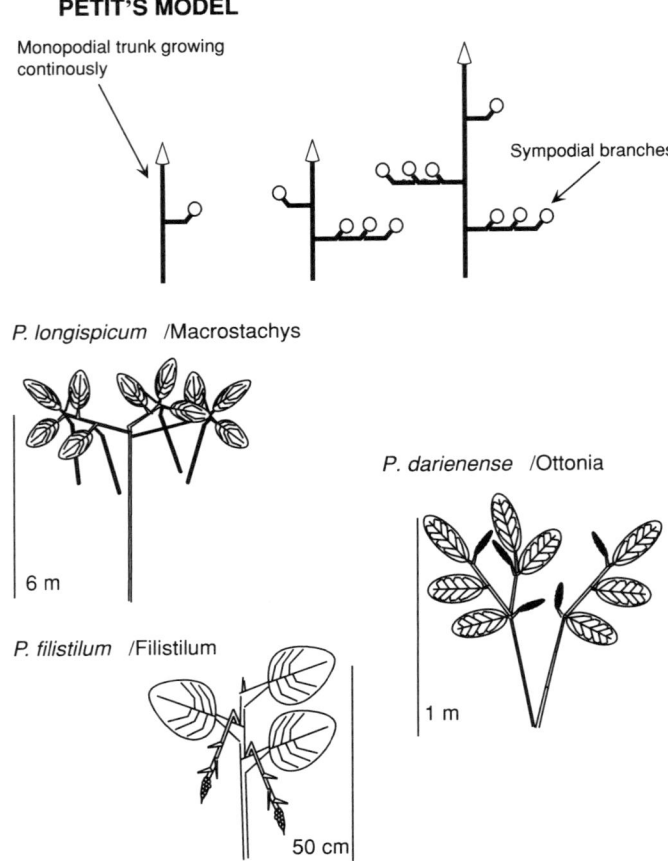

FIGURE 10.5. Petit's model of plant architecture (Hallé et al. 1978) and three examples of this architectural model in the genus *Piper*: upright appearance, *P. longispicum*; rounded appearance, *P. darienense*; dwarf habit, *P. filistilum*.

number of branches, number of internodes in reproductive branches, length of the internodes, and size of the leaves in flowering shoots. All of these elements and the dwarf habit in particular are positively correlated with low light levels (typical of wet tropical forest understory) and dwarf appearance (Jaramillo and Callejas 2004). Dwarfism seems to be an adaptation of some tropical angiosperm taxa to the low light levels of the understory of wet tropical forests, this phenomenon has also been reported in the palm genus *Geonoma* (Chazdon 1991).

10.5. ACKNOWLEDGMENTS

The authors thank Lee Dyer for the invitation to write a chapter in this book, field companions in Colombia, Philippines, Viet Nam, Mexico, and Ecuador, especially to

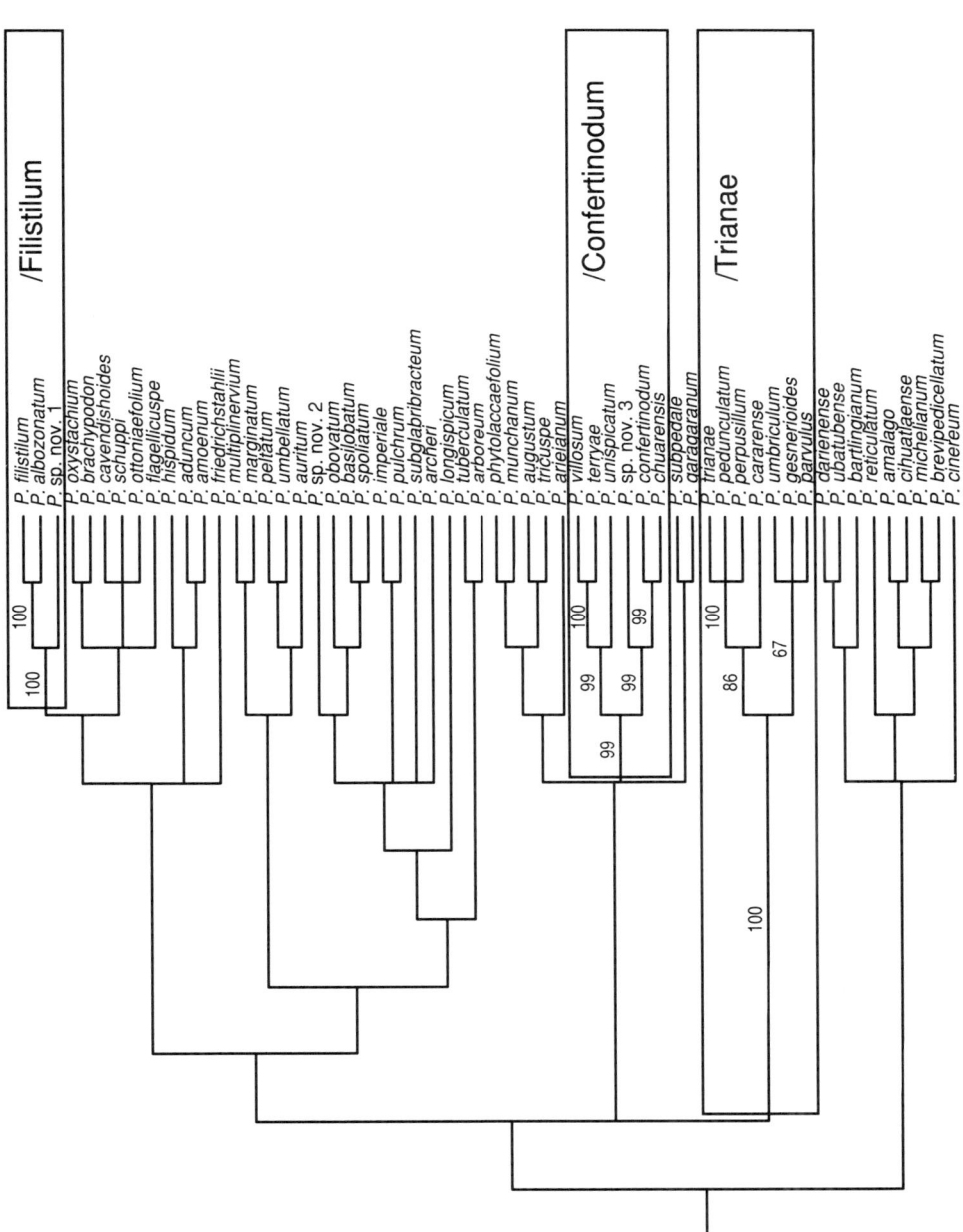

FIGURE 10.6. Phylogeny of Neotropical *Piper* indicating the three clades of dwarf pipers (modified from Fig. 10.2).

Fernando Alzate, Felipe Cardona, M. Cristina Lopez, and F. Javier Roldán in Colombia, and N. T. Hiep and P. K. Lôc in Viet Nam. Financial support for our studies were provided by World Wildlife Fund, National Geographic Research Committee, and U.S. National Science Foundation.

APPENDIX 10.1

Phylogenetic Analysis of the Genus Piper Using ITS

In order to reconstruct the phylogenetic relationships within the genus *Piper* we used an exemplar sampling from Central and South America as well as the Asian tropics. The sampling was more intensive in the American tropics (58 species) given that the residence of the authors facilitated the collections. Thirty-three species from the Paleotropics (Asia, Africa, and the South Pacific) were included.

Plant material was collected from natural populations; a list of species sampled with collection localities, vouchers, and accession numbers is provided in Table 10.1. Total DNA was extracted from silica gel–dried material using the Dneasy Plant Mini kit (Qiagen Corporation, Valencia, California). The ITS region was amplified using the universal primers

TABLE 10.1
Species Included in the Phylogenetic Analyses

Species	Source	Voucher[1]	Genbank No.
Macropiper hooglandii Hutton & Green	Cultivation, Auckland Museum	ROG 8496 (AK)	AF 275192
Macropiper mechilor Sykes	Cultivation, Auckland Museum	ROG 8495 (AK)	AF 275191
Piper aduncum Linn	Depto. Valle del Cauca, Colombia	MAJ 76	AF 275157
Piper albispicum C.DC.	Prov. Ha Tinh, Viet Nam	MAJ 388	AY 572317
Piper albozonatum C.DC.	Prov. Esmeraldas, Ecuador	MAJ 697	AY 326195
Piper amalago Linn.	Edo. Veracruz, Mexico	MAJ 561	AF 275186
Piper amoenum Yuncker	Depto. Chocó, Colombia	MAJ 116	AF 275160
Piper appendiculatum C.DC.	Depto. Cauca, Colombia	MAJ 740	AY 326196
Piper arborescens Roxb.	Prov. Albay, Philippines	MAJ 200	AF 275158
Piper arboreum Aub.	Depto. Antioquia, Colombia	MAJ 112	AF 275180
Piper arboricola C.DC.	Prov. Ha Tinh, Viet Nam	MAJ 467	AY 572319
Piper arieianum C.DC.	Depto. Chocó, Colombia	MAJ 69	AF 275163
Piper atrospicum C.DC.	Prov. Lagunas, Philippines	MAJ 168	AY 572318
Piper augustum Rudge	Depto. Chocó, Colombia	MAJ 122	AF 275165
Piper auritum H. B. & K.	Depto. Chocó, Colombia	MAJ 63	AF 275175
Piper bartlingianum C.DC.	Guyana	RE 1267B	AF 275183
Piper basilobatum Trel. & Yuncker.	Depto. Chocó, Colombia	MAJ 596	AY 326197
Piper bavinum C.DC.	Prov. Ha Tinh, Viet Nam	MAJ 392	AF 275199
Piper betle Linn.	Cultivation, Duke U. Green Houses	DU82–298	AF 275201
Piper boehmeriaefolium Wall.	Prov. Ha Tinh, Viet Nam	MAJ 235	AF 275204
Piper brachypodon C.DC.	Depto. Valle del Cauca, Colombia	MAJ 757	AY 326198
Piper brevipedicellatum A. J. Bornstein	Estado de Colima, Mexico	MAJ 544	AF 275189
Piper brevicuspe (Miq.) Merr.	Prov. Leyte, Philippines	MAJ 211	AY 572321
Piper caballense C.DC.	Depto. Antioquia, Colombia	MAJ 87	AF 275178
Piper cajambrense Treal. & Yunck.	Dep. Valle del Cauca, Colombia	MAJ 768	AY 326199

TABLE 10.1
(cont.)

Species	Source	Voucher[1]	Genbank No.
Piper caninum Blume	Prov. Surigao del Norte, Philippines	MAJ 218	AY 326195
Piper capensis Linn.	Uluguru Mtns, Tanzania	Faden 96-75	AY 326200
Piper cararense Trel. & Yunck.	Depto. Chocó, Colombia	MAJ 601	AY 326201
Piper cavendishoides Trel. & Yunck.	Depto. Chocó, Colombia	MAJ 70	AF 275153
Piper celtidiforme Opiz	Prov. Lagunas, Philippines	MAJ 171	AF 275205
Piper chuarense M.A. Jaramillo & R. Callejas	Depto. Valle del Cauca, Colombia	MAJ 721	AY 326202
Piper cihuatlanense A. J. Bornstein	Edo. Jalisco, Mexico	MAJ 543	AF 275187
Piper cinereum C.DC.	Depto. Chocó, Colombia	MAJ 66	AF 275190
Piper colligatispicum Trel. & Yunck.	Depto. Valle del Cauca, Colombia	MAJ 780	AY 326204
Piper confertinodum (Trel. & Yunck.) M.A. Jaramillo & R. Callejas	Depto. Chocó, Colombia	MAJ 54	AF 275166
Piper cordatilumbum Quisumbing	Prov. Lagunas, Philippines	MAJ 178	AY 572323
Piper darienense (Miq.) C.DC.	Depto. Antioquia, Colombia	MAJ 103	AF 275181
Piper decumanum Linn.	Prov. Leyte, Philippines	MAJ 210	AF 275203
Piper densum Blume	Prov. Nim Binh, Viet Nam	MAJ 508	AY 615963
Piper excelsum Forst. F.	Cultivation, Auckland Museum	ROG 8494 (AK)	AF 275193
Piper filistilum C.DC.	Depto. Nariño, Colombia	MAJ 157	AF 275155
Piper flagellicuspe Trel. & Yunck.	Depto. Chocó, Colombia	MAJ 65	AF 275154
Piper friedrichsthalii.C.DC.	Depto. Chocó, Colombia	MAJ 584	AY 326205
Piper garagaranum C.DC.	Depto. Chocó, Colombia	MAJ 73	AF 275162
Piper gesnerioides R. Callejas	Depto Chocó, Colombia	MAJ 694	AY 326206
Piper gymnostachyum C.DC.	Prov. Ha Tinh, Viet Nam	MAJ 389	AY 572325
Piper harwegianum (Bentham) C.DC.	Depto. Valle del Cauca, Colombia	MAJ 781	AY 326207
Piper hernandii, C.DC.	Prov. Ha Tinh, Viet Nam	MAJ 475	AY 572324
Piper hispidum Sw.	Depto. Chocó, Colombia	MAJ 53	AF 275156
Piper hymenophyllum Miq.	Prov. Nim Binh, Viet Nam	MAJ 505	AY 572327
Piper imperiale C.DC.	Depto. Chocó, Colombia	MAJ 61	AF 275176
Piper korthalsii Miq.	Prov. Quezón, Philippines	MAJ 184	AF 275208
Piper laosanum C.DC.	Prov. Ha Tinh, Viet Nam	MAJ 467	AY 572326
Piper lolot C.DC.	Prov. Ha Tinh, Viet Nam	MAJ 234	AY 326208
Piper longispicum C.DC.	Depto. Cauca, Colombia	MAJ 739	AY 326209
Piper longivillosum Trel. & Yunck.	Dept. Chocó, Colombia	MAJ 605	AY 326221
Piper marginatum Jacq.	Prov. Esmeraldas, Ecuador	MAJ 713	AY 326211
Piper medellinifolium Quisumbing	Prov. Sorsogon, Philippines	MAJ 196	AY 667455
Piper methysticum Forst. f .	Cultivation, National Tropical Botanical Garden, NTBG	NTBG-950585	AF 275194
Pipe michelianum C.DC.	Edo. Jalisco, Mexico	MAJ 537	AF 275188
Piper multiplinervium C.DC.	Depto. Chocó, Colombia	MAJ 139	AF 275168
Piper munchanum C.DC.	Depto. Chocó, Colombia	MAJ 120	AF 275164
Piper myrmecophyllum C.DC.	Prov. Samar, Philippines	MAJ 205	AY 572328
Piper nigrum Linn.	Cultivation, Duke U. Greenhouses	DU94-006	AF 275197
Piper obovatum Vahl	Depto. Chocó, Colombia	MAJ 641	AY 326212
Piper ottoniaefolium C.DC.	Depto. Valle del Cauca, Colombia	MAJ 759	AY 326213
Piper oxystachyum C.DC.	Depto. Chocó, Colombia	MAJ 140	AF 275142
Piper parvulum M.A. Jaramillo & R.Callejas	Depto. Chocó, Colombia	MAJ 55	AF 275167
Piper pedunculatum C.DC.	Depto. Chocó, Colombia	MAJ 691	AY 326214
Piper peltatum Linn.	Depto. Chocó, Colombia	MAJ 142	AF 275169

(cont.)

TABLE 10.1
(cont.)

Species	Source	Voucher[1]	Genbank No.
Piper penninerve C.DC.	Prov. Surigao del Norte, Philippines	MAJ 213	AF 275206
Piper perpusillum R. Callejas	Prov. Esmeraldas, Ecuador	MAJ 699	AY 326215
Piper pierrei C.DC.	Prov. Ha Tinh, Viet Nam	MAJ 394	AF 275200
Piper phyttolaccaefolium Opiz	Depto. Chocó, Colombia	MAJ 599	AY 326216
Piper pulchrum C.DC.	Depto. Chocó, Colombia	MAJ 100	AF 275177
Piper reticulatum Linn.	Depto. Chocó, Colombia	MAJ 128	AF 275184
Piper retrofractum Vahl	Prov. Ha Tinh, Viet Nam	MAJ 395	AF 275196
Piper sabaletasanum Trel. & Yunck.	Depto. Chocó, Colombia	MAJ 623	AY 326217
Piper schuppii A.H. Gentry	Prov. Esmeraldas, Ecuador	MAJ 687	AY 326218
Piper sorsogonum C.DC.	Prov. Quezón, Philippines	MAJ 185	AY 572320
Piper sp. nov. 1	Prov. Esmeraldas, Ecuador	MAJ 674	AY 326219
Piper sp. nov. 2	Prov. Esmeraldas, Ecuador	MAJ 689	AY 326230
Piper spoliatum T. & Y.	Depto. Chocó, Colombia	MAJ 60	AF 275179
Piper subglabribracteatum C.DC.	Depto. Cauca, Colombia	MAJ 747	AY 326220
Piper subpedale T. & Y.	Depto. Chocó, Colombia	MAJ 57	AF 275161
Piper terryae Standley	Depto. Chocó, Colombia	MAJ 605	AY 326221
Piper tomas-albertoi T. & Y.	Depto. Antioquia, Colombia	RC-s.n.	AY 326222
Piper topingii C.DC.	Prov. Quezón, Philippines	MAJ 186	AY 572322
Piper trianae C.DC.	Depto. Nariño, Colombia	MAJ 165	AY 326223
Piper tricuspe C.DC.	Depto. Valle del Cauca, Colombia	AG 41 (CUVC)	AY 326224
Piper tuberculatum Jacq.	Prov. Esmeraldas, Ecuador	MAJ 710	AY 326225
Piper ubatubense R. Callejas	Edo. São Paulo, Brasil	CC 2	AF 275182
Piper umbellatum Linn.	Cultivation, Fairchild Bot. Gard.	78-211B	AF 275174
Piper umbriculum M.A. Jaramillo & R. Callejas	Depto. Chocó, Colombia	MAJ 602	AY 326226
Piper unispicatum R. Callejas	Depto. Chocó, Colombia	RC 11854 (HUA)	AY 326227
Piper urdanetanum C.DC.	Prov. Mindanao, Philippines	MAJ 232	AF 275207
Piper villosum C.DC.	Prov. Esmeraldas, Ecuador	MAJ 667	AY 326228

[1] Vouchers deposited in DUKE (or otherwise noted). Collectors: AG: A. Giraldo; CC: C. Castro; MAJ: M. A. Jaramillo; RC: R. Callejas, RE: R. Elke, ROG: R. O. Gardner, DU: Collected in the greenhouses at Duke University.

LEU1 or ITS 5 and ITS 4, sequencing primers were LEU1 or ITS 5, ITS 4, ITS 3, and ITS 2. Sequencing reactions were prepared using the ABI Prism Dye terminator Cycle Sequencing Reaction Kit, according to the protocols provided by the manufacturer. Double-stranded sequences were obtained in every case, sequences were assembled using Sequencher 4.1 (GeneCodes, Ann Arbor, Michigan), and aligned by eye using previous ITS data sets for *Piper* as a template (Jaramillo and Manos 2001).

A Bayesian phylogenetic analysis was carried out using MrBayes 3.0 (Ronquist and Huelsenbeck 2003). The best model of molecular evolution was selected using Modeltest (Posada and Crandall 1998). The model preferred was HKY + I + G (HKY model of substitution, Hasegawa *et al.* 1985; I = proportion of invariable sites; G = gamma shape parameter). Four Markov Chain Monte Carlo chains were run, 1 topology was sampled every 100 generations for 1,000,000 generations, starting from a random tree. The first 9,100 generations, before convergence, were considered the burn-in period and were not included in generating the majority-rule consensus phylogeny. The forward slash is used here as a "clade-mark" as suggested by Baum *et al.* (1998) to distinguish provisional clade names from ranked taxonomic classification.

Phylogenetic Analysis of the /Ottonia Using Morphological Characters

The morphological data set used by Callejas (1986) was reexamined here. Phylogenetic analyses were carried out using PAUP* (Swofford 2002) and the maximum parsimony (MP) algorithm. The parsimony analysis was performed using heuristic searches with 100 random addition replicates, TBR swapping, MULPARS, steepest descent and AMB- options as implemented in PAUP* version 4.10 b. Relative branch support was determined using 1,000 bootstrap (BS) replicates (Felsenstein 1985) and a complete heuristic search.

REFERENCES

APG. (1998). An ordinal classification of flowering plants. *Annals of the Missouri Botanical Garden* 85:531–553.
APG. (2003). An update of the Angiosperm Phylogeny Group classification for the orders and families of flowering plants: APG II. *Botanical Journal of the Linnean Society* 141:399–436.
Backer, C. A., and van den Brink, R. C. B. (1963). Piperaceae. In: Backer, C. A., and van den Brink, R. C. B. (eds.), *Flora of Java*. N. V. P. Noordhoff, Groningen.
Baum, D. A., Alverson, W. A., and Nyffeler, R. (1998). A Durian by any other name: Taxonomy and nomenclature of the core Malvales. *Harvard Papers in Botany* 3:315–330.
Blanc, P., and Andraos, K. (1983). Remarques sur la dynamique de crossance dans le genre *Piper* L. (Piperaceae) et les genres affines. *Bulletin du Muséum national d'histoire naturelle. Section B. Adansonia.* 23:259–282.
Bornstein, A. J. (1989). Taxonomic studies in the Piperaceae—I. The pedicellate pipers of Mexico and Central America (*Piper* subg. arctottonia). *Journal of the Arnold Arboretum* 70:1–55.
Burger, W. C. (1971). Piperaceae. In: Burger, W. C. (ed.), *Flora Costaricensis. Fieldiana, Bot.* 35:1–227.
Callejas, R. (1986). *Taxonomic revision of* Piper *subgenus* Ottonia *(Piperaceae)*. City University of New York, New York.
Chase, M. W., Soltis, D. E., Olmstead, R. G., Morgan, D., Les, D. H., Mishler, B. D., Duvall, M. R., Price, R. A., Hills, H. G., Qiu, Y.-L., Kron, K. A., Rettig, J. H., Conti, E., Palmer, J. D., Manhart, J. R., Systma, K. J., Michaels, H. J., Kress, W. J., Karol, K. G., Clark, W. D., Hedrén, M., Gaut, B. S., Jansen, R. K., Kim, K.-J., Wimpee, C. F., Smith, J. F., Fournier, G. F., Strauss, S. H., Xiang, Q.-Y., Plunkett, G. M., Soltis, P. S., Sweensen, S. W., Williams, S. E., Gadek, P. A., Quinn, C. J., Eguiarte, L. E., Golenberg, E., Learn, G. H., Jr., Graham, S. W., Barrett, S. C. H., Dayanandan, S., and Albert, V. A. (1993). Phylogenetics of seed plants: An analysis of nucleotide sequences from the plastid gene rbcL. *Annals of the Missouri Botanical Garden* 80:528–580.
Chazdon, R. L. (1991). Plant size and form in the understory palm genus *Geonoma*: Are species variations on a theme. *American Journal of Botany* 78:680–694.
Chew, W.-L. (1953). The genus *Piper* (Piperaceae) in New Guinea, Solomon Islands and Australia, 1. *Journal of the Arnold Arboretum* 53:1–25.
Cronquist, A. (1981). *An Integrated System of Classification of Flowering Plants*. Columbia University Press, New York.
de Candolle, C. (1866). Piperaceae. *Prodromus* 16:235–471.
de Candolle, C. (1923). Piperacearum clavis analytica. *Candollea* 1:65–415.
Ehringhaus, C. (1997). *Medicinal Uses of* Piper *spp. (Piperaceae) in an Indigenous Kaxinawa Community in Acre, Brazil*. Department of Anthropology, Florida International University, Miami.
Endress, P. K. (1987). The Chloranthaceae: Reproductive structures and phylogenetic position. *Botanische Jahrbücher für Systematik* 109:153–226.
Felsenstein, J. (1985). Confidence limits on phylogenies: An approach using the bootstrap. *Evolution* 39:783–791.
Fleming, T. H. (1981). Fecundity, fruiting pattern, and seed dispersal in *Piper amalago* (Piperaceae), a bat dispersed tropical shrub. *Oecologia* 51:42–46.
Fleming, T. H. (1985). Coexistence of five sympatric *Piper* (Piperaceae) species in a tropical dry forest. *Ecology* 66:688–700.
Gentry, A. H. (1990). Floristic similarities and differences between southern Central America and upper Central Amazonia. In: Gentry, A. H. (ed.), *Four Neotropical Rainforests*. Yale University Press, New Haven, Connecticut, pp.141–157.
Hallé, F., Oldeman, R. A., and Tomlinson, P. B. (1978). *Tropical Trees and Forests. An Architectural Analysis*. Springer-Verlag, Berlin.
Hasegawa, M., Kishino, H. and Yano, T. 1985. Dating the human-ape splitting by a molecular clock of mitochondrial DNA. *Journal of Molecular Evolution* 22: 160–174.

Jaramillo, M. A., and Callejas, R. (2004). A reappraisal of *Trianaeopiper* Trelease: Convergence of dwarf habit in some *Piper* species of the Chocó. *Taxon* 53: 269–278.

Jaramillo, M. A., and Manos, P. S. (2001). Phylogeny and the patterns of floral diversity in the genus *Piper* (Piperaceae). *American Journal of Botany* 88:706–716.

Jaramillo, M. A., Manos, P. S., and Zimmer, E. A. (2004). Phylogenetic relationships of the perianth-less Piperales: Reconstructing the evolution of floral development. *International Journal of Plant Sciences* 165.

Kunth, K. (1839). Bemerkungenuber die Familie der Piperaceen. *Linnaea* 13:562–726.

Lebot, V., and Lèvesque, J. (1989). The origin and distribution of Kava (*Piper methysticum* Forst. f., Piperaceae): A phytochemical approach. *Allertonia* 5:223–281.

Lei, L.-G., and Liang, H.-X., (1998). Floral development of dioecious species and trends of floral evolution in *Piper sensu lato*. *Botanical Journal of the Linnean Society* 127:225–237.

Lei, L.-G., Wu, Z.-Y., and Liang, H.-X., (2002). Embryology of *Zippelia begoniaefolia* (Piperaceae) and its systematic relationships. *Botanical Journal of the Linnean Society* 140:49–64.

Liang, H.-X., and Tucker, S. C. (1995). Floral ontogeny of Zippelia begoniaefolia and its familial affinity: Saururaceae or Piperaceae? *American Journal of Botany* 82:681–689.

Liu, T.-S., and Wang, F.-L. (1975). Piperaceae. In: Li, H.-L. (ed.), *Flora of Taiwan*, vol. 2. Taipei, pp. 556–565.

Miquel, F. A. G. (1843–1844). *Systema Piper Acearum*. Kramers, Rotterdam.

Nickrent, D. L., Blarer, A., Qiu, Y.-L., Soltis, D. E., Soltis, P. S., and Zanis, M. J. (2002). Molecular data place Hydnoraceae with Aristolochiaceae. *American Journal of Botany* 89:1809–1817.

Quisumbing, E. (1930). Philippine Piperaceae. *The Philippine Journal of Science* 43:1–233.

Ridley, H. N. (1924). Piperaceae. In: Ridley, H. N. (ed.), *The Flora of the Malay Peninsula*, Apetalae, vol. 3. L. Reeve & Co., Ltd., London, pp. 25–51.

Rogers, H. M., and Hartemink, A. E. (2000). Soil seed bank and growth rates of an invasive species, *Piper aduncum*, in the lowlands of Papua New Guinea. *Journal of Tropical Ecology* 16:243–251.

Ronquist, F., and Huelsenbeck, J. P. (2003). MrBayes 3: Bayesian phylogenetic inference under mixed models. *Bioinformatics* 19:1572–1574.

Smith, S. A. (1975). The genus *Macropiper* (Piperaceae). *Botanical Journal of the Linnean Society* 71:1–38.

Soltis, P. S., Soltis, D. E., Zanis, M. J., and Kim, S. (2000). Basal lineages of angiosperms: Relationships and implications for floral evolution. *International Journal of Plant Science* 161:S97–S107.

Steyemark, J. A. (1984). Piperaceae. In: Instituto Nacional de Parques (eds.), *Flora of Venezuela*, Part 2. Ediciones Fundación, Caracas, Venezuela.

Swofford, D. L. (2002). *PAUP*. Phylogenetic Analysis Using Parsimony (*and Other Methods). Version 4*. Sinauer Associates, Sunderland, Massachusetts.

Tebbs, M. C. (1989). Revision of *Piper* (Piperaceae) in the New World. 1. Review of characters and taxonomy of *Piper* section *Macrostachys*. *Bulletin of the Natural History Museum of London* 19:117–158.

Tebbs, M. C. (1993). Revision of *Piper* (Piperaceae) in the New World. 3. The taxonomy of *Piper* sections *Lepianthes* and *Radula*. *Bulletin of the Natural History Museum of London* 23:1–50.

Trelease, W. (1927). The Piperaceae of Panama. *Contributions of the U.S. National Herbarium* 26:15–50.

Trelease, W. (1928). *Trianaeopiper*, a new genus of Piperaceae. *Proceedings of the American Philosophical Society* 67:47–50.

Trelease, W. (1929). *Lindeniopiper*, a generic segregate from *Piper*. *Proceedings of the American Philosophical Society* 68:53–54.

Trelease, W. (1930). The geography of American peppers. *Proceedings of the American Philosophical Society* 69:309–327.

Trelease, W. (1934). New amplifications of the North American Piperaceae. *Proceedings of the American Philosophical Society* 73:327–330.

Trelease, W. (1940). Piperaceae. In: MacBride, W. (ed.), *Flora of Peru*, vol. 2. Publications of the Field Museum of Natural History (Botany), Chicago, pp. 3–253.

Trelease, W., and Yuncker, T. G. (1950). *The Piperaceae of Northern South America*. University of Illinois Press, Indiana.

Tucker, S. C. (1982). Inflorescence and flower development in the Piperaceae. III. Floral ontogeny of *Piper*. *American Journal of Botany* 69:1389–1401.

Tucker, S. C., Douglas, A. W., and Han-Xiang, L. (1993). Utility of ontogenetic and conventional characters in determining phylogenetic relationships of Saururaceae and Piperaceae (Piperales). *Systematic Botany* 18:614–641.

Wanke, S. D.-T. (2003). Phylogenie der Piperales basierend auf Sequenzen des Chloroplastengens matK und angrenzender nicht-codierender Bereiche. Diplom-Thesis, Botany Institute, Universität of Bonn. Bonn, Germany.

Yuncker, G. (1972). The Piperaceae of Brazil, I. *Piper*-group I, II, III and IV. *Hoehnea* 19:336.

Yuncker, G. (1973). The Piperaceae of Brazil, II. *Piper*-group V, *Ottonia*, *Potomorphe* and *Sarcorhachis*. *Hoehnea* 3:29–284.

11
Future Research in *Piper* Biology

M. Alejandra Jaramillo[1] and Robert Marquis[2]

[1] *Departmento de Bioquímica, Médica, Universidade Federal do Rio de Janieiro, Rio de Janiero, Brazil*

[2] *Department of Biology, University of Missouri—St. Louis, St. Louis, Missouri*

11.1. INTRODUCTION

The very high diversity in the genus *Piper* (1,000+ species) cries out for research into the evolutionary processes that have led to such variety. What have been the driving forces in the diversification of the genus? What has been the relative role of biotic versus abiotic factors in diversification? Of the potential biotic factors, what have been the influences of dispersers, pollinators, pathogens, herbivores, and natural enemies of those herbivores? Is diversification a result of geological events? Are diversification processes different in the American, African, and Asian tropics? By answering these questions, we can begin to learn what factors have led to diversification in tropical plant communities.

Below, we discuss each of these questions and associated issues as possible foci for future research. A major difficulty in answering the above questions is the rather chaotic classification of the genus. A start has been made toward improving this situation with recent monographs focusing on natural groups, supplemented with a few phylogenetic analyses. Nevertheless, these monographs and phylogenies have focused mainly on Tropical American taxa and have sampled only an exemplar number of species (less than 10% of the total). An effort should be made to mount a systematic revision of the genus, especially for Tropical American species, where description of new species has been rampant (Jones 1985).

11.2. PLANT–ANIMAL INTERACTIONS

Animals interact with *Piper* plants in at least four ways: as dispersers, pollinators, herbivores, and as biotic defense agents. Both bats and birds are known to disperse the seeds of many Tropical American *Piper* species. Detailed studies of dispersal have been carried out in a few species of *Piper* (Fleming 1981, 1985, Loiselle 1990, Bizerril and Raw 1998, Mori 2002; Chapter 4). Seed dispersal and its probable effects on plant recruitment have been evaluated in shade-tolerant and shade-intolerant species (Greig 1993). Bats, in particular a few species of the genus *Carollia*, are thought to be the main dispersers of *Piper* seeds. These plant–animal interaction studies include *Piper* species representing a variety of growth habits, species from several geographic regions in South and Central America, as well as taxa from various clades. Future research on *Piper* dispersal should focus on comparing species of different growth habits and also species from different clades. Such studies will shed light on the overall importance of dispersers in the evolution and diversification of *Piper*, and the role of the genus as an important component of tropical ecosystems.

Pollen-collecting bees are thought to be the main pollinators of *Piper* plants. However there are reports of visits by flies, in particular many syrphid species, and there is evidence for wind pollination (Chapter 3). Future research on the pollination of *Piper* should address the genetic consequences of pollination by different types of pollinators, especially in comparison with gene flow provided by seed dispersal. Studies of phenology in relation to pollination success would reveal whether temporal partitioning of the pollinator community, both among *Piper* species and between the understory and canopy, has been necessary in maintaining coexistence of many species at individual sites.

Piper plants experience damage by a large variety of generalist and specialist herbivores (Marquis 1991), and defense against these herbivores includes production of toxins or protection by symbionts. In *P. cenocladum*, toxins are more effective against generalist herbivores (Chapter 7), whereas mutualistic ants are effective against specialist herbivores and pathogens, with these two types of defense acting in concert to increase the plant's overall fitness (Chapter 2). Mutualistic interactions seem to be common antiherbivore defense strategy in Central American *Piper* in general, as reflected by the research of Gastriech and Gentry (Chapter 6). Mutualisms in the genus *Piper* deserve to be studied in greater detail, including taxa (species and clades) from a larger distribution range. In addition, there is great potential for using both the endophytic and exophytic communities of organisms associated with pipers as mesocosms for addressing basic questions in community ecology.

Plant–ant interactions have been the focus of several detailed studies in Central America. Several species of *Piper*, especially some belonging to the clade Macrostachys, house ant colonies in their hollow stems (see Chapter 9). Ants are part of a complex endophytic community, including at least four different levels of interaction: plants, herbivores, ants, and ant predators (Chapter 2). One important addition to this field is the recent comprehensive study by E. Tepe, studying the evolution of myrmecophytism in *Piper* taxa from Central America. Tepe used a phylogenetic framework for the clade *Macrostachys* to depict the multiple origins of plant–ant interactions in this group. Tepe studied the stem morphology of several *Piper* species closely related to known ant-plants and was able to predict possible interactions that should be tested with ecological studies. It is important to note that plant–ant interactions are not limited to the clade *Macrostachys*, and are not present in all species of this clade. Ant colonies have been observed in several Asian species of *Piper* that, coincidentially or not, form part of a monophyletic group (see Chapter 10). Of special interest is *P. myrmecophyllum* C.DC., a species that has secondary domatia formed by the leaf auricles and host ants of the genus *Tetramorium*. Nothing is known about the ecology of this interaction.

Piper species have a large number of secondary compounds of known and potential medical interest (Chapters 7 and 8). Not surprisingly, in a literature search on "Piperaceae" and "Piperaceae-pharmacology," up to 65% of the publications in the last 5 years were related to pharmacology. Most of these studies have been conducted in species selected without regard to systematic position or ecology and, in many cases, the surveys are not exhaustive (as pointed out in Chapter 7). Thus, little can be derived from these studies regarding the evolution and ecological importance of *Piper* compounds or their significance in the diversification of the genus. Only a few studies have focused on the relationship between *Piper* phytochemistry and herbivory and none has specifically studied the evolution of phytochemically active compounds. Further research should focus on at least two topics: evolution of secondary compounds and their synthesis pathways, and coevolution between *Piper* species chemistry and the ability of herbivores to detoxify these compounds.

11.3. ABIOTIC FACTORS

Data available to date regarding reproductive ecology (dispersal and pollination), and herbivore interactions, do not seem to provide sufficient explanation for the large diversity of species in the genus *Piper*. Marquis (Chapter 5) proposes that diversification has been driven by adaptation to unique habitats and through the evolution of nonshrub growth forms, rather than through biotic interactions. His analyses reveal that species number is highest in humid lowland rain forests, in part, because both shrubs and other life forms (lianas, vines, herbs, and small trees) occur in these high-rainfall habitats, whereas only shrub species occur in more seasonal climates.

In general, studies of the distribution of species with respect to abiotic factors, including soil nutrients, soil pH, soil texture, light availability, temperature, and water table, are few. For example, the distribution of *Piper* species relative to soil types has not been quantified for any site, other than as part of species descriptions in local florulas. Some researchers have evaluated the ecological importance of crown architecture and photosynthetic capacity (Chazdon and Field 1982, Nicotra *et al.* 1997) in understory and gap species, demonstrating that species that grow preferentially in open areas are more plastic. However, the potential contribution of these factors to local *Piper* diversity has not been addressed. Plant architecture has received little attention in the study of *Piper* ecology. Future research should address the importance of architectural variation in the diversification of *Piper*.

11.4. GEOGRAPHICAL DISTRIBUTION

One major issue, which is quite apparent from the topics covered in this volume, is a lack of knowledge concerning the biology of tropical Asian and African *Piper* species. There are at least 300 species occurring in Asia and only two species native to Africa. Why diversity is lower in Africa is not clear, but it parallels the lower diversity seen there for trees compared with that in Asia and the Americas. Has diversity always been low in Africa, or has it been reduced because of climatic events?

Piper plants grow in tropical ecosystems around the world, and yet the most recent studies of *Piper* biology outside of the Americas are taxonomic monographs 30–70 years old (Ridley 1924, Quisumbing 1930, Chew 1953, Smith 1975). One peculiarity of tropical Asian taxa is that they are easily recognized at the species level but difficult to place into morphologically similar groups, a condition that may reflect a totally different pattern of speciation or different divergence time relative to tropical American taxa. In addition, many Old World species are dioecious, suggesting that their reproductive biology and population dynamics may be distinctly different from that of tropical American species. The two African *Piper* species are very different from each other: *P. guieenense* is a dioecious vine, similar to Asian taxa, and *P. capensis* is a shrub with hermaphroditic flowers similar to American taxa. The phylogenetic relationships of the African species remain a puzzle, and nothing is known about the ecology of these plants.

Even our understanding of the distribution of species in the Tropical Americas is based on spotty sampling. Using information published in local florulas from Central and South America, Marquis (Chapter 5) has confirmed the occurrence of three major biogeographic provinces in the region: Mexico–Central America, the Amazon and the Atlantic Forest of Brazil. This study shows that the large diversity of species in the Andes is the result of a confluence of elements from these three provinces with local diversification of several groups. Future research should systematically sample local sites and regions throughout each of these tropical realms, and analyze the results in a phylogenetic context. This approach would allow determination of the history of the genus as it diversified across the landscape in the context of geologic events (e.g., uprise of the Andes, or the drying of the African climate).

11.5. SUMMARY

This book presents data on mutualistic relationships with ants, coevolution with bat dispersers, phytochemical compounds, phylogenetic relationships, and biogeography. Little has been published regarding *Piper* demography (but see Greig 1993), habitat specialization, genetics (Heywood and Fleming 1986), and speciation processes. Similarly, little is known regarding *Piper* ecophysiology beyond light control of photosynthesis (Chazdon *et al.* 1988) and the ground-breaking work of Vázquez-Yanes (e.g., Vázquez-Yanes and Orozco-Segovia 1982) on the environmental control of seed germination. Research efforts should be designed toward using multiple approaches to study *Piper* biology in a more inclusive manner. For example, the methods of population genetics can be used to distinguish the relative importance of dispersal vs. pollination for the genetic structure of the population. In turn, phylogenetic reconstructions can be used to guide studies of ecological specialization and biogeography. In so doing, we may learn much about the forces that have generated one of the most diverse tropical plant genera.

REFERENCES

Bizerril, M. X. A., and Raw, A. (1998). Feeding behaviour of bats and dispersal of *Piper arboreum* seeds in Brazil. *Journal of Tropical Ecology* 14:109–114.

Chazdon, R. L., and Field, C. B. (1987). Determinants of photosynthetic capacity in six rainforest *Piper* species. *Oecologia* 73:222–230.

Chazdon, R. L., Williams, K., and Field, C. B. (1988). Interactions between crown structure and light environment in five rain forest *Piper* species. *American Journal of Botany* 75:1459–1471.

Chew, W.-L. (1953). The genus *Piper* (Piperaceae) in New Guinea, Solomon Islands and Australia, 1. *Journal of the Arnold Arboretum* 53:1–25.

Fleming, T. H. (1981). Fecundity, fruiting pattern, and seed dispersal in *Piper amalago* (Piperaceae), a bat dispersed tropical shrub. *Oecologia* 51:42–46.

Fleming, T. H. (1985). Coexistence of five sympatric *Piper* (Piperaceae) species in a tropical dry forest. *Ecology* 66:688–700.

Greig, N. (1993). Regeneration mode in Neotropical *Piper*: Habitat and species comparisons. *Ecology* 74:2125–2135.

Heywood, J. S., and Fleming, T. H. (1986). Patterns of allozyme variation in three Costa Rican species of Piper. *Biotropica* 18:208–213.

Jones, A. G. (1985). An annotated catalogue of type specimens in the University of Illinois Herbarium (ILL)—I. Piperaceae, except *Peperomia*. *Phytologia* 58:1–102.

Loiselle, B. A. (1990). Seeds in droppings of tropical fruit-eating birds: Importance of considering seed composition. *Oecologia* 82:494–500.

Marquis, R. J. (1991). Herbivore fauna of *Piper* (Piperaceae) in a Costa Rican wet forest: Diversity, specificity and impact. In: Price, P. W., Lewinsohn, T. M., Fernandes, G. W., and Benson, W. W. (eds.), *Plant–Animal Interactions, Evolutionary Ecology in Tropical and Temperate Regions*. John Wiley & Sons, New York, pp. 179–208.

Mori, S. (2002). *Bat/Plant Interactions in the Neotropics*. New York Botanical Garden, New York.

Nicotra, A. B., Chazdon, R. L., and Schlichting, C. D. (1997). Patterns of genotypic variation and phenotypic plasticity of light response in two tropical *Piper* (Piperaceae) species. *American Journal of Botany* 84:1542–1552.

Quisumbing, E. (1930). Philippine Piperaceae. *The Philippine Journal of Science* 43:1–233.

Ridley, H. N. (1924). Piperaceae. In: Ridley, H. N. *The Flora of the Malay Peninsula*, Apetalae, Vol. 3. L. Reeve & Co., Ltd., London, pp. 25–51.

Smith, S. A. (1975). The genus *Macropiper* (Piperaceae). *Botanical Journal of the Linnean Society* 71:1–38.

Vázquez-Yanes, C., and Orozco-Segovia, A. (1982). Phytochrome control of seed germination in two tropical rain forest pioneer trees: *Cecropia obtusifolia* and *Piper auritum* and its ecological significance. *New Phytologist* 92:477–485.

Index

Abiotic factors, 201
Abiotic pollination, 44
Acacia, 13, 15, 17, 173
Acarina, 13
Acromyrmex, 128
Acuyo, 179
Aduncamide, 132
Aedes aegypti, 133
Africa, 1, 180, 189, 194, 201–202
Agrobacterium rhizogenes, 151
Agua Buena, Coto Brus Province, 8
Alkaloids, 118
Alpha-pinene, 118
Amazon, 52, 84, 180, 185, 187, 188, 202
   biogeography of, 85, 89, 90, 91, 93
   flower morphology in, 191
   phenology displacement in, 38
Ambates, 9
Ambates melanops, 21
Ambates scutiger, 21
Ambophily, 45, 51
American tropics, 183, 185, 189, 191, 194, 202
Amides, 3, 19, 20–21, 24, 34, 118–133
   ecology of, 128–130
   evolution of, 131–132
   isolation and quantification of, 120–121
   synthesis of, 121–127
Ananeae, 11
Andes, 164, 180, 185, 188, 202
   biogeography of, 84, 85, 86, 89, 90–91, 93
   dispersal ecology in, 74, 75
   Macrostachys in, 187
Anemophily, 41, 44, 45, 47, 49–51
Angiosperms, 51, 181

Annelida, 12–13, 24
Annelids, 12–13, 25
Anticarcinogens, 132, 133
Antiherbivore defenses, 3, 200. *See also* Mutualism
   artificial, 17
   tritrophic interactions and, 18–22
   trophic cascades and, 22–27
Ant lions, 12
Ants, 5–29, 114. *See also Piper* ant-plants; specific species
   amides and, 128–129, 130
   dispersal ecology and, 65–66
   leafcutter, 10, 21, 128, 129
   mutualism and, 2, 3, 5–7, 10–11, 13–18, 23, 28, 98, 99–102, 105, 112–113, 130, 168, 169–174, 200
   origin of plant structures associated with, 173–174
   Piper sect. *Macrostachys* and, 156–157, 159–162, 166, 168–174
   predators of, 107
Anyphaenidae, 106
Anyphaeniids, 107
Apatelodids, 27
Apidae, 43, 45
Apomixis, 42, 43
Aporcelaimium, 13
Aprovechado, 28
Arachnida, 11
Araneae, 11
Arboreumine, 132
Archipteridae, 13
Arctottonia, 180, 181, 183, 185
Argentina, 90
Aristolochiaceae, 182
Artanthe, 188

Arthropoda, 12–13, 24
Arthropods, 128, 129. *See also* specific types
    Piper ant-plants and, 12–13, 25
    Piper urostachyum and, 102, 103–107
Asca, 13
Ascalaphidae, 12
Ascidae, 13
Asebogenin, 118, 119f
Asexual reproduction, 37, 38, 51
Asia, 179, 182, 183, 186, 188–189, 191, 194, 201–202
Atlantic Forest, Brazil, 52, 180, 185, 188, 202
    biogeography of, 84, 88, 89, 90, 93
    Macrostachys in, 187
Atta, 10, 128
Aurantiamide benzoate, 123
Australia, 1, 43, 44, 52, 141
Automeris postalbida, 10
Awke, 146, 148t, 149, 150–151
Azteca, 17, 19, 171

Bana, 84
Barro Colorado Island, Panama, 58
Barteria, 16
Bats, 2, 3, 34, 128, 179
    characteristics of, 59–62
    coevolutionary aspects of *Piper* interactions, 72–74
    dispersal ecology and, 58–62, 64–67, 71–74, 199, 200
    mating system of, 61
Bayesian analysis, 162, 164, 175, 183, 196
Bees, 42, 43–44, 45, 47, 48, 51, 58, 99, 200
Beetles, 113, 128, 129
    clerid, 25, 26–27, 28, 106
    coccinellid, 106, 108
    leaf-feeding flea, 9
    Piper ant-plants and, 9, 12, 18, 23, 25, 26–27, 28
    Piper urostachyum and, 105, 106, 108, 110
    pollination by, 43
Belboidea, 13
Belize, 83
Beltian bodies, 173
Betle leaf. *See Piper betle*
Bignoniaceae, 93
Biogeographic affinity, 84–85
Biogeography, 78–94. *See also* Geographical distribution
    growth form variation and, 86–89
    regional species pools and, 84–85
    species richness and, 78–79, 84, 86, 91–93
    study site characteristics, 80–82t
Biotic defenses, 19–22, 199
Biotic pollination, 44

Birds, 2, 64, 65, 66, 67, 72, 199
Bisexual flowers, 2, 185, 189
Black pepper, 2, 42, 52, 145, 179. *See also Piper negrum*
Bolivia, 90, 187
Bottom-up trophic cascades, 22–23, 24–26, 28
Branch fragmentation, 37, 38
Branch prostration, 37–38
Branch repositioning, 37, 38
Brasília, Brazil, 40
Braullio Carillo National Park, 8
Brazil, 3, 180. *See also* specific locations
    biogeography of, 84
    dispersal ecology in, 63
    pollination and resource partitioning in, 33–52
Broadcast dispersal, 67, 72
Bull's Horn *Acacia*, 13, 15
Burger's hypotheses (revisited), 156, 157, 165–167, 174
Butterflies, 45

Caatinga, 84
Cadenas project, 22
Camphor, 118
Campo rupestre, 84
Carabodoidea, 13
Carara Biological Reserve, Puntarenas Province, 8
Carbohydrates, 20, 160, 173
Caribbean, 5, 8, 73, 183
Carollia, 65, 66, 67, 71–74, 200
    characteristics of, 59–62
Carollia brevicauda, 71
Carollia castanea, 71, 72, 73
Carollia perspicillata, 60, 61, 65, 66, 71, 72
Carolliinae, 59, 74, 179
Ceccidomyiids, 103, 107
Cecidomyiidae, 12, 13, 44
Cecropia, 17, 19, 170, 173
Cecropiaceae, 65
Cecropia peltata, 65, 66
Cenocladamide, 20, 120, 121t, 124, 132
Cenozoic era, 73
Central America, 38, 52, 128, 164, 167, 180, 183, 186, 194, 200, 202
    biogeography of, 84–85, 86, 87f, 89, 91, 93
    dispersal ecology in, 58, 59, 73, 75
    flower morphology in, 189
    Macrostachys in, 187
Central Cordillera, Costa Rica, 12
Cepharadione, 119f
Cerradão, 84
Cerro El Hormiguero, 83
Chalcones, 118
Chemical defenses, 19–22, 130
Chiapas, Mexico, 65

Chironomidae, 13
Chloranthaceae, 51, 181
Chlorophora tinctoria, 66
Chocó Region, 180, 191
Cholesterol, 118
Chrysomelidae, 9
Ciccadellidae, 107
Cinnamic acid, 118
Cinnamylidone butenolides, 118
Classification of *Piper*, 180–181
Cleridae, 11
Clerid beetles, 25, 26–27, 28, 106
Climate, 89–90, 201
Clonal growth, 3, 37, 93
Coccidae, 103
Coccinellid beetles, 106, 108
Coccinellidae, 12
Coevolutionary theory
  on amide susceptibility, 132
  on bat-*Piper* interactions, 72–74
Coleoptera, 9, 11, 43, 103
Coleopterans, 9, 25, 129
Collembola, 42
Collembolans, 12, 25
Colombia, 84, 85, 86, 88, 89, 93
Combretaceae, 88
Competition
  dispersal ecology and, 66
  resource partitioning and, 33, 34, 36, 49–52
Conservation ecology, 34, 49–52
Continual flowering strategies, 39
Convolvulaceae, 88
Copestylum, 45
Corcovado National Park, Osa Peninsula, 5, 8
Cosmolaelaps, 13
Costa Rica, 99, 180. *See also* specific locations
  dispersal ecology in, 58, 61, 62, 63, 64, 65, 66, 72, 74
  Piper ant-plants of, 5–29
  Piper sect. *Macrostachys* in, 156, 157, 159, 164, 165, 166, 167, 174, 175
  pollination in, 37, 40, 43, 48
Cotton, 114
Coumaperine, 121
Crab spiders, 99
Crematogaster, 105, 107, 171, 172
Ctenidae, 106
Cubeb pepper, 2
Curculionidae, 9, 103
Cynopterus brachyotis, 72

Damaeoidea, 13
Dehiscence, 167
Dehydropipernonaline, 123, 124f
Density. *See* Species density

Density-mediated indirect interactions (DMIIs), 97–99, 113
Department of Antioquia, Colombia, 83–84, 88, 89f
Desmethoxyyangonin, 149t, 150f
4'-Desmethylpiplartine, 120, 121t, 124, 126f
Dichapetalaceae, 88
Dietary Supplement Health and Education Act (DSHEA), 140
Dihydrochalcones, 118
Dihydrokavain, 149, 150f, 151
(+)-Dihydrokawain, 127
(+)-Dihydrokawain–5–ol, 127
Dihydromethysticin, 149, 150f, 151
Dihydropipercide, 121
8, 9-Dihydropiplartine, 132
5, 6-Dihydropyridin-2(1H)-one, 126
Dihydropyridone, 132
Dioecious species, 2, 182, 189, 191, 202
Diplopods, 12
Dipoena, 11
Dipoena schmidti (banksii), 11, 16, 99–102, 107, 113
Diptera, 12, 43, 44, 48, 51
Dipterans, 25, 28
Dispersal ecology, 58–75, 93, 199–200
  coevolutionary aspects of, 72–74
  fates of seeds, 65–67
  fruiting phenology and, 62–64, 72
  patterns of, 64–65
  postdispersal patterns of, 67–72
Diversity. *See* Species diversity
DNA techniques, 73, 162, 174, 182, 194
Dominica, 73
Dry season, 63, 65
Dwarf pipers, 191, 193f

East Indian islands, 1
Ecuador, 61, 79
Eleocarpaceae, 65
Elevation, 79, 83–84, 86, 89f, 91, 180
Enchytraeid worms, 13
Enckea, 183–185, 191
Endophytic animals, 24–26
Entomophily, 45, 51
Eocene era, 73
Eois, 103, 105, 106, 129, 130, 132
Eois apyraria, 9, 129
Eois dibapha, 9
Epidorylaimus, 13
Epimecis, 9
Episodic flowering strategies, 39
Erophylla bombifrons, 73
Eucephalobus, 13
Euclea plungma, 10
Eudorylaimus, 13
Eugenol, 118

Euphtheracaroidea, 13
Evenness, of visitor insects, 44, 48t
Evolutionary aspects. *See also* Coevolutionary theory
 of ant-plant associations, 156–157, 168–174
 of flower morphology, 189–191
 of *Piper* chemistry, 131–132
 of *Piper* sect. *Macrostachys*, 156–157, 167, 168–174
 of plant architecture, 191–192
 of pollination and resource partitioning, 33, 34, 49–52
 of species richness, 93

Fabaceae, 168
Facultative associations, 3, 107, 161, 162, 168, 172, 173, 174
Fertilizer, 20, 24–26
Ficus, 41, 65, 72
Fijian archipelago, 188
Flavanones, 118
Flavones, 118
Flies, 12–13, 43–44, 45, 58, 118, 200
Flora Costaricensis, 156
Flowering phenology, 38–40, 41f, 42, 42f, 43f, 44f
 biogeography and, 93
 dispersal ecology and, 72
Flower morphology, 189–191
Flowers, 62
 bisexual, 2, 185, 189
 dispersal ecology and, 58
 hermaphroditic (*see* Hermaphroditic flowers)
 of *Piper* ant-plants, 8
 unisexual, 188, 189
Folivores, 99, 113
 amides and, 129
 on *Piper* ant-plants, 10, 16, 17, 26–27
 on *Piper obliquum*, 100, 101–102
 on *Piper urostachyum*, 109–112
Food and Drug Administration, U.S. (FDA), 143–144
Food bodies, 98, 99, 101–102, 103, 104f, 106, 107, 112, 114, 173
 of *Cecropia* trees, 19
 of *Piper* ant-plants, 11, 13, 18–19, 20
Food webs, 13, 14t, 24–26
Formicidae, 10
N-Formyl aporphines, 118
French Guiana, 59, 61, 64, 84, 87
Fruiting phenology, 38, 39f, 40–41, 42f, 43f, 44f
 dispersal ecology and, 62–64, 72
Fruits, 2, 42, 58, 185
Fulgoridae, 107

Gall making ceccidomyiids, 103
Galumnoidea, 13
Gamasida, 13

Gap species, 63, 67, 70–71, 72, 201
Geocoridae, 114
Geographical distribution, 201–202. *See also* Biogeography
Geometridae, 103
Geometrids, 9, 27, 130
Geonoma, 192
Germination, seed, 66–67, 71
Glossophaga soricina, 61–62
Glossophaginae, 61
Golfo de Uraba, Costa Rica, 84
Golfo Dulce Forest Reserve, Costa Rica, 99
Gondwana, 73, 74
Greater Antilles, 73
Green desert model, 22–23, 24
Grenada, 73
Guadeloupe, 73
Guaiol, 118
Guineensine, 132

Habitat affinity, 79, 86–89
Habitats, 36–37, 144
Hacienda Loma Linda, Costa Rica, 8
Halictidae, 43, 47
Haplozetidae, 13
Harem-polygynous mating system, 61
Hawaii, 141
Height of plants, 87, 92f, 93, 141
Heliconiaceae, 93
Heliotropin, 118
Hemiepiphytes, 88, 93
Hemipodium, 162
Hemiptera, 114
Hepatotoxicity, 143–144
Herbivores, 2. *See also* Antiherbivore defenses
 amides and, 128–130
 generalist, 129, 130, 132, 200
 Piper ant-plants and, 9–10
 Piper urostachyum and, 103–105
 specialist, 129, 130, 132, 200
Hermaphroditic flowers, 2, 41, 58, 182, 183, 185, 187, 191, 202
Hesperiids, 27
Heteroptera, 103, 105
Hoja santa, 2
Hollow stems, 169–171, 173, 200
Honeybees, 99
Hoverflies, 45, 47, 48–49, 51
Hydnoraceae, 182
N-5-(4-Hydroxy-3-methoxyphenyl)-2E, 4E, pentadienoyl piperidine, 121
N-5-(4-Hydroxy-3-methoxyphenyl)-2E–pentenoyl piperidine, 121
Hymanea courbaril, 28
Hymenoptera, 10, 43, 48, 51

INDEX

Hymenopterans, 9
Hypericaceae, 59
Hypoaspididae, 13
Hypoponera, 11

India, 2, 42
Indirect interactions, 26–27, 97–99, 111–112. *See also* Density-mediated indirect interactions; Trait-mediated indirect interactions
Inflorescences, 2, 38, 41, 51, 160, 183
   insects on, 43–44, 45, 46–47t, 48
   of *Macrostachys*, 187
   of *Piper* ant-plants, 18
   stamen number and, 189, 191
Infrageneric relationships, 183
Infructescences, 2, 38, 40, 51, 58, 160
Inga, 19
Insecticidal chemicals, 120, 121, 133
Insect pollination, 39–40, 43–44, 49, 51, 58, 93
Internal transcribed spacers (ITS), 162, 165, 166, 170, 172, 174, 183, 194–196
Isobutyl, 132
Isopods, 12

Jaborandi, 185
Java, 180
Jug nests, 105, 107

Kadsurin, 118, 119f
Kava, 2, 3, 133, 140–153, 179, 188
   cell cultures of, 146–151
   description and growth, 141–142
   discovery and origins of use, 141
   hepatotoxic potential of, 143–144
   phytochemical research, 144–145
   phytochemicals in extracts of, 142–143
   tissue culture growth of, 144–145
Kavain, 144, 149, 150, 151
Kavapyrones, 118, 142–143, 144, 146–153
   in vitro production of, 146–151
Kawain, 127

Lactoridaceae, 182
Laelaspis, 13
La Llorona Point, 5
La Selva Biological Station, Costa Rica, 1, 2, 102, 128
   biogeography of, 89
   dispersal ecology in, 58, 64, 66, 67–72
   Piper ant-plants of, 8–9, 11, 12, 18, 22
Latitude, 86, 87f, 88f
Leafcutter ants, 10, 21, 128, 129
Leaf-feeding flea beetles, 9
Leaf miners, 103
Leonardoxa africana, 168

Lepidoptera, 9, 103
Lepidopterans, 9, 10, 15, 21, 27, 129
Lesser Antilles, 73
Li-Cor model 3100 area meter, 108, 110
Light, 23, 24–26, 67, 129, 191, 201
Light-demanding species, 37, 38
Lignans, 118
Limacodids, 27
Linalool, 118
Lindeniopiper, 180
Lipids, 20, 128, 160, 173
Lohmanniid mites, 12
Lolot leaves, 179. *See also Piper lolot*
Long pepper, 2;. See also *Piper lomgum*

Macaranga, 161, 171, 172, 173
Macropiper, 183, 188
Macrostachys, 3, 183, 187, 191, 200
Makea, 146, 147, 148t, 149, 150
Malphigiaceae, 88
Manakins, 64
Manantlan, Mexico, 84
Manaus, Brazil, 58
Maximum Likelihood (ML) tree, 163, 164, 165, 175
Maximum parsimony (MP) method, 162, 164, 169, 171, 175, 197
Malay Peninsula, 180
Melanesia, 141
Melastomataceae, 58
Menispermaceae, 88
Mesocosm, 5–29. *See also Piper* ant-plants
Mesostigmata, 13
5-Methoxy-5, 6-dihydromethysticin, 143
Methysticin, 119f, 127, 133, 149, 150, 151
Mevalonic acid pathway, 118
Mexico, 179, 180, 202. *See also* specific locations
   biogeography of, 84, 85, 86, 87f, 89, 91, 93
   dispersal ecology in, 64, 67, 73
Microdon, 12
Micronesia, 141
Miocene era, 73
Miridae, 105
Mirids, 105, 107, 108, 109, 110, 111, 112, 113
Missouri Botanical Garden, 83
Mites, 12, 13, 25, 103, 105–106
Monkeys, 65
Monophyly, 181–183
Monoterpenes, 118
Monteverde, 1
Montserrat, 73
Morphological characters, 196
Morphological variation, 1, 2, 78
Mosquitoes, 128, 133
Müllerian bodies, 19, 114, 173
Muntingia calabura, 65, 66

Mutualism, 2, 3, 128, 130, 200
    ants and (*see under* Ants)
    bats and, 34
    evolution of, 169–174
    facultative (*see* Facultative associations)
    in model *Piper* systems, 97–114
    obligate (*see* Obligate associations)
    parasites of, 98, 106
    Piper ant-plants and, 5–7, 10–11, 13–18, 23, 28
    Piper sect. *Macrostachys* and, 168, 169–174
    plant characteristics encouraging, 103
    spiders and, 11, 28, 97–114
Mycetophilidae, 13
Myristicine, 118
Myrmicinae, 10

Nematoda, 12–13, 24
Nematodes, 12–13, 25
Neolignans, 21, 118
Neostruma myllorhapha, 11
Neotropical *Pipers*, 1, 2, 3
    biogeography of, 78–94
    dispersal ecology of, 58–75
    phylogenetics of, 183–188
    pollination and resource partitioning of, 33–52
Neuruopterans, 28
Neuroptera, 12
New Guinea, 180
New World pipers, 2, 72, 78, 79, 132
Nicaragua, 83
Niche overlap, 43, 49
Niches, 78–79, 92, 93
Nigeria, 16
Notonycteris, 73
Nouragues Reserve, French Guiana, 84, 87

Obligate associations, 3, 160, 161, 166, 168, 173, 174
    hollow stems and, 169–171
    number of species with, 162
Ocyptamus, 45, 49
Ocyptamus propinqua, 51
Old World pipers, 2, 72, 120, 132, 202
Oligocene era, 73
Ornidia obesa, 45
Orthoptera, 10
Orthopterans, 9, 10, 21, 129
Osa Peninsula, Costa Rica, 2, 99, 100, 113. *See also* specific sites
Ottonia, 35, 183, 185, 186f, 191, 197. *For species, see also Piper*
Oxoaporphine, 118

Pachysima, 16
Panama, 38, 63, 165, 180. *See also* specific locations
Papua New Guinea, 15
Parasites of mutualism, 98, 106
Paratrechina, 107
Passiflora, 93
Passifloraceae, 16
Pearl bodies, 2, 114, 159–160, 169, 171, 173–174
Penninervia, 180–181, 189
Peperomia, 1, 180, 182
Pepper, 34
    black, 2, 42, 52, 145, 179. *See also Piper nigrum*
    cubeb, 2
    long, 2. *See also Piper longum*
Peppercorns, 34, 52
Peru, 59, 64, 187
Pesticidal agents, 2, 132. *See also* Insecticidal chemicals
Petiole chambers, 98, 99, 100, 101, 103
    of *Piper* ant-plants, 11, 13, 18, 24, 25
    of *Piper* sect. *Macrostachys*, 159, 160, 161, 162, 169–170, 172
Pheidole, 3, 66, 102, 105, 107, 187
Pheidole bicornis, 5, 8, 10–11, 13–18, 28, 98, 160, 166, 170, 171, 172
    amides and, 128–129
    evidence for anti–folivore defense, 16
    evidence for higher plant fitness via, 17, 18
    nutrient procurement by, 15–16
    on *Piper obliquum*, 99–102, 112, 113
    tritrophic interactions and, 18–22
    trophic cascades and, 22–27
Pheidole campanae, 11
Pheidole ruida, 11
Pheidole specularis, 11
Pheidole susannae, 11
Phenology, 38–41, 51, 52. *See also* Flowering phenology; Fruiting phenology
N-(3-Phenylpropranoyl)pyrrole, 123
Philippines, 180, 181, 189, 191
Pholcidae, 106
Photoblastic response, 67, 72
Phthiracaroidea, 13
Phyllobaenus. *See Tarsobaenus letourneau*
Phyllostomidae, 59, 73, 74
Phylogenetics, 3, 181–189, 194–196
    of ants, 168, 170
    morphological characters in, 196
    of Neotropical taxa, 183–188
    of Piperales, 181–183
    of *Piper* sect. *Macrostachys*, 156–157, 159, 162–165
Phylogeny. *See* Phylogenjitics
Physimera, 9
Piperaceae, 19, 27, 34, 51, 58, 141, 181, 182, 201
    genera of, 1
    stamens of, 189

INDEX

Piper aduncum, 70, 167, 173, 186
  amides of, 132, 133
  biogeography and, 84
  dispersal ecology and, 67, 72
  pollination of, 37, 39, 40, 43, 45, 49, 52
Piper aequale, 63, 65, 67
Piper aereum, 162, 166
Piper alatabaccum, 185
Piper albispicum, 189
Piperales, 181–183
Piper amalago
  amides of, 132
  dispersal ecology and, 61, 63, 65, 66, 67, 72, 74
  pollination of, 37, 38, 39, 40, 42, 43, 45, 48, 49
  stamens of, 189
Piper amazonicum, 84
Piperamide-C9, 123
Piper amides. *See* Amides
Piper ant-plants, 5–29
  herbivores of, 9–10
  mutualism and, 5–7, 10–11, 13–18, 23, 28
  species studied, 8–9
  study sites, 8
  tritrophic interactions and, 18–22
Piper arborescens, 189
Piper arboreum, 162, 166, 174
  amides of, 132
  biogeography and, 87
  dispersal ecology and, 60, 63
  pollination of, 37, 38, 39–40, 45, 49
Piper archeri, 164, 167
Piper arieianum, 2
  biogeography and, 84
  dispersal ecology and, 63, 67, 68
  pollination of, 43–44
Piper aurantiacum, 123
Piper auritum, 2, 72, 143, 173, 179, 187
  amides of, 118, 133
  biogeography and, 89
  dispersal ecology and, 64, 65, 66, 67, 70
  pollination of, 43
Piper bartlingianum, 185
Piper begoniicolor, 164
Piper betle, 2, 118, 132, 145, 179
Piper biseriatum, 70, 161, 165, 172
Piper blattarum, 42–43
Piper boehmeriaefolium, 189
Piper brachypodon, 88
Piper brevicuspe, 189
Piper calcariformis, 156, 160, 166, 169, 170
Piper capensis, 189, 202
Piper caracasanum, 167
Piper carpunya, 88
Piper carrilloanum, 63
Piper castroanum, 88

Piper cenocladum, 5, 8, 9, 10f, 11, 12, 15, 99, 103, 105, 106, 112, 113, 169, 170, 174
  amides of, 120, 121t, 124, 128–130, 132
  antiherbivore defenses of, 18, 19–20, 200
  dispersal ecology and, 69
  food web of, 13, 14t
  leaf lost to herbivores, 21t
  trophic cascades and, 23–27
Piper cernuum, 165
Piper chemistry, 117–134
  applied, 132–133
  ecology of, 128–130
  evolution of, 131–132
  future research on, 133–134
*Pipercide, 121, 123
Piper citrifolium, 42
Piper cordulatum, 63, 166
Piper coruscans, 84
Piper crassinervium, 37, 38, 39, 40, 45, 48, 51
Piper cubeba, 2
Piper culebranum, 63, 67
Piper dactylostigmum, 88
Piperdardine, 124
Piper darienense, 192f
Piper dariense, 63
Piper decumanum, 189
Piper demeraranum, 84
Piper dilatatum, 63
Piperene, 119f
Piper euryphyllum, 164, 165–166
Piper filistilum, 192
Piper fimbriulatum, 8, 15, 18, 21, 157, 164, 169, 170
Piper francovilleanum, 185
Piper friedrichsthalii, 43
Piper garagaranum, 2, 67, 69
Piper gaudichaudianum, 37, 39, 40, 45
Piper gibbosum, 165–166, 170
Piper gigantifolium, 164
Piper glabratum, 37, 38, 39, 40, 45, 49
Piper grande, 63
Piper guineense, 121, 133, 189, 202
Piper gymnostachyum, 189
Piper hebetifolium, 166, 170
Piper hermannii, 84
Piper heterophyllum, 84
Piper heterotrichum, 88
Piper heydi, 187
Piper hispidinervium, 118, 133
Piper hispidum, 63, 64, 65, 67, 167
Piper holdridgeianum, 69
Piper holtii, 84
Piperic acid, 121, 126
*Pipericide, 120, 132
Piperidine, 121, 132, 133
Piper imperiale, 105, 120, 130, 165, 172

Piperine, 2, 120, 121, 123, 124f, 132
Piper jacquemontianum, 43, 63, 66
Piper kegelii, 191
Piper khorthalsii, 191
Piper klotzchianum, 185
Piper laevigatum, 183
Piper laosanum, 189
Piper lenticellosum, 88
Piper lhotzkyanum, 84
Piper lolot, 179, 189
Piper longispicum, 164, 192f
Piper longum, 2, 128, 133, 145
Piperlongumine, 123, 124f
Piper macedoi, 37, 39, 40, 45, 49
Piper macrotrichum, 84
Piper marginatum, 43, 63, 185, 187
Piper marsupiatum, 164
Piper martiana, 37, 38, 39, 40, 45, 49
Piper melanocladum, 67, 162, 166
Piper methysticum, 2, 3, 118, 127, 133, 140–153, 179, 188. *See also* Kava
Piper mikanianum, 37, 38, 39, 40, 45, 49
Piper mollicomum, 37, 38, 39, 40, 45, 84
**Piper myrmecophilum*, 189
**Piper myrmecophylum*, 200
Piper nicoyanum, 183
Piper nigrum, 2, 34, 145, 165, 173, 179
  amides of, 121, 133
  chemistry of, 118
  pollination of, 42, 43, 51, 52
Piper nitidum, 65
Piper nobile, 167
Pipernonaline, 123, 124f, 133
Piper novae-hollandiae, 44
Piper nubigenum, 84
Piper nudilimbum, 183
Piper obliquum, 8, 18, 21, 164, 169, 170, 187
  mutualism and, 99–102, 103, 107, 111, 112, 113
Piper obliquum complex, 8, 156, 157, 174
  evolutionary trends in, 167
  systematic relationships in, 165–167
Piper oblongum, 88
Piper obtusilimbum, 164
Piperolein A, 123
Piperolides, 118
Piperonyl, 123
Piperonyl butoxide, 118
Piper ottoniifolium, 88
Piper ottonoides, 185
Piper peltata, 43, 72
Piper peltatum, 187
Piper perlasense, 63
Piper pinaresanum, 88
Piper propinqua, 37, 38, 39, 40, 45, 48
Piper pseudobumbratum, 89

Piper pseudofuligineum, 2
  dispersal ecology and, 61, 63, 65, 66, 74
  pollination of, 43
Piper pseudonobile, 167, 174
Piper regnelli, 37, 38, 39, 40, 45, 48, 49, 51
Piper reticulatum, 63, 132, 183
Piper retrofractum, 123
Piper rothiana, 44
Piper sagittifolium, 8, 18, 21, 157, 160f, 166, 167, 169, 170, 173, 174
Piper sancti-felicis, 64, 66, 67, 70
Piper sanctum. *See Piper auritum*
Piper sarmentosum, 123
Piper scrabrum, 42
Piper scutifolium, 185
Piper sect. *Macrostachys*, 156–175
  Burger's hypotheses (revisited) and, 156, 157, 165–167, 174
  natural history of, 159–162
  phylogenetics of, 156–157, 159, 162–165
  species included in study, 158t
  taxonomic history of, 158–159
Piper spoliatum, 174
Piperstachine, 121, 123
Piper subglabribracteatum, 174
Piper sylvaticum, 123
Piper toppingii, 189
Piper trichostacyon, 121
Piper tuberculatum, 43, 84, 121, 124, 166, 173
Piper ubatubense, 185
Piper umbellatum, 67, 187, 189
Piper urostachyum, 69, 99, 102–112, 113, 114
  characteristics encouraging mutualism, 103
  resident arthropods of, 103–107
Piper villiramulum, 43
Piper wichmannii, 141, 143
Piper xanthostachyum sect. *Churumayu*, 167
Piper yzabalanum, 65
Piplartine, 120, 121t, 124, 126f, 132
Pipridae, 64
Plant-animal interactions, 199–201. *See also* Antiherbivore defenses; Dispersal ecology; Pollination
Plant architecture, 191–192, 201
Plant Preservative Mixture (PPM), 146, 147
Plectus, 13
Pleiostachyopiper, 180, 181
Podothrombium, 13
Pollination, 3, 33–52, 58, 99, 199, 200
  abiotic, 44
  amides and, 128
  biotic, 44
  in evolutionary and conservation ecology, 33, 34, 49–52
  guidelines for future research, 52

habit and habitat utilization in, 36–37
insect (*see* Insect pollination)
site and species studied, 35
visitors and, 41–49
wind (*see* Wind pollination)
Polyaspidoidea, 13
Polynesia, 2, 141
Pothomorphe, 183, 187. *For species, see also Piper*
Pothomorphe peltata, 84
Pothomorphet umbellata, 84
Predators, 200
 generalist, 100, 107
 mutualist, 98, 105–106
 postdispersal, 65–66, 67
 specialist, 100, 107
 top (*see* Top predators)
Primary forests, 67, 70, 71
Propenylphenols, 118
Proteins, 20, 72, 128, 129, 160, 173
Protoangiosperms, 51
Pseudomyrmex, 17, 171
Pseudomyrmicinae, 16
Pteropodidae, 72
Puerto Rico, 42, 73
Pyrethrins, 118
Pyrrolidine, 132

Quadrus cerealis, 9

Radula, 167, 183, 185–186, 187, 191
Rainfall, 8, 42, 43, 58, 79, 84, 85, 86, 87, 88f, 89–90, 91–93, 201
Reproduction
 asexual, 37, 38, 51
 vegetative, 37–38, 51, 52, 67
Reproductive phenology. *See* Phenology
Reserve Ducke, Brazil, 88
Resource partitioning, 33–52
 in evolutionary and conservation ecology, 33, 34, 49–52
 habit and habitat utilization in, 36–37
 site and species studied, 35
Retrofractamide C, 123
Rhabditidae, 13
Rhinophylla, 59–60
Rhinophylla pumilio, 59, 61
Rhizome fragmentation, 37, 38
Richness. *See* Species richness
Rio Magdalena River valley, 83
Rubiaceae, 58

Safrole, 34, 118, 119f, 132–133
St. John's wort, 140
Salpingogaster nigra, 45, 48
Salticidae, 106

Salticids, 107, 113
Sandstone mesas, 84
Santa Genebra Reserve, Brazil, 40, 50t
Santa Rosa National Park, Costa Rica, 1, 58, 65, 66
Sapindaceae, 88
Sarcorhachis, 180, 182, 183
Saururaceae, 181, 182
Saururus, 167
Schilleria, 187–188, 191
Sciaridae, 13
Scytodidae, 106
Seasonal flowering strategies, 39
Secondary forests, 67, 71
Seed rain, 64–65, 71
Seeds. *See* Dispersal ecology
Seed traps, 65
Semideciduous forests, 35, 36–37, 39, 49
Serra da Chapadinha, Brazil, 84
Serra do Japi Reserve, Brazil, 40
Sesamin, 118, 119
Sesquiterpenes, 118
Shadehouse experiments, 129
Shade-intolerant species, 67, 200
Shade-tolerant species, 37, 38, 67, 200
Shimodaira-Hasegawa test, 171, 175
Sierra Madre, 185
Sirena, Osa Peninsula, 1
Sitosterol, 118
Skippers, 9
Snake bite antidotes, 185
Soil, 23, 24, 58, 129, 201
Solanaceae, 58, 59
Solanum, 59
Solenopsis, 172
Sørensen similarity index, 70, 84
South America, 52, 128, 164, 165, 168, 180, 183, 194, 202
 biogeography of, 86, 88f, 90, 91
 dispersal ecology in, 58, 73, 75
 Macrostachys in, 187
 phenology in, 38
Southeast Asia, 1, 2, 180
South Pacific Islands, 141, 144, 179, 183, 186, 188–189, 194
Species density, 67, 69t, 70, 71, 87f
Species diversity, 1, 2
 biogeography and, 78, 86, 91, 92, 93
 dispersal ecology and, 67, 69–70, 71, 73, 74
 geographical distribution and, 201–202
 trophic cascades and, 24–26
 of visitor insects, 44, 48t
Species richness
 biogeography and, 78–79, 84, 86, 91–93
 dispersal ecology and, 58, 67, 70
 of visitor insects, 44, 45, 48t

Spiders, 3, 51, 97–114. *See also* individual species
   crab, 99
   on *Piper* ant-plants, 11, 28
   on *Piper obliquum*, 99–102, 107, 111, 112, 113
   on *Piper urostachyum*, 102, 103–112, 113
   Thomisid, 51
   as top predators, 11, 106–107, 114
Spikes, 41
Stamen number, 189–191
Steffensia, 188
Stem borers, 25, 160
Stem-boring weevils, 9, 18
Sternodermatinae, 59, 61, 179
Steroids, 118
Stigmaidae, 13
Strychnine, 142
Sturnira lilium, 61–62, 65, 73
Sturnira ludovici, 62
Sturnira thomasi, 73
Swamp forests, 35, 36–37, 84
Syconycteris australis, 72
Sylvamide, 123
Synergistic defenses, 118, 134
Syrphidae, 12, 43, 45, 48
Syrphids, 12, 47–48, 200

Taiwan, 180
Tanagers, 64, 66
Tarchon, 10
Tarsobaenus, 26, 113
Tarsobaenus letourneau, 11, 106, 128
Tectocepheidae, 13
Tegeozetes, 13
Temperature, 79, 201
Tenthrenidae, 106
Tents, 60, 106, 108, 110
Terpenes, 118, 132
Terpinolene, 119f
Tetramorium, 200
Tettigoniidae, 10
Theridiidae, 11, 107
Theridiids, 99, 107
Thermodynamics model, 22–23, 24
Thomisid spiders, 51
Thonus, 13
Thraupinae, 64

Tococa, 168, 171
Top-down trophic cascades, 22, 23–26, 27, 28
Top predators, 114
   Piper ant-plants and, 11–12, 23–27
   Piper urostachyum and, 106–107
Trait-mediated indirect interactions (TMIIs), 97–99, 100–102, 111–112, 113
N-Trans-feruloyl pipertidine, 121
N-Trans-feruloyl tyramine, 121
Transphytol, 119f
Trianaeopiper, 180, 181, 191
Trigona, 48
Tritrophic interactions, 18–22
Trombidiidae, 13
Trophic cascades, 3, 7, 22–27, 28, 99
Tropical forests, 79
   dry, 58, 62, 63, 65, 66, 71, 74
   wet, 58, 62, 66, 67–72
TROPICOS database, 83

Unisexual flowers, 188, 189
Uropodina, 13

Varzea forests, 84
Vegetative reproduction, 37–38, 51, 52, 67
Venezuela, 79, 180
Vera Cruz, Mexico, 65
Viet Nam, 179
Visitor insects, 41–49
Vismia, 59

Wassmannia, 172
Weevils, 9, 18, 103
Wetlands, 84, 87
Wet season, 63, 65, 66
Wind pollination, 39–40, 42–43, 49, 51, 200
Winteraceae, 51
Wisanidine, 121
Wisanine, 121

Xylobates, 13

Yangonin, 149t, 150f
Yellow fever, 133

Zippelia, 182–183
Zippeliae, 182

DATE DUE

DUE DATE SUBJECT TO CHANGE
IF A RECALL IS REQUESTED